Occupational Health in the Construction Industry

Editors

Mariusz Szóstak
Marek Sawicki

MDPI

Basel • Beijing • Wuhan • Barcelona • Belgrade • Novi Sad • Cluj • Manchester

Editors

Mariusz Szóstak
Department of Building Engineering, Faculty of Civil Engineering
Wroclaw University of Science and Technology
Wroclaw
Poland

Marek Sawicki
Department of Building Engineering, Faculty of Civil Engineering
Wroclaw University of Science and Technology
Wroclaw
Poland

Editorial Office
MDPI
St. Alban-Anlage 66
4052 Basel, Switzerland

This is a reprint of articles from the Special Issue published online in the open access journal *Buildings* (ISSN 2075-5309) (available at: https://www.mdpi.com/journal/buildings/special_issues/5LSH03FZ4G).

For citation purposes, cite each article independently as indicated on the article page online and as indicated below:

Lastname, A.A.; Lastname, B.B. Article Title. *Journal Name* **Year**, *Volume Number*, Page Range.

ISBN 978-3-7258-1055-0 (Hbk)
ISBN 978-3-7258-1056-7 (PDF)
doi.org/10.3390/books978-3-7258-1056-7

Contents

About the Editors

Mariusz Szóstak

Mariusz Szostak is employed at the Department of Building Engineering, Faculty of Civil Engineering, at Wroclaw University of Science and Technology (WUST), Poland. He obtained an MSc in Civil Engineering from WUST in 2013 and a PhD (Hons) in Civil Engineering in April 2018. Since 2021, he has been the Deputy Head of the Department of Building Engineering. He is the author of more than 90 scientific papers, including papers on work safety in construction and articles in journals in the Journal Citation Reports database. He developed several dozens of reviews of articles in journals from the JCR list such as *Applied Sciences, Automation in Construction, Buildings, International Journal of Environmental Research and Public Health, Sustainability, Safety Science*, etc. His research interests concern issues related to safety and health protection in construction processes, particularly the modeling of accidents in the construction industry, and an analysis of the causes of accidents in construction and construction project management, such as a cost analysis of construction projects, value engineering, and BIM technology. He has been a Project Manager of the Wroclaw University of Science and Technology of the project Erasmus+ Strategic Partnerships for vocational educational and training, Innovation "SafeCRobot Virtual Reality Immersive Safety Training Environment for Robotised and Automated Construction Sites" 2020-1-UK01-KA202-079176, 2020–2022, and is a member of the research team of the project "Model of the assessment of risk of the occurrence of building catastrophes, accidents, and dangerous events at workplaces with the use of scaffolding" supported by Polish National Centre of Research and Development within Project PBS3/A02/19/2015. Since 2022, he has been a Member of the Academy of Young Scholars and Artists.

Marek Sawicki

Marek Sawicki is employed at the Department of Building Engineering, Faculty of Civil Engineering, at Wroclaw University of Science and Technology (WUST), Poland. He received his master's degree in civil engineering from WUST in 1988 and his Ph.D. (Hons) in Civil Engineering in July 1997. He is the author of more than 130 scientific papers, including works on building renovation, organization of works and work safety in the construction industry, and articles in journals listed in the Journal Citation Reports database. He is a member of the Program Council and the Board of Reviewers of Builder journal. He has actively participated in two research grants on knowledge maps and scaffolding safety. He was a member and the Chairman of the Supervisory Board of a housing cooperative from 2001 to 2022. His research interests include issues regarding the recycling of construction materials, renovation of balconies in residential buildings, organization and technology of works, assembly and issues related to safety and health in construction processes, particularly the modeling of accident phenomena in construction, analysis of causes of accidents in construction, and management of construction projects with a cost analysis of construction projects.

Preface

Welcome to this Special Issue of *Buildings* titled "Occupational Health in the Construction Industry". The construction industry, with its dynamic nature and diverse activities, presents unique challenges concerning occupational health and safety. This Special Issue aims to shed light on the most recent advances in multidisciplinary research related to occupational safety in the construction sector, emphasizing the enhancement of safety protocols and practices.

The papers compiled in this Special Issue cover a broad spectrum of topics, from technological innovations like near-miss detection metrics using sensing technologies to proactive construction safety management, to psychological aspects such as the impact of the COVID-19 pandemic on construction workers' mental health. We delve into the roles and motivations of safety officers, explore safety risks from overlapping construction activities using Building Information Modeling (BIM), and identify critical factors leading to fatal accidents in small-scale construction sites.

The contributions gathered here are a testament to the ongoing efforts to improve occupational health and safety in the construction industry. They provide valuable insights for researchers, safety managers, policymakers, and practitioners alike, offering practical solutions and innovative approaches to address the complex challenges faced by construction workers.

We would like to express our sincere gratitude to all the authors who have contributed their research to this Special Issue. Their dedication and commitment to advancing the field of construction safety have made this publication possible. We also extend our appreciation to the reviewers for their invaluable feedback and constructive criticism, which have helped to enhance the quality of the articles.

This Special Issue is dedicated to everyone involved in ensuring the health, safety, and well-being of construction workers around the world. May the insights shared in these pages contribute to creating safer and healthier construction sites for all.

Mariusz Szóstak and Marek Sawicki
Editors

buildings

Article

Health and Safety Protocol for the Management of Building Demolition Waste with High Mercury Contamination

Rafael Rodríguez [1,*], Hector Garcia-Gonzalez [2], Ángel Pastrana [3] and Zenaida Hernández [4]

1 Department of Mining Exploitation and Prospecting, School of Mining, Energy and Materials Engineering, University of Oviedo, C/Independencia 13, 33004 Oviedo, Spain
2 Instituto Nacional de Silicosis, C/La Minería 1, 33011 Oviedo, Spain; hectorg@ins.es
3 Global Service, C/Suero de Quiñones 24, 24002 León, Spain; angelpastra91@gmail.com
4 Recuperación y Renovación SL, C/Santa Susana 29, 33007 Oviedo, Spain; zenaida.hernandez@recuperacionrenovacion.com
* Correspondence: rrodrifer@uniovi.es; Tel.: +34-985104253

Abstract: The LIFE-funded European research project SUBproducts4LIFE seeks to demonstrate the use of industrial subproducts for the large-scale remediation of contaminated soils and industrial building debris connected to Hg mining. The main purpose of the present research was to ensure worker health and safety by creating a protocol for working in a highly mercury-contaminated demolition debris. A methodology consisting of sampling campaigns with a Lumex RA-915 mercury analyser, evaluating the accuracy of an empirical Hg emission model, evaluating each working task, providing recommendations for minimising the workers' exposure and calculating the maximum work period in each area was proposed. It was also shown to forecast Hg biological markers. As a result, a work protocol was developed with three scenarios which allow planning the work and forecasting the workers' mercury exposure as a function of the daily temperature, ensuring that the workers' mercury exposure is below occupational mercury levels. The working protocol allows planning the works safely with minimum exposure to gaseous mercury and working fulfilling standard requirements. Plans for restoration or new use of industrial mercury-contaminated sites have increased in recent years, and the research improves the knowledge of Hg gas distribution and worker Hg exposure.

Keywords: industrial building; mercury mining; mercury airborne; working conditions; remediation; working protocol; workers; health and safety; contamination

Citation: Rodríguez, R.; Garcia-Gonzalez, H.; Pastrana, Á.; Hernández, Z. Health and Safety Protocol for the Management of Building Demolition Waste with High Mercury Contamination. *Buildings* **2023**, *13*, 1310. https://doi.org/10.3390/buildings13051310

Academic Editors: Mariusz Szóstak and Marek Sawicki

Received: 19 April 2023
Revised: 12 May 2023
Accepted: 14 May 2023
Published: 18 May 2023

1. Introduction

Mercury has been a well-known metal since ancient times; in the past 500 years, 992,812 tons of mercury have been produced worldwide [1], with the most significant producers being Spain, the USA, Slovenia, and Italy. Since 1971, the use of mercury has been declining, and it is now prohibited in many countries. However, certain nations still produce mercury, including China, Tajikistan, and Mexico, and in other cases, mercury is produced as a subproduct. In most cases, mercury mining facilities were abandoned without any intentions for restoration, similar to other mercury-related buildings such as mercury chlor-alkali plants [2], scrap metal processing facilities [3], copper and zinc smelting factories, and weapon production buildings [4] generating tons of hazardous waste which nowadays is a significant public health issue. There are plans to decommissioning these facilities, and it is crucial to ensure the health and safety of the workers and all the people involved. This research evaluates the working conditions in heavily mercury-contaminated facilities.

1.1. Brief State of the Art

Heavy metal contamination in buildings is a major environmental problem threatening public health. Because of the issue's importance, this topic has been thoroughly researched in the scientific literature. Several categories can be used for group research. The preservation of the environment comes first. Understanding mercury gas emissions from places with high mercury concentrations has advanced significantly in the recent decades, e.g., Wang et al. [5]; Gustin [6]; Feng et al. [7] demonstrated in their findings that the Hg gas emission rates from mercury-enriched soils are significantly higher than the values observed in the background area and that the amount of gaseous mercury contributed by Hg-enriched soil in the mercuriferous belt to the atmosphere has been substantially underestimated [6,7]. As a result, many topics have been thoroughly studied, such as the concentration of mercury (Hg) in the soil and water near contaminated sites [8–10], as well as the rates of emission into the atmosphere and the distribution of the contamination around the sites affected by mercury mining [5,11–16].

Another area of study is the creation of emission models, such as those by Lindberg et al. [17] and Llanos et al. [18], which aid in analysing potential dangers associated with Hg pollution. Matanzas et al. [19] have conducted other particular investigations, such as the transfer of Hg to plants.

The impact of pollution on human health is a different area of study [20]. Both broad investigations, such as those by Kim et al. [21] and Wu et al. [22], and more focused ones, such as those by Phelps et al. [23] or Koeningsmark [24], have been conducted. The World Health Organization (WHO) states that breathing in mercury can have catastrophic consequences for the lungs, kidneys, and immunological, neurological, and digestive systems. Memory loss, neuromuscular effects, headaches, cognitive and motor problems, tremors, and insomnia are a few adverse effects of the exposure. When mercury is present in exceptionally high amounts, it has been demonstrated to produce a variety of malignancies in rats and mice [25,26]. The most poisonous forms of mercury are methylmercury and metallic mercury vapour, which can irreversibly damage the kidneys, the developing foetus, and the brain at large doses. Other effects of mercury include diminished fertility, abdominal pain, inflammatory bowel disease, ulcers, bloody diarrhoea, and intestinal flora loss. The liver, brain, and kidneys are the body's primary locations for mercury bioaccumulation [27].

The commerce and use of mercury are currently prohibited under the Minamata Convention on Mercury. Some traditional products have disappeared, such as mercury-based measurement equipment (thermometers, barometers and sphygmomanometers), light bulbs, neon sign producers, etc. On the contrary, the production has increased in other industries where workers can be exposed to mercury, such as waste and recycling or companies involved with contaminated land and bioremediation. Attempts to diminish the risk of worker exposure to mercury have made the Hg levels in workers' urine in the UK decrease from 90% (P90) of 24.7 μmol/mol creatinine in 1997 to 2.1 μmol/mol creatinine in 2019, as it has been studied by Morton et al. [28].

The remediation of contaminated sites and specifically the occupational dangers of working in these locations are the subject of another category of research. Despite the subject's significance, there are not many studies in this area. However, there have been recent studies that are pretty pertinent; for instance, Wcislo et al. [29] evaluated the human health risk assessment in restoring safe and beneficial use of contaminated areas that have been abandoned. There are some studies about biological parameters such as the level of Hg in blood, urine or hair [30–32]. The assessment and clean-up of mercury-contaminated sites and human and ecological Hg exposure and risk are covered in depth by Eckley et al. [33]. Due to the complexity of the pollution situations, González-Valoys et al. [34] suggest risk evaluations of an abandoned gold mine in Panama using combinations of indices. There is also some research on managing hazardous mercury waste [4].

Similarly, Wcislo et al. [35] assess the health risk of working in post-mining HgAs-contaminated soils. Other studies, including those by Wongsasuluk et al. [36], analyse the health risks associated with exposure to heavy metals such as arsenic in operating mines. This latter class may incorporate the current piece of work.

1.2. SUBproducts4LIFE Project

A research project called SUBproducts4LIFE is being co-funded by the European Union as a part of the LIFE initiative. By reusing industrial waste (coal ash, gypsum, blast furnace slag, and steelmaking slag) to restore polluted soils, brownfield sites and building debris connected to Hg mining in the decommissioned site of La Soterraña, the initiative seeks to demonstrate cutting-edge circular economy concepts (Figure 1).

Figure 1. Facilities of the decommissioned Hg mine, La Soterraña [37].

One of the project's goals is to ensure the safety of the workers by creating guidelines and a manual of best practices for working in these highly hazardous locations.

This study is the continuation of the previous studies on the characterisation of Hg contamination in the air:

Preliminary analysis by Garcia-Gonzalez et al. [25] was conducted, in which it was demonstrated that particle concentrations of As and Hg in the air were minimal.

Then, an empirical model was developed to predict the Hg gaseous concentration emissions at any temperature in highly contaminated areas (Rodríguez et al. [38]).

Finally, a chemical–physical model to explain the emission and diffusion of Hg at a short distance from the metallurgical plant demolition was established by Rodríguez et al. [39].

1.3. Empirical Model: Description, Planning, Risk Assessment

The starting point of the present work is the emission and diffusion empirical models determined previously. It is based on the solid relationship between emissions and soil temperature, as stated by Scholtz et al. [11] (referencing Siegel and Siegel [40], Zhang et al. [41] and Lindberg et al. [17]). Mercury concentration measurements were carried out under different climatic conditions (temperatures between 4 °C and 30 °C and without rain or wind) to evaluate the influence of temperature on the potential release of Hg in the area with demolition debris from the metallurgical plant building. All the measurements were carried out at 1.0–1.5 m above ground level because it is the recommended height for airborne pollution environmental values [42]. Figure 2A displays these values and shows that the correlation between temperature θ (°C) and concentration over debris area C_{10} (ng/m^3) can be explained well by Equation (1) [38].

$$C_{max} = C_{10} = 6759\ e^{0.0704 \cdot \theta}. \quad (1)$$

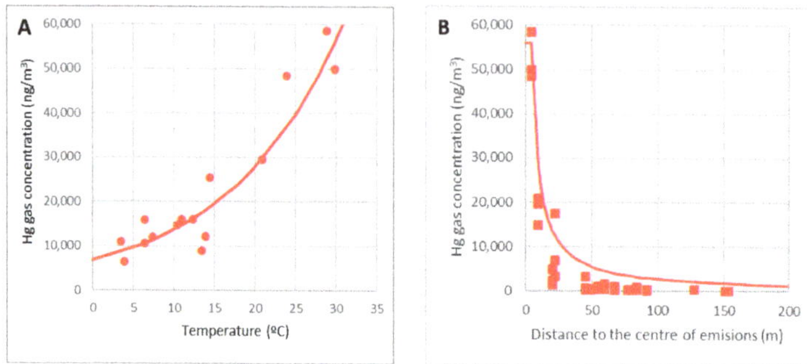

Figure 2. Relationship between temperature and concentration at point 10 (**A**) and concentration variation with the distance to the focus for θ = 25–30 °C (**B**).

As previously mentioned, measurements were also performed at various distances from the centre of the focus. Radial diffusion can explain the existence of gaseous mercury as we move away from the source of contamination. This diffusion follows a hyperbolic curve that varies inversely with distance from the source. Figure 2B represents Hg concentration between 24 °C and 30 °C. Therefore, according to the following Equation (2), also established empirically, the fluctuation of the concentration varies with the distance to the focus of the contamination:

$$C(r) = 2550\ e^{0.0704\ \theta} \left(\frac{10}{r}\right). \quad (2)$$

1.4. Objectives of the Study

The principal purpose of the research is to evaluate the work managing demolition waste highly contaminated with mercury in terms of gaseous mercury in the environment and to design a plan for health and safety at work. The steps taken to achieve this objective are the following:

- A gaseous Hg measurement campaign was carried out in the area, with a sampling period of 2 hours following the EN-689:2018 standard [43]. The workplace was differentiated based on the Hg gas concentration in the environment. The accuracy of the empirical model proposed in previous work was verified, and it was shown that temperature is the crucial variable.
- A risk prevention criterion for exposure to gaseous Hg was defined following EN-689:2018, which limits working hours in the most contaminated areas based on ambient temperature; based on it, a protocol was developed to work in health and safety conditions for workers. On the other hand, it is stated that covering the contaminated material is an effective safety measure for worker protection.
- Analysis of the actual work completed was conducted, detailing the actual activities in the riskiest areas (something that is not typically published in the specialised bibliography); based on the empirical model and the risk prevention criterion, an appropriate way to respond was defined, first during the planning of the tasks, and then during their execution, so that the quality standards in terms of health and safety are met at all times.

2. Materials and Methods

2.1. Measuring Device and Sampling Procedure

The LUMEX RA-915 sampling device had an analytical gaseous Hg range of 1–100,000 ng/m^3 (Figure 3, left). The instrument collected 10 litres of air every minute and recorded the data in an internal data logger while reporting analysis every second as well as the 10 s average. This instrument is widely used in scientific research and by reference organisations to measure gaseous mercury in various circumstances and conditions. Mercury particles removed via a filter at the equipment's inlet are not included in the LUMEX RA-915's analysis of gaseous mercury.

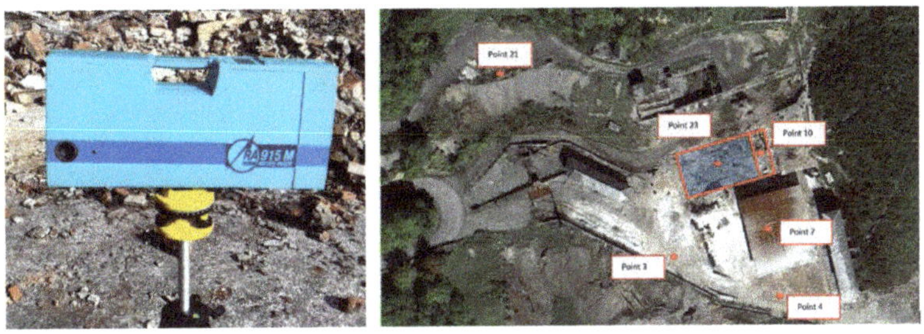

Figure 3. The Lumex analyser (**left**) and areas represented by points 10 and 23 (**right**).

In order to create work protocols, a sample procedure was designed; sampling locations and systematic monitoring campaigns were planned. A route with 22 control stations spread out along the area where the Hg gas concentrations were measured under various circumstances to establish the empirical model. Nevertheless, this study only monitored the points at which work was carried out. The study's main objective is to evaluate the location to carry out restoration and remediation plans and avoid Hg gas occupational risks.

Two points are related to the areas where demolition debris from the old metallurgical plant accumulate (Figure 3, right):

- Point 10: area with demolition debris in its original location;
- Point 23: area with demolition rubble covered with slag and ash (next to the previous one).

It is noticeable from point 10 that the concentration of gaseous Hg rises with temperature, according to the exponential pattern determined by Rodríguez et al. [38]. Since this is the site's most critical location, a thorough statistical analysis using at least six measurements—consisting of continuous monitoring of the Hg gaseous for at least two hours—must be performed to determine the conditions under which work can be conducted at point 10 according to EN-689:2018. In this instance, 24 measurement campaigns were conducted under various temperature circumstances, and the 20% occupational exposure limit value (OELV) (4000 ng/m^3) was surpassed in each.

Other points were outside the demolition debris zone, where different activities occur.

- Point 3 is an area where the store and welfare facilities are located; it is a rest area for workers.
- Point 4 is where ancillary tasks such as repair and machinery maintenance are carried out.
- Point 7 is a covered work area where the filter channels are located.
- Point 21 is an earthwork work area in the furnace slag heap.

Gaseous Hg content at these points was very low, less than 10% of the OELV, according to the preliminary analysis using the empirical model. Therefore, following EN-689:2018 using three measurements, it is possible to conclude that the gaseous Hg exposure is below the legal limits for the scenario investigated.

The study is focused on points 10 and 23 because these are the most critical areas (Figure 4). Due to Hg gas emissions varying with temperature, it was decided to test Hg gas concentrations from low to high temperatures to identify a range of emission values for the location. The campaigns had to be conducted throughout the year and the four seasons to obtain measurements at various temperatures. The campaigns continued as a result of warm days when no work was conducted and cooler days when some work was carried out, as will be discussed further below.

Figure 4. Sampling points 10 (**left**) and 23 (**right**).

The averages of Hg gas measured while workers were on site might be used to evaluate working conditions.

2.2. EN-689:2018 Standard Application

EN-689:2018 provided guidelines for an accurate sampling process [43]. A group of employees exposed to a chemical agent at a similar level while carrying out their duties is known as an SEG (similar exposure group). The goal of the standard is to ascertain whether work in a similar exposure group SEG is consistent with the OELV created for jobs requiring chemical agent exposure.

In this instance, the SEG is the workers who conduct their work in the area with metallurgical plant demolition debris. The degree of exposure is determined by the average concentration of Hg in the atmosphere over eight hours. The chemical agent is gaseous Hg, with an established OELV of 20,000 ng/m^3.

The standard specifies that measurements of exposure to the chemical agent must be conducted to ascertain whether the work is compatible with or adheres to that OELV. The standard allows carrying out only three exposure measurements when exposure is less than 10%, four measurements when exposure is less than 15% or five when exposure is less than 20% of the OELV. However, it is necessary that a minimum of 6 measurements and a statistical analysis must be undertaken when the exposure is anticipated to be greater than 20% of the OELV; this case must be applied to the analysis in this situation. According to the guideline, a measurement must monitor the exposure for a minimum of 2 h to accurately represent an entire 8 h day. The weighted average concentration of Hg gas C_k acquired from the Hg gas measurements at a specific site for two hours is considered one sample, assuming there are n samples.

For each sample, the probability p_k that the concentration is less than that of the sample C_k is calculated:

$$p_k = \frac{k - \frac{3}{8}}{n + \frac{1}{4}}. \quad (3)$$

On a log-probability paper, the observed exposure C_k values are organised in ascending order and plotted on the horizontal axis against the corresponding probabilities p_k on the vertical axis. These findings are distributed lognormally, as evidenced by the solid fit for the straight line [43].

These data must be transformed into the geometric mean GM and geometric standard deviation GSD using the following formula:

$$\ln(\text{GM.}) = \frac{\sum_1^n \ln(C_k)}{n} \rightarrow \text{GM} = \exp\left(\frac{\sum_1^n \ln(C_k)}{n}\right), \tag{4}$$

$$\ln(\text{GSD}) = \sqrt{\frac{\sum_1^n (\ln(C_k) - \ln(\text{GM.}))^2}{n-1}} \rightarrow \text{GSD} = \exp\left(\sqrt{\frac{\sum_1^n (\ln(C_k) - \ln(\text{GM.}))^2}{n-1}}\right). \tag{5}$$

This test compares the 95th percentile distribution of the results with the 70% upper confidence limit (UCL). Geometric mean (GM) and standard deviation (GSD) determine UCL. Then, the variable UR is calculated:

$$U_R = \frac{\ln(\text{OELV}) - \ln(\text{GM.})}{\ln(\text{GSD})}. \tag{6}$$

The UT variable, tabulated under n (Table 1), must be used to verify the UR value.
If UR $\geq$ UT, then the conclusion is compliant with the OELV.
If UR < UT, then the conclusion is non-compliant with OELV.

Table 1. UT variable tabulation according to UNE-EN-689:2018 standard.

n	U_T	n	U_T	n	U_T	n	U_T	n	U_T
6	2.187	11	1.981	16	1.905	21	1.863	26	1.836
7	2.12	12	1.961	17	1.895	22	1.857	27	1.832
8	2.072	13	1.944	18	1.886	23	1.851	28	1.828
9	2.035	14	1.929	19	1.878	24	1.846	29	1.824
10	2.005	15	1.917	20	1.870	25	1.841	30	1.820

Statistical analysis results can be used to decide the interval from the initial compliance test before periodic measurements. Assuming that the results indicate that the SEG complies with a fraction j of the OELV, the procedure is as follows. First, we need to calculate the value of j and then the interval T (months):

$$j = \exp(U_T \ln(\text{GSD}) + \ln(\text{GM}) - \ln(\text{OELV})). \tag{7}$$

- If j < 0.25, then T = 36 months.
- If 0.25 < j < 0.5, then T = 30 months.
- If j > 0.5, then T = 24 months.

Nevertheless, as demonstrated later, this procedure cannot be applied in the case study.

2.3. Biological Markers

Mercury is present in the body in three different forms. The first is as a pure element, without combining ("elemental mercury"), the second is by forming salts or inorganic compounds such as chlorides, sulfides, sulfates, and nitrates ("inorganic mercury"), and the third is by combination with Carbon to form organic compounds such as methylmercury ("organic mercury"). A study performed in Asturias in 2013 [44] showed that the relationship between inorganic Hg and total Hg in the blood is, on average, 50–55%.

Biological marker and regulated biological marker levels are fundamental tools in the control of worker health. In Spain, the legally accepted biological markers are the concentration of total inorganic Hg in blood $C_{Hg\text{-}B}$ (µg/L) and the concentration of total inorganic Hg in urine related to creatinine $C_{Hg\text{-}U}$ (µg/g). The limit values established by the National Institute for working safety and health (Instituto Nacional de Seguridad y Salud en el Trabajo, INSST) are recorded in Table 2.

Table 2. Biological mercury levels according to INSST [45].

Fluid	Biological Indicator	Moment of Sampling	Year of Update
Urine	Total inorganic mercury 30 μg/gcreatinine	Before working hours	2013
Blood	Total inorganic mercury 10 μg/L	End of the workweek	2013

Drake et al. [46] state, using data from workers exposed to mercury in gold mining operations, that creatinine is present in a concentration between 0.5 and 3 g per litre of urine. On the other hand, standard medical information shows that a typical value for creatinine production per day is between 14 and 26 mg/day of creatinine per kg of body mass. Taking the average of 20 mg/day × kg and assuming a 75 kg worker drinks 1.5 L of water daily, we obtain 20 × 75/1.5 = 1000 mg/L = 1 g/L, that is, 1 g of creatinine per litre of urine. Therefore, regarding average workers, the expression $C_{Hg\text{-}U}$ (μg/gCre) ≈ $C_{Hg\text{-}U}$ (μg/L) can be used, and the limit for Hg in the urine of 30 μg/gCre is equivalent to the limit of 30 μg/L.

Due to data protection issues, this work does not include any actual results related to biological markers. However, estimating the value of biological markers is very useful in the planning phase, and this is included in the work protocol. The extensive experience obtained in the Almadén mines, Spain, over hundreds of years of mercury exploitation is used for this.

Regarding the relationship between the concentration of Hg in urine and Hg in the work environment, Español Cano [47], citing the WHO, said that there is a proportionality between the inorganic Hg concentration in urine $C_{Hg\text{-}U}$ (μg/L) and the Hg concentration in the air $C_{Hg\text{-}A}$ (μg/m^3).

$$C_{HgU} = k\, C_{Hg\text{-}A}, \tag{8}$$

where the proportionality factor k takes the value 0.7, which is in accordance with the results of other researchers, although the value of k is slightly different. For example, Yoshida [31] provided data from which it can be deduced that the concentration of inorganic Hg in urine $C_{Hg\text{-}U}$ is a function of individual exposure, weighted in time $C_{Hg\text{-}A}$ with the coefficient k = 0.13. In the same way, from the data of Iden et al. [48], in a study among workers of a fluorescent lamp factory, the proportional factor is 0.11. Then, we can assume an average value of k = 0.10 and, using the units ng/m^3 for the concentration in the air, for inorganic mercury, we obtain:

$$C_{Hg\text{-}U} = 0.0010\, C_{Hg\text{-}A}. \tag{9}$$

Drake et al. [46] proposed a linear relationship between neper logarithms ln($C_{Hg\text{-}U}$) and ln($C_{Hg\text{-}A}$), useful even for very high Hg concentration in the environment. In the range from 5000 to 60,000 ng/m^3, both Equations (8) and (9) provide similar results, and both can be used as an approach.

On the other hand, some authors have found a relationship between Hg in urine and Hg in blood. The works of Almadén, Español Cano [47] and Tejero Manzanares [49] established the relationship between the concentrations of inorganic Hg in urine $C_{Hg\text{-}U}$ and in blood $C_{Hg\text{-}B}$ (both in μg/L):

$$C_{Hg\text{-}U} = \frac{1}{30}\left(C_{Hg-B}\right)^2 + 1.75\, C_{Hg\text{-}B}, \tag{10}$$

from which follows

$$C_{Hg\text{-}B} = \frac{-1.75 + \sqrt{1.75^2 + 4\, C_{Hg\text{-}U}}}{2}. \tag{11}$$

According to Español Cano [47] citing the WHO, there were no specific symptoms below 35 μg/L mercury in the blood (when ambient concentration is about 0.050 mg/m^3 or 50,000 ng/m^3). As seen, the legal limit for Hg blood concentration is 10 μg/L.

In a study by Yoshida [31] with a sample of workers in a mercury thermometer factory, the relationship between the concentrations of inorganic Hg in urine $C_{Hg\text{-}U}$ and blood $C_{Hg\text{-}B}$ (both in nmol/L) was

$$C_{Hg\text{-}U} \approx 3.0\ C_{Hg\text{-}B}. \tag{12}$$

When the exposure is below 100 μg/m^3, it fits with Almadén ambient mercury concentration values of less than 100 μg/m^3, which implies blood concentration values below 400 nmol/L or 80 μg/L (80 micrograms of inorganic Hg per litre of blood). In this case, the following formula can be used:

$$C_{Hg\text{-}B} \approx\ 0.3\ C_{Hg\text{-}U}. \tag{13}$$

A very relevant topic is the frequency of biomonitoring controls. Some references can be found in the literature related to mining. A typical one is related to very hard working conditions in mercury mining. For example, Kobal and Dizdarevic [50] described that the miners' biomonitoring in 1997 was performed every month. There are also some references related to gold mining operations, such those of as Drake et al. [46] or Ramírez [51]; in both cases, the recommendation is to control the biomarkers every six months.

Related to the chlor-alkali industry, Lovejoy and Bell [52], in the year 1973, recommended urine analysis every four months, assuming Hg concentration in urine between 100 and 250 μg/L, increasing the frequency if the concentration increases. More recently, in 2010, Euro Chlor [53] recommended a frequency of sampling urinary mercury concentration of 2 controls/year (every six months) when it is under 20 μg/g creatinine and higher than 4 controls/year (every three months) when it is higher than this value.

Manson [54] analysed different industries in the UK. They recommended a urinary sampling interval for urinary Hg between 1 and 3 months, consistent with the reported toxicokinetics of Hg excretion, according to the medical guidance from the Health and Safety Executive on Hg Exposure MS 12 (1996) [54].

Another practical empirical law is the decrease along the time of the urinary Hg concentration after cessation of exposure. After mercury exposure, Hg in the body decreases. Starting from the initial concentration of Hg in urine $C_{Hg\text{-}U}(0)$, that is, the one it has when exposure ceases, the concentration $C_{Hg\text{-}U}(t)$ can be estimated after a time t (days) using the following Ellingesen [55] expression:

$$C_{Hg\text{-}U}(t) =\ C_{Hg\text{-}U}(0) \times 10^{-\beta\ t}. \tag{14}$$

The analysis of several independent individuals provides a mean value β = 0.0046. This formula allows determining the concentration in urine that will be present when workers exposed to mercury return to work after a break of t days $C_{Hg\text{-}U}(t)$, if, at the time the holiday began, the concentration was $C_{Hg\text{-}U}(0)$. On the other hand, taking a limit for $C_{Hg\text{-}U}$ =30 μg/L also allows for estimating the total rehabilitation time, that is, the time necessary to eliminate Hg from the body.

Finally, it is interesting to comment on Hg concentration in hair. As many authors have already verified [47], the determination of mercury concentration in hair is not suitable for controlling occupational exposure due to external contamination and because mercury can stably bind to the -SH groups of keratin. In addition, hair analysis provides the value of mercury accumulated over several weeks or even months, not from day to day. However, carrying out Hg measurements in hair can be very useful since they can be conducted with the same environmental control equipment, faster than urine or blood analysis.

The concentration of Hg in hair is not a biological marker accepted by the authorities; nevertheless, it is interesting to observe relationships between Hg in hair and blood or urine. Español Cano [47] points out that a linear correlation has been established between the levels of mercury in hair $C_{Hg\text{-}H}$ and the total mercury in blood $C_{Hg\text{-}B}$, with relationships varying from 300 to 500. According to Diez et al. [56], citing Phelps et al. [23] and the World Health Organisation (WHO), the Hg ratio in hair to total Hg in the blood is 250 to 1. Assuming a blood density of 1.06 kg/L, there are

$$C_{Hg\text{-}B} \approx 4\, C_{Hg\text{-}H}, \tag{15}$$

where $C_{Hg\text{-}B}$ is in μg/L, and $C_{Hg\text{-}H}$ is in mg/kg (or what is the same ppm).

After Packull-McCormick et al. [57], using the 250:1 ratio derived by the World Health Organisation to estimate blood mercury concentrations from hair mercury concentrations substantially overestimates blood mercury concentrations. However, geometric mean site-specific hair-to-blood mercury ratios can estimate central tendency measures for blood mercury concentrations from hair mercury concentrations at a population level.

Diez et al. [56] cited Harada et al. [58] to establish a limit of 10 ppm for the concentration of Hg in hair in workers exposed to Hg. The WHO sets no specific symptoms below 35 μg/L in the blood (ambient concentrations of 0.050 mg/m^3 or 50,000 ng/m^3). As can be seen, there is consistency with the limits since 10 ppm in hair corresponds to approximately 40 μg/L, which is of the same order.

3. Results and Discussion

3.1. Results in Areas with Demolition Debris from the Metallurgical Plant Building

There are two areas where demolition debris accumulates:

- Point 10: area with demolition debris in its original location;
- Point 23: area with demolition rubble covered with slag and ash (next to the previous one).

In previous studies, the following conclusion was reached: the demolition rubble from the metallurgical plant, point 10, behaved as an emitting source. At this point, the emissions of Hg gas and its concentration in the air strongly depended on the temperature. With temperatures around 30 °C, the average Hg gas concentration in the rubble air reached almost 60,000 ng/m^3. Figure 5 shows the rubble area and a Hg gas record obtained at 15.5 °C; the average concentration reached 13,680 ng/m^3.

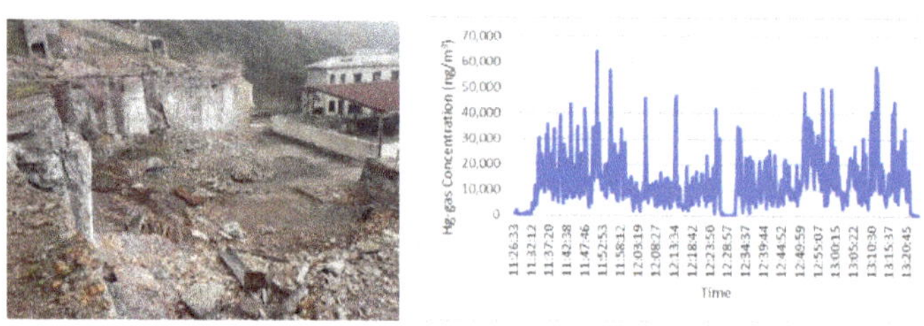

Figure 5. The area at point 10 and a record of Hg gas concentration.

One of the tasks to be carried out in the project is to use industrial waste (blast furnace slag and coal ashes) to improve the environment of the old mining–metallurgical facilities. For this reason, a large part of the demolition rubble was covered with slag and ashes, creating a new work zone named point 23.

A photo of the new area is shown in Figure 6. It is verified that the layer of slag and ash stops the emission of the rubble that is covered. For this reason, the area begins behaving like other surrounding points in which the Hg gas concentration in the air results from the dispersion of Hg gas from the rubble that remains uncovered. Figure 6 shows a record of the monitoring of gaseous Hg at point 23 on the same day. With a temperature of 15.5 °C, the average concentration of Hg in the air is only 814 ng/m^3 (less than 10% of the OELV), utterly different from point 10 previously analysed.

Figure 6. The area at point 23 and a record of Hg gas concentration.

Table 3 and Figure 7 summarise the analysis for points 10 and 23. In the case of point 10, it is seen that the concentration of gaseous Hg increases with temperature, following the same exponential pattern previously determined by Rodríguez et al. [38]. Since this is the critical point of the site, 24 measurement campaigns were carried out in different temperature conditions, and in all cases, the 20% OELV (4000 ng/m^3) was exceeded. Therefore, to determine under which requirements works can be carried out in point 10 following EN-689:2018, a complete statistical analysis must be carried out (see the next section).

Table 3. Measurements in points 10 and 23.

Point 10		Point 23	
Temp. (°C)	**C_{10} (ng/m^3)**	**Temp. (°C)**	**C_{23} (ng/m^3)**
3	8505	16.2	2394
4	7420	15.5	814
13.2	9915	22	445
11.1	12,738	19.2	641
11.5	11,195	28	3912
8.1	10,800	30	3317
21.2	35,883	22	1383
14.5	14,481	-	-
23	31,404	-	-
14	13,680	-	-
15.5	13,570	-	-
15.8	17,000	-	-
17.6	15,805	-	-
22	19,606	-	-
22.5	16,600	-	-
19.2	17,315	-	-
21.1	22,323	-	-
30	52,313	-	-
30.6	44,308	-	-
28	48,334	-	-
30	53,095	-	-
22	32,327	-	-
27.4	56,746	-	-
20	10,082	-	-

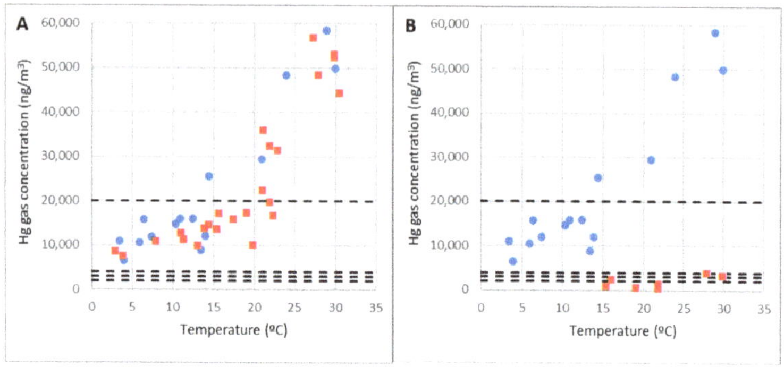

Figure 7. Gaseous Hg concentration at point 10 (**A**) and point 23 (**B**), current (red) and previous measurements (blue).

Figure 7A shows that below 20 °C, the average readings are below the OELV (20,000 ng/m^3). Likewise, it is observed that above 25 °C, the temperatures are higher than the OELV. However, there is a range of temperatures between 20 °C and 25 °C in which it can be above or below, and we believe that the explanation for this is related to the weather.

Temperatures around 30 °C are relatively infrequent in Asturias and occur with very stable conditions, with clear skies and no wind. The high solar radiation favours the emission of Hg gas, and the absence of wind prevents its dispersion, thus increasing its concentration in the air. The same happens with the coldest days in Asturias, less than 5 °C, and typical winter days with clear skies that favour frost and little wind. However, in the more normal temperature range of 10 °C to 15 °C, rain, wind, clouds, or clearings may cause Hg gas emission to fluctuate.

Wind influence is evident. With temperatures around 20 °C and more or less stable conditions, concentrations around 20,000 ng/m^3 have been found. However, on one day with a temperature of 20 °C but with strong gusts of wind, the concentration drastically dropped to 10,000 ng/m^3.

It should be noted that the work with the rubble was always carried out at temperatures below 15 °C, in conditions compatible with the regulations. The measurements obtained with temperatures above 15 °C were recorded in the absence of any other work and only to determine how the temperature influenced the Hg gas concentration. For the measurements carried out with very high concentrations (always less than 60,000 ng/m^3), the presence of personnel was reduced to about 2–3 min that it took to place the measurement equipment on top of the rubble (always less than the 15 min allowed by the legislation).

Another critical aspect, typical in underground coal mining, is that no worker carries out a task alone in this area; to work in this area, there must be at least two workers together.

At point 23, it is verified that with temperatures around 20 °C, the concentration of gaseous Hg in the environment is moderate, similar to other site points. Therefore, five measurements were carried out to analyse the exposure to gaseous mercury. All of them were below 15% of the OELV, showing that the work at point 23, on the rubble covered with ash and slag, is compatible with the mercury legislation on air, and it is up to temperatures of the order of 20 °C.

Since work on the rubble was never carried out at temperatures above 15 °C, the standard does not impose more samples to be carried out to analyse the OELV limit. However, two more measurements were conducted to see the effect of slag and ash, reducing and even eliminating gaseous mercury emissions at the highest temperatures in Asturias (around 30 °C).

In conclusion, the concentration of gaseous Hg dropped drastically after covering the rubble with slag and ashes. It should be noted that covering the contaminated material is an effective safety measure for protecting workers.

3.2. Application of the EN-689:2018 Standard and Development of a Risk Prevention Criterion

As previously stated, to determine under what conditions work can be carried out in point 10 following the EN-689:2018 standard, the statistical analysis defined in the same standard must be carried out.

There are 24 samples of Hg gas in the air in the area of demolition debris $C_k = C_{10}$ (ng/m^3) taken in the temperature range 5 °C to 30 °C. A first statistical analysis of all of them already provides precious information. Following the standard, the graph of Figure 8A is represented, which indicates that conditions cannot be considered homogeneous in the temperature range of 5 °C to 30 °C. The calculation of the parameters UR = 0.028 < UT = 1.846 shows no compliance with the OELV.

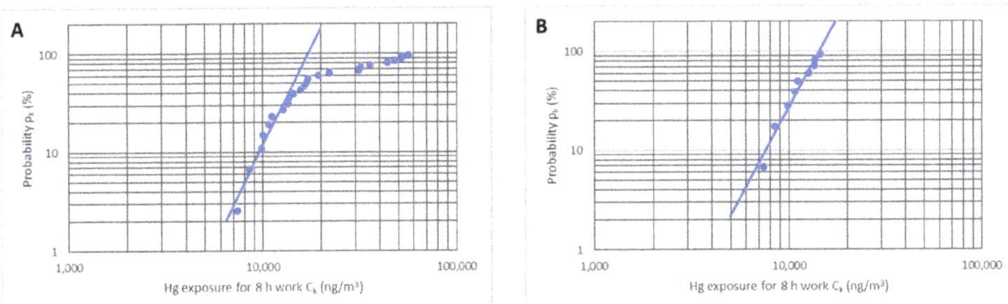

Figure 8. Analysis of the log-normal distribution for the ranges 5–30 °C (**A**) and 5–15 °C (**B**).

It is also noted that the graph indicates that there are different sets of conditions or trends and that temperature may be the critical variable that defines the difference between them. Up to $C_k \approx$ 14,000 ng/m^3, a line representing more or less homogeneous conditions can be fitted. However, the data do not follow that trend from that concentration, which could indicate other situations. The concentration of 14,000 ng/m^3 is reached at temperatures around 15 °C. Therefore, homogeneous working conditions could be working at temperatures below 15 °C.

Repeating the analysis with the nine data obtained in approximate conditions with θ < 15 °C and C_k < 14,000 ng/m^3, we see that they conform to a log-normal distribution (Figure 8B) with a correlation coefficient r^2 = 0.94. The calculation of the parameters UR = 2.562 > UT = 2.035 indicates compliance with the OELV. The work of 8 h/day and 40 h/week follows the OELV = 20,000 ng/m^3.

On the other hand, the calculation provides j = 0.89, meaning that the period until a new reassessment should be T = 24 months. However, the dependence of the mercury concentration on the ambient temperature makes the working conditions change quickly, and then this interval must be diminished. Due to this, the calculation of T based on j will not be carried out in the following.

It is clear that when C_k > OELV = 20,000 ng/m^3, the conditions for working 8 h/day and 40 h/week are not met. As seen in Figure 7A, this concentration is reached at 20 °C; below that temperature, the concentration is always lower than the OELV.

It is necessary to analyse the conditions with temperatures between 15 °C and 20 °C to check for limitations. Using the six available data obtained between 15.5 °C and 21.1 °C, it is concluded that in this temperature range, there is no compliance with the OELV. Therefore, working 8 h/day and 40 h/week is unacceptable regarding occupational mercury limits. In addition, there is also no conformity if the six available data are used with temperatures between 14.5 °C and 20 °C. This is because, as we choose a range closer to 20 °C, the concentration becomes closer to the OELV, and non-conformance is more likely than at lower temperatures. The solution to this problem is to limit the number of working hours per day so that the equivalent dose is less than the OELV.

Let us assume that 6 h are worked every day at point 10, 1 h at point 3 (we assume a concentration of $C_3 = 1000$ ng/m^3) and 1 h at point 7 ($C_7 = 4000$ ng/m^3). If $C_{k1} \ldots C_{k6}$ are the six values used for the analysis, assuming that 8 h are worked, the new representative values to carry out the analysis are

$$C'_{ki} = \frac{6 \times C_{ki} + 1 \times 1000 + 1 \times 4000}{8}. \quad (16)$$

If the six values are taken between 14.5 °C and 20 °C, it can be found that there is compliance with the OELV (UR = 2.940 > UT = 2.187). However, if the six representative points are taken between 15.5 °C and 21.1 °C, working 6 h per day is not possible, even if the remaining two hours are rested at point 3. In this case, the values to be used are

$$C'_{ki} = \frac{6 \times C_{ki} + 2 \times 1000}{8}. \quad (17)$$

The analysis result is UR = 1.978 < UT = 2.187; that is, there is no compliance with the OELV.

Only if we reduce the number of working hours in point 10 to 5 is compliance fulfilled, even assuming that the remaining three hours are worked in point 7 ($C_7 = 4000$ ng/m^3):

$$C'_{ki} = \frac{5 \times C_{ki} + 3 \times 4000}{8}. \quad (18)$$

Figure 9A represents the log-normal distribution for this case. The calculation provides UR = 2.502 > UT = 2.187, which means compliance with the OELV = 20,000 ng/m^3.

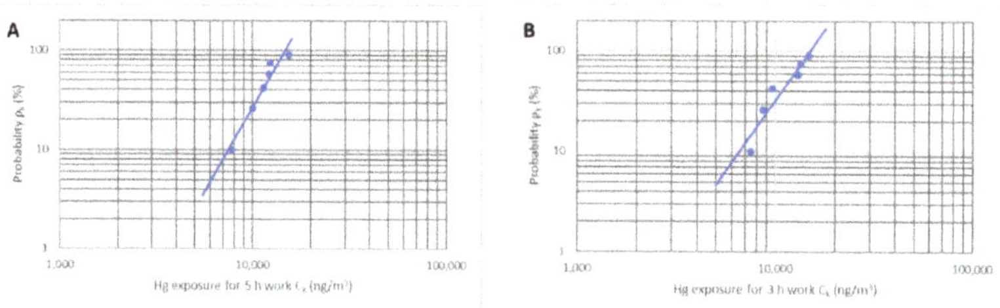

Figure 9. Analysis of the log-normal distribution for the ranges 15 °C to 20 °C (**A**) and 20 °C to 25 °C (**B**).

Therefore, the best solution to be able to carry out work at point 10 when the temperature is between 15 °C and 20 °C is to limit the work at that point to only 5 h/day, with a possiobility to carry out work at any other point in the remaining three hours.

The analysis can now be repeated in the interval of 20 °C to 25 °C. In that interval, the concentration fluctuations can be huge; thus, the six available data vary between 16,600 and 35,883 ng/m^3.

Repeating the previous analysis, it is found that, with temperatures between 20 °C and 25 °C, at point 10, work can be carried out for up to a maximum of 3 h, provided that the remaining 5 h were worked at points with a concentration of less than 2000 ng/m^3 (points 4 or 21). The values to be used are calculated as follows:

$$C'_{ki} = \frac{3 \times C_{ki} + 5 \times 2000}{8}. \quad (19)$$

The log-normal distribution is depicted in Figure 9B. Considering that UR = 2.245 > UT = 2.187, it is concluded that there is compliance with the OELV = 20,000 ng/m^3.

Based on the results of the previous analysis, it is possible to define a Risk Prevention Criterion that appears in Table 4. Three work scenarios are defined with which the classic green–orange–red colours are associated.

Table 4. Risk prevention criterion for exposure to Hg according to the EN-689:2018 standard.

Hg Gas Concentration on the Rubble C_{10} (ng/m^3)	Scenario	Hg Gas Concentration Ranges C_{10} (ng/m^3)	Temperature Ranges θ (°C)	Permissible Working Time EN689 Standard (h/day)	Working Time Assumed in the Protocol (h/day)	Hg Gas Monitoring Interval	Additional Measures
≤10,000	Green	5000–7500	0–5 °C	8	6 *	1 month	No
		7500–10,000	5–10 °C	8	6 *	1 month	No
10,000–20,000	Orange	10,000–15,000	10–15 °C	8	6 *	1 month	No
		15,000–20,000	15–20 °C	5	5	2 weeks	Recommended
≥20,000	Red	20,000–40,000	20–25 °C	3	3	2 weeks	Recommended
		40,000–60,000	25–30 °C	1 **	1 **	Continuous	Necessary

(*) 1 h breaks are mandatory after 2 h work for using specific Hg gas masks; (**) 1 h distributed into 4 periods of 15 min/h for 4 h.

The green stage occurs when the Hg concentration is below 10,000 ng/m^3 (typically with the temperature below 10 °C). In this case, working in the demolition rubble area for 8 h/day (standard shift) is possible. However, it must be said that gas semi-masks are always used at work, and the authority recommends a 0.5 h break after continuous work for 1 h (although due to operational reasons, in our case, a 1 h break is adopted after 2 h of continuous working). Therefore, even if the conditions are favourable, the maximum number of working hours per day is 6 h/day.

The orange scenario is defined in the Hg concentration range of 10,000 to 20,000 ng/m^3 (temperature range of 10 °C to 20 °C). If it can be ensured that the Hg in the air is below 15,000 ng/m^3, the considerations of the green procedure are valid. However, if the Hg concentration exceeds 15,000 ng/m^3 (typically above 15 °C), the maximum working hours are 5 h daily. Considering it is possible to work 6 h/day in the most favourable conditions, the orange scenario does not represent a drastic change in routine work.

Above 20,000 ng/m^3 (typically above 20 °C), we enter the red stage. Some work in rubble areas is allowed only in the range of 20,000 to 40,000 ng/m^3 (20 °C to 25 °C), but with some extreme restrictions since it is not possible to work more than 3 h/day. From above 40,000 ng/m^3 (typically above 25 °C), work with demolition rubble is prohibited without taking additional measures. Only particular jobs (such as sampling) could be carried out with a maximum of 1 h/day, distributed into four periods of 15 min/hour for 4 h.

Work with temperatures between 20 °C and 25 °C, with concentrations that could reach 40,000 ng/m^3, represents the worst conditions in which practical work can be carried out following the legislation. The limitation to 3 h/day of work, or 60 h/month, is the same order as the 48 h/month of normal conditions in Spanish mercury mining, with average concentrations of 121,000 ng/m^3 [47]. It is seen that this criterion follows the experience in the mercury mines, although, by the requirement of the legislation, it is even more conservative.

This risk prevention criterion has the advantage of being simple and mnemonic: the mercury concentrations in the environment at typical working temperatures, 10 °C, 15 °C and 20 °C, are 10,000, 15,000 and 20,000 ng/m^3, respectively.

Finally, from all of the above, it is deduced that a fundamental variable, temperature, deterministically controls gaseous Hg emissions. For this reason, setting a temperature range is equivalent to defining homogeneous working conditions. The fluctuations in the concentration of Hg gas within that temperature range would be determined by other variables related to the climate, such as wind, solar radiation or rain.

It must be kept in mind that the work has to be planned and carried out according to the Hg gaseous concentration. The temperature variable allows the quick estimate of the Hg concentration and decision on the work during the planning and execution phases. Nevertheless, if continuous monitoring enables us to determine that the Hg concentration is less than 10,000 ng/m^3, the work could be carried out according to the green scenario, although the temperature is higher than 10 °C. It can occur, for example, if the strong wind makes the concentration under the expected Hg concentration for this temperature.

Respecting the frequency of the monitoring of Hg in the air, we must remember that at our latitude, the temperature can be over 15 °C in any period of the year, even during the winter. Consequently, the works for the next days must be planned based on the Spanish Meteorological Agency's (AEMET) forecast. Simultaneously, it is recommended to carry out control monitoring every month, according to Spanish Mercury Technological Center [59]. It is also recommended to diminish the interval to 2 weeks with medium and high temperatures.

When the temperature is over 15 °C, additional measures can be recommended, especially if we want to increase the working time to 6 h. Some are not expensive or difficult to implement, such as covering the contaminated rubble (in our case, with ash and slag), using a long-arm excavator that allows the driver to be far from the rubble, and using a full mask with HgP3 filters instead of a semi mask with HgP3 filters, etc. In the case of taking measures that diminish the concentrations of Hg in the working place, the working time could be increased.

The Spanish law does not define the frequency of the workers' biomonitoring, and the law only requires that all workers must be checked at least once a year. For specific risks, the number of biomonitoring controls must be defined by the company's occupational health service physician after assessing the workplace risks. The Mercury Technological Center has established for its employees who work in exposed places to carry out urine analysis every month and blood analysis every 3 months, which can be taken as a first reference.

3.3. *Analysis of the Actual Work with the Demolition Rubble of the Metallurgical Plant*

As can be deduced from the previous analyses, from the point of view of worker health and safety, the work with demolition rubble is the most critical. It requires careful planning and strict onsite control measures. Typically, only three operators work, and in some cases four (Table 5). External personnel only participate in placing the HDPE waterproof sheet (another four people), although they only work one day. Several illustrative photographs of these tasks are shown in Figures 10–12.

Table 5. Tasks carried out in the rubble area (point 10 and surroundings).

Task	Work to Be Carried Out	Plant Machinery	Number of Workers
1	Debris removal, demolition of benches and scaling the work area	Backhoe, excavator with a hydraulic hammer, diesel generator with irrigation equipment	3
2	Recessing and spreading of the concrete slab	Backhoe, concrete tank, diesel generator, compressor	3
3	Construction of perimeter retaining wall and placement of drainage pipe	Backhoe, diesel generator, compressor, concrete mixer	4
4	Waterproofing with HDPE sheet	Backhoe, diesel generator, compressor, HDPE welding equipment	3 (+4)
5	Application slag surface layer on an HDPE sheet	Backhoe, track chain excavator, tipper truck	3
6	Filling of the treatment area with demolition rubble	Backhoe, track chain excavator	3
7	Filling of the treatment area with furnace slag	Backhoe, track chain excavator, tipper truck	3
8	Ash coating	Backhoe, track chain excavator, tipper truck	3

Figure 10 (left) shows the development of task 1, where the rubble is removed with a backhoe excavator to carry out task 2, building a concrete slab, Figure 10 (right). The photograph of Figure 10 (right) shows the personal protective equipment used by the workers. In addition to the standard personal protective equipment PPE (hard hat, safety boots), the specific protection necessary to work in the area can be seen: integral protection cover, gloves protecting against chemical risks, safety glasses and a half mask filtering gases and mercury vapours HgP3.

Figure 10. (**Left**): task 1 (debris removal); (**Right**): task 2 (concrete slab construction).

Figure 11 (left) shows the perimeter wall construction, and Figure 11 (right) shows the placement of the HDPE sheet. It is the task that requires the most personnel. Firstly, the Hg gaseous concentration meter in the environment is part of the real-time monitoring frequently carried out as a control measure on site.

Figure 11. (**Left**): task 3 (perimeter wall construction); (**Right**): task 4 (HDPE sheet placement).

Figure 12 (left) illustrates task 6 of filling with demolition rubble, and the same Figure 12 (right) shows the work of tasks 5, 7 and 8 of filling with by-products (slag and ashes transported in tip trucks).

Figure 12. (**Left**): task 6 (excavator in rubble), (**Right**): tasks 7, 8 (dump truck).

As seen below, the knowledge provided by the previous studies makes it possible to carry out the planning and control of the work systematically and relatively simply. The procedure for planning the work with the rubble (point 10) is as follows:

- The dates on which the work will be carried out are defined.
- The temperature θ (°C) is defined with the average monthly value of the last 10 years according to the Spanish Agency for Meteorology (Agencia Española de Meteorología, AEMET).
- The allowed work period n_{max} = 6 h/day is established as a first approximation.

- The concentration at point 10 and the concentration at other alternative points are estimated (for simplicity, it is assumed that they are at least 50 m from point 10):

$$C_{max} = C_{10} = 6759\ e^{0.0704\ \theta}, \tag{20}$$

$$C_i = C(50) = 2550\ e^{0.0704\ \theta} \left(\frac{10}{50}\right). \tag{21}$$

- The weighted average exposure to gaseous Hg in an 8 h workday is estimated as

$$C_{eq} = \frac{n_{max} C_{10} + (8 - n_{max})\ C_i}{8}. \tag{22}$$

- The average concentration of inorganic Hg in urine and blood is estimated:

$$C_{Hg\text{-}U} = 0.001\ C_{eq}, \tag{23}$$

$$C_{Hg\text{-}B} = \frac{-1.75 + \sqrt{1.75^2 + 4\ C_{Hg\text{-}U}}}{2}. \tag{24}$$

- It is checked if the C_{eq} weighted average exposure is in accordance with the OELV = 20,000 ng/m^3.
- The concentrations in urine $C_{Hg\text{-}U}$ and blood $C_{Hg\text{-}B}$ are compared with the OELV 30 µg/L and 10 µg/L.

Table 6 summarises the calculations. In all cases, the control variables are below the OELV limits imposed by the legislation; the work planned this way would follow the legislation.

Table 6. Analysis of the work to be carried out in the planning phase.

Task	Year	Month	Temp. (°C)	Working Hours (h/day)	C_{10} (ng/m^3)	C_i (ng/m^3)	C_{eq} (ng/m^3)	$C_{Hg\text{-}U}$ (µg/L)	$C_{Hg\text{-}B}$ (µg/L)	C_{eq}/20,000	$C_{Hg\text{-}U}$/30	$C_{Hg\text{-}B}$/10
1	1	October	14.8	6	19,159	1446	14,731	14.7	3.1	0.74	0.49	0.31
2	2	February	8.1	6	11,955	902	9191	9.2	2.3	0.46	0.31	0.23
3	2	February	8.1	6	11,955	902	9191	9.2	2.3	0.46	0.31	0.23
4	2	March	9.7	6	13,380	1010	10,287	10.3	2.4	0.51	0.34	0.24
5	2	October	14.8	6	19,159	1446	14,731	14.7	3.1	0.74	0.49	0.31
6	2	October	14.8	6	19,159	1446	14,731	14.7	3.1	0.74	0.49	0.31
7	3	February	8.1	6	11,955	902	9191	9.2	2.3	0.46	0.31	0.23
8	3	February	8.1	6	11,955	902	9191	9.2	2.3	0.46	0.31	0.23

On the other hand, using the same procedure described, a work control can be carried out simultaneously with the development of the works or retrospective analysis of the work once completed, verifying that the safety and health parameters are met.

In the actual case of the SUBproducts4LIFE project, the work was carried out in two phases. The first half of the initial area was worked on, and then the other half, so the tasks were repeated, and their number increased. Part of the monitoring and control of Hg gas was carried out while the work was carried out, although it is now presented as a retrospective analysis. In this case, the data used are the real data, and the risk prevention criteria were used as a guide since they were defined with the legal OELV values of the INSST measuring the mercury in the air according to the EN-689:2018 standard. This demonstrates that the work is compatible with that OELV.

The process was the following:

- The best months to work were chosen: those in which the average temperature is below 15 °C according to the prevention criteria (in Asturias, all except those in summer, June, July, August and September).

- The daily temperature was recorded, or the average daily temperature in the area provided by the meteorological services (AEMET) was used. The graph of Figure 13A shows the daily variation of the real temperature, as well as the average temperature of the period.
- The allowed work time was established according to the prevention criteria: n_{max} = 6 h/day as the temperature is always $\theta < 15$ °C.
- The concentration in point 10 was estimated based on the prevention criteria. In Figure 13B, it is seen that also the curve of the empirical model (dotted) could be used to define the concentration at point 10 at a given temperature; however, as said, it is easier to use the concentration specified by the risk prevention criteria (solid line) according to EN-689:2018.
- The equivalent exposure C_{eq} was calculated assuming that, with the regulatory breaks, 6 h/day was spent at point 10, 1 h/day at point 3 and 1 h/day at point 7 (the most unfavourable work point). Assuming $C_3 \leq 1000$ ng/m^3 and $C_7 \leq 4000$ ng/m^3, the weighted average exposure to gaseous Hg in an 8 h shift was estimated.

$$C_{eq} = \frac{6\,C_{eq} + C_3 + C_7}{8}. \tag{25}$$

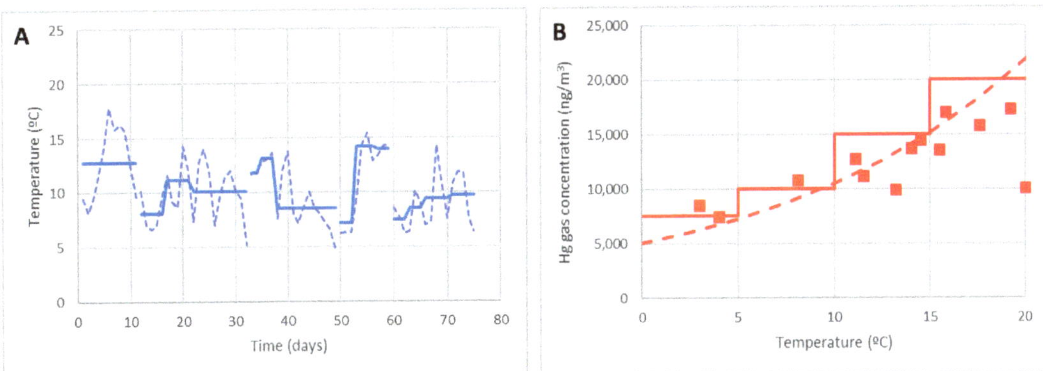

Figure 13. (**A**): Daily temperature (dotted line) and the average for each period (continuous line). (**B**): Estimation of the concentration with the empirical model (dotted line) and with the risk prevention criterion (continuous line).

It was checked whether the C_{eq} weighted average exposure was in accordance with the OELV = 20,000 ng/m^3.

Table 7 shows the results. In addition to the indicated data, those significant periods without work concerning the body's capacity to eliminate Hg were also noted. The general result is that the average exposure of a worker is approximately 50% of the OELV maximum permitted dose, so it must be concluded that the work was carried out in conditions in compliance with the legal limits.

Table 7. Retrospective analysis of the work performed.

Phase	Task	Year	Month	Duration (Days)	Days between Works	Temp. (°C)	Working Hours (h/day)	C_{10} (ng/m^3)	C_{eq} (ng/m^3)	C_{eq}/OELV
1	1	1	October	11	104	12.8	6	15,000	11,875	0.59
1	2	2	February	5	-	8.1	5, 5	10,000	8125	0.41
1	3	2	February	5	-	11.1	6	15,000	11,875	0.59
1	4	2	March	11	>365	10.1	5, 5	10,000	8125	0.41
1	5	2	October	2	-	11.8	6	15,000	11,875	0.59
1	6	2	October	3	-	13.1	6	15,000	11,875	0.59
2	1	3	November	12	36	8.5	6	10,000	8125	0.41
2	3	3	December	3	-	7.1	6	10,000	8125	0.41
2	2	3	December	4	-	14.2	6	15,000	11,875	0.59
2	4	3	December	3	42	14.0	6	15,000	11,875	0.59
2	5	4	February	3	-	7.5	6	10,000	8125	0.41
2	6	4	February	3	-	8.5	6	10,000	8125	0.41
1/2	7	4	February	5	-	9.4	4	10,000	8125	0.41
1/2	8	4	February	5	>365	9.7	6	10,000	8125	0.41

It must be said that the calculation carried out is conservative, since it is considered that all the tasks carried out involve the handling of the rubble, which is not true. The tasks related to the construction of the walls, the concreting of the floor, etc., were carried out at the limits of the rubble zone, where the concentration drops to half. On the other hand, it is assumed that working with rubble implies being on the rubble 100% of the time, something that is not true; some of the work can be conducted near the debris boundary where concentrations are already half those in the centre. In addition, it is assumed that no measure is carried out to diminish the Hg gaseous emission, which is also conservative. If necessary, covering the rubble with subproducts diminishes the Hg concentration, significantly improving the working conditions.

As can be deduced from Table 6, the work should be carried out during different months over 1.5 years. Nevertheless, due to a significant delay in obtaining permits from authorities and mainly due to the COVID-19 pandemic, the work was carried out for 2.5 years, as shown in Table 7. This means that there were short periods of continuous working (maximum 16 days) and long periods without working, which is very positive from a health and safety point of view.

Regarding controlling worker health based on biological markers, the results are not available for the actual case. Therefore, they were not included in the retrospective analysis. As the number of workers involved is minimal, the anonymity required by the data protection law cannot be guaranteed, which is why it was left out of the scope of this study.

The work in mining tailings contaminated by metals, specifically by Hg, is very topical because administrations must restore decommissioning mining facilities and industrial buildings highly contaminated with mercury. In addition, nowadays, tailings are an emerging market opportunity.

In this context, all research aimed at the health and safety of workers is very important, so having a replicable methodology is very useful.

On the other hand, the information is sensitive, and companies are not inclined to publish the results. The results of our work, with valuable field data and a new perspective, could benefit other engineers and/or technicians in charge of Health and Safety working in highly contaminated areas.

The work conducted above was to illustrate or provide an example of how the methodology is used at two different and very interesting moments in any project:

(a) During planning (prediction);
(b) After executing the jobs (back-analysis).

To be sure that predictions can be offered and used during planning, it is important to demonstrate the causal relationship between the variables evolved.

The fact that the actual final results are similar to those predicted is a consequence of several underlying causal phenomena: the physical law that determines the temperature as a function of the time of year, the physical law that defines the Hg emission as a function of the temperature and the rules that determine the concentration of Hg in blood and urine in a person who breathes air in which there is gaseous Hg. Based on underlying physical laws, causality is assured.

It is important to note that with the data of the last 10 years, the average value of the average, minimum and maximum daily temperatures are θ_{medM} = 13.2 °C, θ_{minM} = 8.18 °C and θ_{maxM} = 18.33 °C, and the average temperature is reached at the end of the month of April (month 4, t = 120 days) and that the temperature varies with a sinusoidal law in which the period is the 365 days of the year T = 365 days:

$$\theta = 13.2 + \left(\frac{18.33 - 8.18}{2}\right) \sin\left[\frac{2\,\pi}{365}\,(t - 120)\right]. \tag{26}$$

This law, together with the law that governs the emission of gaseous Hg and its concentration on the source, Equation (1), and those that govern the concentration of Hg in urine and blood, Equations (10) and (11), ensure that, ultimately, conducting work at one time or another of the year is the cause of the varying Hg levels in a worker's body.

3.4. Brief Analysis under ESG Criteria

The sustainable development of the global economy and society calls for the practice of the environmental, social and governance (ESG) principle. The ESG principle has been developing for almost 20 years following its formal proposal in 2004. Over the past decades, ESG factors have become increasingly relevant for investors and stakeholders (creditors, regulators, etc.). In this way, it is helpful to contextualise our findings in light of this growing trend (increasing relevance of ESG), especially considering that the results speak to the "E" (environmental) and the "S" (social, as far as worker safety is a societal problem) of ESG criteria. As Dantas (2021) [60] pointed out, institutional investors increasingly use ESG factors in their portfolio when making decisions.

Taking different proposals for the ESG criteria as a reference, what is described in this work can positively assess various topics related to pillar (E) and pillar (S), as presented in Table 8.

Table 8. Topics in which the research would be positively assessed according to different ESG criteria proposals.

Reference	Source	Environmental (E)		Social (S)	
		Dimension	Factor	Dimension	Factor
[60]	BlackRock FMA analysis	Natural Resources and Pollution	Waste Management/Toxic Emissions	Internal Stakeholder Management	Worker's rights: safe working conditions
[61]	EBA report on ESG risk management and supervision	Environmental	Waste production and management (water, solid, hazardous)	Social	Workplace health and safety
[62]	Thomson Reuters	Environmental	Emissions	Social	Workforce
[63]	Refinitiv	Emissions reduction	Emissions; Waste; Environmental management systems	Workforce	Working conditions; Health and safety

4. Conclusions

Managing demolition waste from a mercury mine facility can be a significant hazard regarding Hg emissions, so it is crucial to plan the work carefully. The research developed a proper sampling procedure following the standard EN-689:2018 to ensure the workers' protection; the sampling campaigns included at least 2 h of sampled periods in different conditions, and all the tasks carried out in the restoration project were evaluated.

A working protocol with three scenarios (green, orange, red) was developed to calculate the maximum working hours in the function of the forecast temperature, a valuable tool for planning the works. In the same way, it was checked that the empirical model developed in la Soterraña by Rodríguez [38] fits in the measurements allowing forecasting of the workers' exposure all over the year. Evaluating the Hg gas concentration in the air, it is possible to estimate the biological markers such as Hg in blood or Hg in urine, essential parameters to protect the workers' health.

Lastly, it was demonstrated that covering high-emission Hg rubble with furnace slag and ashes reduces Hg gas emissions is a vital engineering control tool to avoid or minimise Hg gas exposition to workers and an important environmental measure to minimise the mercury emissions in la Soterraña.

All the aspects defined in this study (mercury gaseous monitoring system, risk prevention criterion or working protocol) were helpful to the performance of the work under safety conditions in this specific place: the rubble from the demolition of the metallurgical plant. These aspects should only be used as a guide for other different sites. Nevertheless, the defined methodology that measures the Hg gas concentration, relates it with other variables such as the temperature, develops a risk prevention criterion and defines a working protocol can be extrapolated to any other site, but it is an empirical investigation, and therefore it would only be applicable in similar conditions unless the model is readjusted.

Researching the emissions of different gases in abandoned tailings or debris and determining emission laws can be a line of research for the future, both from an environmental point of view, especially if the temperatures will rise, and from a health and safety point of view, if mining dumps become of beneficial interest.

The main practical implication is to use this research as a working guide for work on tailings contaminated with metals, which is becoming increasingly frequent.

Author Contributions: Conceptualisation, R.R. and H.G.-G.; methodology, R.R. and H.G.-G.; validation, R.R., H.G.-G., Á.P. and Z.H.; investigation, R.R., H.G.-G. and Z.H.; data curation, R.R. and H.G.-G.; writing—original draft preparation, R.R., H.G.-G. and Z.H.; writing—review and editing, H.G.-G.; supervision, R.R.; project administration, R.R., Á.P. and Z.H.; funding acquisition, R.R. and Á.P. All authors have read and agreed to the published version of the manuscript.

Funding: The authors would like to thank the program LIFE of the European Commission for the funding received for the project SUBproducts4LIFE (reference LIFE16 ENV/ES/000481).

Institutional Review Board Statement: Not applicable.

Informed Consent Statement: Not applicable.

Data Availability Statement: The data presented in this study are available within the manuscript.

Acknowledgments: Authors would like to thank the collaboration of the institutions and private companies that participated in the project SUBproducts4LIFE: Biosfera consultoría Medioambiental (BIOSFERA), Escorias y Derivados (EDERSA), Global Service (GService), Hidroeléctrica del Cantábrico (EDP), Instituto Asturiano de Prevención de Riesgos Laborales (IAPRL), Recuperación y Renovación (R&R) and Universidad de Oviedo (UNIOVI). Finally, the collaboration of sponsors Arcelor Mittal, Ingeniería de Montajes Norte SA (IMSA), Asturbelga de Minas, Lena Council, and the Instituto Nacional de Silicosis (INS) is also greatly appreciated.

Conflicts of Interest: The authors declare no conflict of interest.

References

1. Hylander, L.D.; Meili, M. 500 Years of Mercury Production: Global Annual Inventory by Region until 2000 and Associated Emissions. *Sci. Total Environ.* **2003**, *304*, 13–27. [CrossRef]
2. UN Environment Programme. Guideline for Decommissioning of Mercury Chlor-Alkali Plants. Available online: http://www.unep.org/resources/report/guideline-decommissioning-mercury-chlor-alkali-plants (accessed on 18 April 2023).
3. Finster, M.E.; Raymond, M.R.; Scofield, M.A.; Smith, K.P. Mercury-Impacted Scrap Metal: Source and Nature of the Mercury. *J. Environ. Manag.* **2015**, *161*, 303–308. [CrossRef]
4. Randall, P.; Chattopadhyay, S. Advances in Encapsulation Technologies for the Management of Mercury-Contaminated Hazardous Wastes. *J. Hazard. Mater.* **2004**, *114*, 211–223. [CrossRef]
5. Wang, S.; Feng, X.; Qiu, G.; Shang, L.; Li, P.; Wei, Z. Mercury Concentrations and Air/Soil Fluxes in Wuchuan Mercury Mining District, Guizhou Province, China. *Atmos. Environ.* **2007**, *41*, 5984–5993. [CrossRef]
6. Gustin, S.M.; Coolbaugh, M.; Engle, M.; Fitzgerald, B.; Keislar, R.; Lindberg, S.; Nacht, D.; Quashnick, J.; Rytuba, J.; Sladek, C.; et al. Atmospheric Mercury Emissions from Mine Wastes and Surrounding Geologically Enriched Terrains. *Environ. Geol.* **2003**, *43*, 339–351. [CrossRef]

7. Feng, X. Mercury Pollution in China—An Overview. In *Dynamics of Mercury Pollution on Regional and Global Scales: Atmospheric Processes and Human Exposures around the World*; Springer: Boston, MA, USA, 2005; pp. 657–678, ISBN 978-0-387-24493-8.
8. Loredo, J.; Ordóñez, A.; Gallego, J.R.; Baldo, C.; García-Iglesias, J. Geochemical Characterisation of Mercury Mining Spoil Heaps in the Area of Mieres (Asturias, Northern Spain). *J. Geochem. Explor.* **1999**, *67*, 377–390. [CrossRef]
9. Lin, Y.; Larssen, T.; Vogt, R.D.; Feng, X. Identification of Fractions of Mercury in Water, Soil and Sediment from a Typical Hg Mining Area in Wanshan, Guizhou Province, China. *Appl. Geochem.* **2010**, *25*, 60–68. [CrossRef]
10. Ordóñez, A.; Álvarez, R.; Charlesworth, S.; Miguel, E.D.; Loredo, J. Risk Assessment of Soils Contaminated by Mercury Mining, Northern Spain. *J. Environ. Monit.* **2011**, *13*, 128–136. [CrossRef]
11. Scholtz, M.T.; Van Heyst, B.J.; Schroeder, W.H. Modelling of Mercury Emissions from Background Soils. *Sci. Total Environ.* **2003**, *304*, 185–207. [CrossRef]
12. Kotnik, J.; Horvat, M.; Dizdarevič, T. Current and Past Mercury Distribution in Air over the Idrija Hg Mine Region, Slovenia. *Atmos. Environ.* **2005**, *39*, 7570–7579. [CrossRef]
13. Loredo, J.; Soto, J.; Alvarez, R.; Ordóñez, A. Atmospheric Monitoring at Abandoned Mercury Mine Sites in Asturias (NW Spain). *Environ. Monit. Assess.* **2007**, *130*, 201–214. [CrossRef]
14. Zhu, J.; Wang, D.; Liu, X.; Zhang, Y. Mercury Fluxes from Air/Surface Interfaces in Paddy Field and Dry Land. *Appl. Geochem.* **2011**, *26*, 249–255. [CrossRef]
15. Qiu, G.; Feng, X.; Meng, B.; Sommar, J.; Gu, C. Environmental Geochemistry of an Active Hg Mine in Xunyang, Shaanxi Province, China. *Appl. Geochem.* **2012**, *27*, 2280–2288. [CrossRef]
16. Cabassi, J.; Tassi, F.; Venturi, S.; Calabrese, S.; Capecchiacci, F.; D'Alessandro, W.; Vaselli, O. A New Approach for the Measurement of Gaseous Elemental Mercury (GEM) and H2S in Air from Anthropogenic and Natural Sources: Examples from Mt. Amiata (Siena, Central Italy) and Solfatara Crater (Campi Flegrei, Southern Italy). *J. Geochem. Explor.* **2017**, *175*, 48–58. [CrossRef]
17. Lindberg, S.E.; Kim, K.H.; Meyers, T.P.; Owens, J.G. Micrometeorological Gradient Approach for Quantifying Air/Surface Exchange of Mercury Vapor: Tests over Contaminated Soils. *Environ. Sci. Technol.* **1995**, *29*, 126–135. [CrossRef]
18. Llanos, W.; Kocman, D.; Higueras, P.; Horvat, M. Mercury Emission and Dispersion Models from Soils Contaminated by Cinnabar Mining and Metallurgy. *J. Environ. Monit.* **2011**, *13*, 3460–3468. [CrossRef]
19. Matanzas, N.; Sierra, M.J.; Afif, E.; Díaz, T.E.; Gallego, J.R.; Millán, R. Geochemical Study of a Mining-Metallurgy Site Polluted with As and Hg and the Transfer of These Contaminants to Equisetum Sp. *J. Geochem. Explor.* **2017**, *182*, 1–9. [CrossRef]
20. McNutt, M. Mercury and Health. *Science* **2013**, *341*, 1430. [CrossRef]
21. Kim, K.-H.; Kabir, E.; Jahan, S.A. A Review on the Distribution of Hg in the Environment and Its Human Health Impacts. *J. Hazard. Mater.* **2016**, *306*, 376–385. [CrossRef]
22. Wu, Z.; Zhang, L.; Xia, T.; Jia, X.; Wang, S. Heavy Metal Pollution and Human Health Risk Assessment at Mercury Smelting Sites in Wanshan District of Guizhou Province, China. *RSC Adv.* **2020**, *10*, 23066–23079. [CrossRef]
23. Phelps, R.W.; Clarkson, T.W.; Kershaw, T.G.; Wheatley, B. Interrelationships of Blood and Hair Mercury Concentrations in a North American Population Exposed to Methylmercury. *Arch. Environ. Health Int. J.* **1980**, *35*, 161–168. [CrossRef]
24. Koenigsmark, F.; Weinhouse, C.; Berky, A.J.; Morales, A.M.; Ortiz, E.J.; Pierce, E.M.; Pan, W.K.; Hsu-Kim, H. Efficacy of Hair Total Mercury Content as a Biomarker of Methylmercury Exposure to Communities in the Area of Artisanal and Small-Scale Gold Mining in Madre de Dios, Peru. *Int. J. Environ. Res. Public Health* **2021**, *18*, 13350. [CrossRef]
25. Garcia Gonzalez, H.; García-Ordiales, E.; Diez, R.R. Analysis of the Airborne Mercury and Particulate Arsenic Levels Close to an Abandoned Waste Dump and Buildings of a Mercury Mine and the Potential Risk of Atmospheric Pollution. *SN Appl. Sci.* **2022**, *4*, 76. [CrossRef]
26. WHO. Mercury and Health. 2017. Available online: https://www.who.int/news-room/fact-sheets/detail/mercury-and-health (accessed on 1 February 2021).
27. Saturday, A. Mercury and Its Associated Impacts on Environment and Human Health: A Review. *J. Environ. Health Sci.* **2018**, *4*, 37–43. [CrossRef]
28. Morton, J.; Sams, C.; Leese, E.; Garner, F.; Iqbal, S.; Jones, K. Biological Monitoring: Evidence for Reductions in Occupational Exposure and Risk. *Front. Toxicol.* **2022**, *4*, 836567. [CrossRef]
29. Wcisło, E.; Bronder, J.; Bubak, A.; Rodríguez-Valdés, E.; Gallego, J.L.R. Human Health Risk Assessment in Restoring Safe and Productive Use of Abandoned Contaminated Sites. *Environ. Int.* **2016**, *94*, 436–448. [CrossRef]
30. Lie, A.; Gundersen, N.; Korsgaard, K.J. Mercury in Urine—Sex, Age and Geographic Differences in a Reference Population. *Scand. J. Work. Environ. Health* **1982**, *8*, 129–133. [CrossRef]
31. Yoshida, M. Relation of Mercury Exposure to Elemental Mercury Levels in the Urine and Blood. *Scand. J. Work. Environ. Health* **1985**, *11*, 33–37. [CrossRef]
32. Rumiantseva, O.Y.; Ivanova, E.; Komov, V. High Variability of Mercury Content in the Hair of Russia Northwest Population: The Role of the Environment and Social Factors. *Int. Arch. Occup. Environ. Health* **2022**, *95*, 1027–1042. [CrossRef]
33. Eckley, C.S.; Gilmour, C.C.; Janssen, S.; Luxton, T.P.; Randall, P.M.; Whalin, L.; Austin, C. The Assessment and Remediation of Mercury Contaminated Sites: A Review of Current Approaches. *Sci. Total Environ.* **2020**, *707*, 136031. [CrossRef]
34. González-Valoys, A.C.; Jiménez Salgado, J.U.; Rodríguez, R.; Monteza-Destro, T.; Vargas-Lombardo, M.; García-Noguero, E.M.; Esbrí, J.M.; Jiménez-Ballesta, R.; García-Navarro, F.J.; Higueras, P. An Approach for Evaluating the Bioavailability and Risk

Assessment of Potentially Toxic Elements Using Edible and Inedible Plants—The Remance (Panama) Mining Area as a Model. *Environ. Geochem. Health* **2021**, *45*, 151–170. [CrossRef]
35. Wcisło, E.; Bronder, J.; Rodríguez-Valdés, E.; Gallego, J.L.R. Health Risk Assessment of Post-Mining Hg-As-Contaminated Soil: Implications for Land Remediation. *Water Air Soil Pollut.* **2022**, *233*, 306. [CrossRef]
36. Wongsasuluk, P.; Tun, A.Z.; Chotpantarat, S.; Siriwong, W. Related Health Risk Assessment of Exposure to Arsenic and Some Heavy Metals in Gold Mines in Banmauk Township, Myanmar. *Sci. Rep.* **2021**, *11*, 22843. [CrossRef]
37. *Mina de La Soterraña (Lena—Asturias) a Vista de Dron*; Ax1 Ahora: La Soterraña, Asturias, 2022.
38. Rodríguez, R.; Garcia-Gonzalez, H.; García-Ordiales, E. Empirical Model of Gaseous Mercury Emissions for the Analysis of Working Conditions in Outdoor Highly Contaminated Sites. *Sustainability* **2022**, *14*, 13951. [CrossRef]
39. Rodríguez, R.; Fernández, B.; Malagón, B.; Garcia-Ordiales, E. Chemical-Physical Model of Gaseous Mercury Emissions from the Demolition Waste of an Abandoned Mercury Metallurgical Plant. *Appl. Sci.* **2023**, *13*, 3149. [CrossRef]
40. Siegel, S.M.; Siegel, B.Z. Temperature Determinants of Plant-Soil-Air Mercury Relationships. *Water Air Soil Pollut.* **1988**, *40*, 443–448. [CrossRef]
41. Zhang, H.; Lindberg, S.E.; Marsik, F.J.; Keeler, G.J. Mercury Air/Surface Exchange Kinetics of Background Soils of the Tahquamenon River Watershed in the Michigan Upper Peninsula. *Water Air Soil Pollut.* **2001**, *126*, 151–169. [CrossRef]
42. Nagl, C.; Spangl, W.; Buxbaum, I. Sampling Points for Air Quality. In *Representativeness and Comparability of Measurement in Accordance with Directive 2008/50/EC on Ambient Air Quality and Cleaner Air for Europe*; European Parliament: Luxembourg, 2019.
43. *EN 689:2018+AC:2019*; Workplace Exposure-Measurement of Exposure by Inhalation to Chemical Agents-Strategy for Testing Compliance with Occupational Exposure Limit Values. European Committee for Standardisation: Brussels, Belgium, 2018.
44. Rodriguez, V. El Estado de Situación Del Mercurio Ambiental En Asturias. 2013. Available online: https://www.astursalud.es/documents/35439/38376/20130613+El+estado+de+situaci%C3%83%C2%B3n+del+mercurio+ambiental+en+Asturias.pdf/c7fecc72-43e4-7b00-1fe5-690e5f8152de?t=1564654288891 (accessed on 1 May 2023).
45. INSST Límites de Exposición Profesional para Agentes Químicos 2022—Portal INSST. Available online: https://www.insst.es/documentacion/catalogo-de-Publicaciones/limites-de-exposicion-profesional-para-agentes-quimicos-2022 (accessed on 26 April 2022).
46. Drake, P.; Rojas, M.; Reh, C.; Mueller, C.; Jenkins, F. Occupational Exposure to Airborne Mercury during Gold Mining Operations near El Callao, Venezuela. *Int. Arch. Occup. Environ. Health* **2001**, *74*, 206–212. [CrossRef]
47. Español Cano, S. Toxicología Del Mercurio. In Actuaciones Preventivas En Sanidad Laboral y Ambiental; Gama Ed., Lima, Perú. 2001. Available online: http://www.gama-peru.org/jornada-hg/espanol.pdf (accessed on 1 May 2023).
48. Iden, M.; Kira, S.; Miyaue, H.; Fukuda, M.; Yamaguchi, K.; Fujiki, Y. Biological Monitoring of Inorganic Mercury in Workers in a Fluorescent Lamp Plant. *J. Occup. Health* **1998**, *40*, 68–72. [CrossRef]
49. Tejero Manzanares, J.; Español Cano, S.; Serrano García, J.J.; de Paula Montes Tubío, F. Niveles de mercurio en ambiente y en fluidos biológicos: Caso de la metalurgia en Almadén, España (1986–2001). *Salud de los Trabajadores* **2011**, *19*, 123–133.
50. Kobal, A.; Dizdarevič, T. The Health Safety Programme for Workers Exposed to Elemental Mercury at the Mercury Mine in Idrija. *Water Air Soil Pollut.* **1997**, *97*, 169–184. [CrossRef]
51. Ramírez, A.V. Mejora de los indicadores biológicos de exposición al mercurio en trabajadores de una refinería de oro. *Anales de la Facultad de Medicina* **2011**, *72*, 177–182. [CrossRef]
52. Lovejoy, H.B.; Bell, Z.G. Mercury Exposure Evaluations and Their Correlation With Urine Mercury Excretions: 6. Recommendations for Medical Evaluation of Mercury Exposed Workers in the Chlor-Alkali Industry. *J. Occup. Med.* **1973**, *15*, 964–966.
53. Euro Chlor Code of Practice: Control of Worker Exposure to Mercury in the Chlor-Alkali Industry | Global Mercury Partnership; Euro Chlor, Brussels. 2010. Available online: https://wedocs.unep.org/bitstream/handle/20.500.11822/13103/Health_2_Edition_6.pdf?sequence=1&isAllowed=y (accessed on 1 May 2023).
54. Mason, H. Biological Monitoring and Exposure to Mercury. *Occup. Med.* **2001**, *51*, 2–11. [CrossRef]
55. Ellingsen, D.G.; Thomassen, Y.; Langård, S.; Kjuus, H. Urinary Mercury Excretion in Chloralkali Workers after the Cessation of Exposure. *Scand. J. Work. Environ. Health* **1993**, *19*, 334–341. [CrossRef]
56. Díez, S.; Esbrí, J.M.; Tobias, A.; Higueras, P.; Martínez-Coronado, A. Determinants of Exposure to Mercury in Hair from Inhabitants of the Largest Mercury Mine in the World. *Chemosphere* **2011**, *84*, 571–577. [CrossRef]
57. Packull-McCormick, S.; Ratelle, M.; Lam, C.; Napenas, J.; Bouchard, M.; Swanson, H.; Laird, B.D. Hair to Blood Mercury Concentration Ratios and a Retrospective Hair Segmental Mercury Analysis in the Northwest Territories, Canada. *Environ. Res.* **2022**, *203*, 111800. [CrossRef]
58. Harada, M.; Nakachi, S.; Cheu, T.; Hamada, H.; Ono, Y.; Tsuda, T.; Yanagida, K.; Kizaki, T.; Ohno, H. Monitoring of Mercury Pollution in Tanzania: Relation between Head Hair Mercury and Health. *Sci. Total Environ.* **1999**, *227*, 249–256. [CrossRef]
59. Ramos, M.; Carrasco, J.; Conde, A.; Delacasa, E.; Pujols, M.; Ballesteros, G.; Malet, A.; Caprino, A.; Chimenos, J.M. Guidelines on Best Environmental Practices for the Environmental Sound Management of Mercury Contaminated Sites in the Mediterranean United Nations Environment Programmes. Athens, Greece. 2015. Available online: https://wedocs.unep.org/handle/20.500.11822/9917 (accessed on 1 May 2023).
60. BlackRock Financial Markets Advisory. *Directorate-General for Financial Stability, F.S. and C.M.U. (European C. Development of Tools and Mechanisms for the Integration of ESG Factors into the EU Banking Prudential Framework and into Banks' Business Strategies and Investment Policies: Final Study*; Publications Office of the European Union: Luxembourg, 2021; ISBN 978-92-76-17252-9.

61. Li, T.-T.; Wang, K.; Sueyoshi, T.; Wang, D.D. ESG: Research Progress and Future Prospects. *Sustainability* **2021**, *13*, 11663. [CrossRef]
62. Seth, R.; Gupta, S.; Gupta, H. ESG Investing: A Critical Overview. *Hans Shodh Sudha* **2021**, *2*, 69–80.
63. Ehlers, T.; Elsenhuber, U.; Jegarasasingam, A.; Jondeau, E. Deconstructing ESG Scores: How to Invest with Your Own Criteria. BIS Work. Pap. Basel, Switzerland. 2022, pp. 1–48. Available online: https://www.bis.org/publ/work1008.pdf (accessed on 1 May 2023).

buildings

MDPI

Article

Exploring the Factors That Influence the Work–Family Interface of Construction Professionals: An Indian Case Study

M. G. Soundarya Priya [1], K. S. Anandh [1,*], K. Prasanna [1], K. Gunasekaran [1], Emmanuel Itodo Daniel [2], Mariusz Szóstak [3] and Della Sunny [4]

1 Department of Civil Engineering, Faculty of Engineering and Technology, SRM Institute of Science and Technology, SRM Nagar, Kattankulathur 603203, Tamil Nadu, India; sm0295@srmist.edu.in (M.G.S.P.); prasannk@srmist.edu.in (K.P.)
2 School of Architecture & Built Environment, University of Wolverhampton, Wolverhampton WV1 1LY, UK; e.daniel2@wlv.ac.uk
3 Department of Building Engineering, Faculty of Civil Engineering, Wroclaw University of Science and Technology, 50-370 Wroclaw, Poland; mariusz.szostak@pwr.edu.pl
4 District Nirmithi Kendra, Palakkad 695043, Kerala, India; dellaann70.da@gmail.com
* Correspondence: anandhk@srmist.edu.in

Abstract: The objective of this paper is to explore the dynamic factors of the work–family interface (WFI) of construction professionals in South India. It also aims to develop a model of the factors that influence the WFI. This study identified seven factors from the literature: work environment, family, stress, personal satisfaction, work culture, top-level acceptance, and work–family balance. Data were collected using questionnaires distributed among the construction professionals in South India. Data analysis was performed using SPSS. The mean standard deviation and shape measures (skewness and kurtosis) revealed that personal satisfaction (3.55) was rated the top factor for WFI, followed by work–family balance (3.15), stress (2.91), and work culture (2.83). The study found that the work culture was not significantly correlated with work environment, family, stress, personal satisfaction, or top-level acceptance. The developed SEM model emphasises the need to pay keen attention to the work environment and the work–family balance among the construction professionals in India. Management should consider these factors to design standard policies to improve the WFI and design work–life balance strategies to create stability in the lives of construction professionals. The current research is limited to only two cities in Kerala and Tamil Nadu in South India. More studies must be carried out for more states in India to better understand the current situation of WFI as there is limited evidence of studies on the WFI of construction professionals in India. Therefore, the findings of this study fill the existing knowledge gap and provide a clear insight into improving the mental and social well-being of construction professionals in the Indian construction industry and construction professionals in other developing countries.

Keywords: work–life balance; work–family interface; construction sector; engineers; quality of life

Citation: Soundarya Priya, M.G.; Anandh, K.S.; Prasanna, K.; Gunasekaran, K.; Daniel, E.I.; Szóstak, M.; Sunny, D. Exploring the Factors That Influence the Work–Family Interface of Construction Professionals: An Indian Case Study. *Buildings* **2023**, *13*, 1511. https://doi.org/10.3390/buildings13061511

Academic Editor: Maxim A. Dulebenets

Received: 23 March 2023
Revised: 4 June 2023
Accepted: 9 June 2023
Published: 12 June 2023

1. Introduction

The Indian construction industry is vast and is the second largest sector contributing to the national GDP. The industry employs many people, however, it is highly fragmented and diversified in nature. As it is a challenging and complex sector requiring a high amount of investment, the productivity of human resources is a key concern [1]. Lingard et al. [2] observed that low productivity is prevalent within the construction industry and it contributes to the low level of work–life balance (WLB) among professionals in this sector. A low level of WLB has adverse effects on individuals, families, and the workplace. According to Townsend et al. [3], stress in the family and workplace negatively impacts family life, sleep dysfunction, and workplace health and safety risks. Given this, Lingard and Francis [4] and

Townsend et al. [3] have suggested that organisations should support their workforce to reduce emotional exhaustion by improving the WLB practices. However, in the management of construction projects, much attention seems to be paid to the arrangements of contracts, legal issues, the adaptation of technology, and the milestones, with little consideration of the emotional and physiological well-being of the employees [5,6]. It is concerning that the construction industry has been characterised by long working hours and overtime, which contributes to stress. In the specific context of developing countries, the construction sectors face many challenges and has low productivity. Studies have identified several factors that contribute to productivity, including human factors, personal factors, work-supporting factors, motivation, change in lifestyle and job satisfaction [7–12]. The improvement of individual performance can be achieved by understanding human nature and the structural nature and indeterminateness of organisations. Top-level management should aim to manage emotions and reduce pressure based on the understanding that employee turnover and productivity are determined by organisational commitment. This will also be reflected in the retention rate of employees in the construction organisation [13–15].

Researchers worldwide are trying to understand WLB in the construction profession, including Australia, UK, Poland, South Africa and US. However, in India, very few studies have attempted to familiarise the various factors that influence the WLB of construction professionals [5,6]. The current study aims to fill this gap and direct future research into the work–family interface (WFI) of construction professionals in India and other parts of the world. This study's objectives are to identify the predominant factors that influence the WFI of construction professionals in South India. In addition, to identify the association between the WFI factors the independent sample t-test, correlation analysis, and confirmatory factor analysis (CFA) were used.

1.1. Work–Life Balance (WLB) and Work–Family Interface (WFI)

Every employee should know what they do, that is, 'live to work' or 'work to live'. In the construction industry, there is no defined work time, and work personnel must be available at any time for work. WLB refers to the equilibrium between work-related activities and personal life responsibilities. It involves effectively managing and integrating work and non-work domains to reduce conflict and enhance well-being. WLB aims to create harmony and satisfaction in both work and personal life. Work–family interface (WFI) refers to the interaction and interplay between an individual's work life and family life. The amount of WFI states the degree of segmentation versus integration among work and family spheres. The quality of WFI refers to negative versus positive interactions between work and family domains and examines how work-related demands and family-related demands can influence each other, leading to various outcomes, such as conflict, enrichment, or segmentation. A positive WLB, where people are able to effectively manage their demands in work and personal life, can contribute to a positive WFI. When individuals have a good WLB, they are better able to meet the demands of both their work and family roles, thereby reducing conflict and improving the compatibility between the two domains. On the other hand, a poor WLB, characterised by excessive work demands, long working hours, and neglect of personal life, can result in a negative WFI. It can lead to work–family conflict (WFC), where the demands and pressures from work interfere with family responsibilities, causing stress and strain in both domains [2]. Every employee works to provide financial support for their family, but when he or she cannot spare personal time for their own family, the WFC arises. WFC also has an unfavourable consequence on work satisfaction, organisational commitment, productivity, etc. On an individual level, it is related to employee burnout and mental health issues. Career development includes the ability to focus on personal and professional values, the appropriate attribution of responsibilities, ethics, roles, knowledge, and mature decision-making; all these aspects must be developed through education at the beginning of the training level so that conflict can be reduced [16]. Work-related pressure and stress will be reflected in the personal lives of the individuals, thus negatively impacting the family life of construction professionals. Most professionals working in the construction industry are exposed to WFC. Therefore, WLB becomes an essential issue in the construction industry in terms of both organisational effectiveness and occupational health. The WFI can be described in terms of amount, quality and context. The context of WFI refers to the perspective from which the WFI is studied. The working environment should be positive and comfortable enough to allow deadlines and goals to be met. Many sub-factors, such as working hours, nature of work, etc., also make it complicated. Stress is also a significant factor, including stress from work, family, and health problems. Additionally, family support is essential to make a person perfect at work. Top-level acceptance is another essential factor. The culture of work of an employee and their co-workers too is a crucial factor affecting the WFI [2].

Lingard and Francis [17] showed a disproportional link between work and family life in the construction industry, providing a significant contribution to the research on WFI, particularly in the construction sector. The factors that they focused on were individualistic, gender, the phase of life, and degree of job or family focus. In summary, WLB and WFI are closely linked. A good WLB can positively influence WFI by reducing conflict and improving the compatibility between work and family domains. In contrast, a poor WLB can lead to WFC and negative outcomes in both work and family life.

Employers should provide a healthy WLB for all employees, as this will enhance the work culture. In Western countries, there is ample evidence of WLB practices; when it comes to developing countries, the implementation of such practices is still in the early stages. Therefore, there is no evidence in the literature for WLB practices in developing countries in the construction industry, which adds significance to the present study [18]. Job satisfaction involves considering various probable work, personal, and cultural incentives or motivations. The nature of the job, the designation of the job, the income rate and the work environment have all been considered; the three elements that they evaluated for WLB were balance of time, participation and satisfaction relevant to work and family [19]. WLB has also been correlated with quality of life, but is limited to assertive conditions [20]. Regarding job satisfaction, self-respect, a fair approach, professional treatment by co-workers, and understanding the human values of employees are all key factors. Specifically, all of this can lead to better job fit while promoting WLB in the construction industry. In particular, women have the strongest influence on this strategy, despite the fact that the construction industry is largely a male-dominated industry. There has been an increasing acceptance of the fact that a healthy WLB can benefit organisations. However, several factors within an organisation must be considered before implementing WLB practises. Management support, employee effectiveness, intentional and behavioural outcomes, work environment, organisational culture, and acceptance by top-level management must all be considered integral to promoting work–family support [21,22].

1.2. Factors Influencing the Work–Family Interface (WFI)

The work environment, stress, and work culture all are interlinked. The work environment comprises the physical and non-physical factors prevailing in that environment. It includes geographic location, technology, communication systems, physical setup of the building, transportation systems, infrastructure, relationships (internal and external), and the work climate. It has a significant impact on the WFI. Recent studies have stated that stress plays an essential role in human behaviour, both mentally and physically [9,23]. As such, mental and physical illnesses caused at work unequivocally affect conflicts in the family.

Hofstede [24] explained that work culture is similar to work climate. Work culture also has several factors: power distance, individualism-collectivism, uncertainty avoidance, masculinity–femininity, and assertiveness. All these factors show mixed results [24]. The influence of family on the WFI generates temporal effects carried out from the family to the workplace. The three main ways in which the family can impact the WFC are balancing time, participation and satisfaction, which means dedicating time equally, participating and satisfying both the organisation and the family [25–27].

Employee satisfaction is a critical factor that impacts their success as a worker, dramatically benefiting the company. Therefore, employee satisfaction is an important driving force for organisations [28]. WFC occurs when there is a demand for targets assigned in work that makes it difficult to fulfil the desires of family life. WFC have been derived in terms of the inter-role conflict wherein role pressures from work and family territory are reciprocally conflicting in some respects, e.g., support in the work role is made more difficult by involvement in the family role. There are three main types of conflict: time-based, strain-based, and behaviour-based. Improper WLB leads to poor WFI [4]. Acceptance, depending on the situation, can change its role. Acceptance may or might not influence the WFI. The study identified seven WFI factors affecting the construction industry from the literature, which are shown in Figure 1.

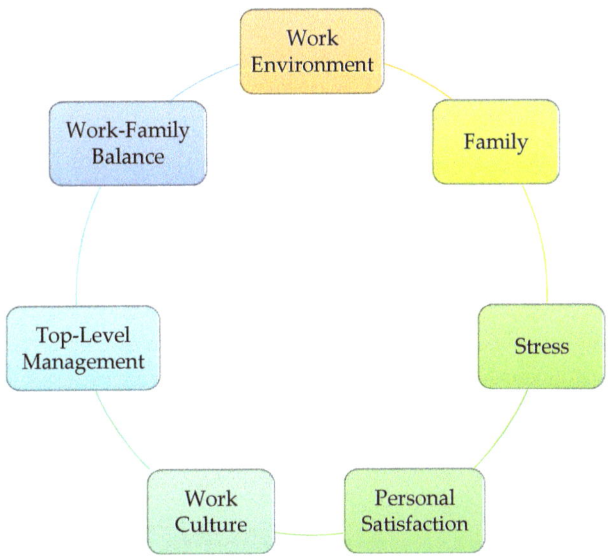

Figure 1. Work–family interface factors.

1.3. Impact of the Work–Family Interface (WFI)

The impact of the WFI has both positive and negative possible outcomes: A good work environment has a positive effect in that it increases productivity, organisational commitment and loyalty, and corporate citizenship, which in turn helps the individual to be motivated, recognise their skill, and gain satisfaction from their well-being. Stress unequivocally has negative impacts on the individual's health. The outcomes of the work culture have shown mixed results: positive (increased commitment); and negative (degradation of mental and physical status and health). However, employees with better personal satisfaction have reported better psychological well-being due to having more control over their family time and thus having more time to spend with their family. A healthy WLB includes the appreciation of expressiveness and emotional sensitivity. Additionally, employees must not take the work burden home to the family. Finally, top-level acceptance (of the need for a healthy WLB for employees), when effectively transmitted throughout the organisation, helps improve employee retention at work, which is positively reflected in an improved WFI [29].

Research in the construction industry has revealed a high level of stress due to higher job demands and long and irregular working hours, leading to burnout of employees [30,31]. The WFI is associated both negatively and positively with the family domain as well the work domain. When it comes to the Indian construction industry, depending on the role, levels of stress vary. Physiological, physiological, and sociological factors significantly affect professionals in India [30–32]. If the employee is satisfied with the quality of work-life (QWL), they may balance their personal and professional life, leading to a healthy WLB. This suggests that critical executives of an organisation should reformulate the firm's policies to improve the QWL and WLB of their employees. Factors that affect the relationship between QWL and WLB have been identified based on the productivity of an organisation [33–35].

2. Methodology

2.1. Data Collection

Figure 2 shows the study methodology flow chart. In this study, a quantitative approach research was adopted using a questionnaire. Many professionals working in private construction companies are invited to take part in the survey. Construction work is nomadic, therefore, the simple random sampling method was used to gather evidence. The questionnaire was distributed to civil engineering professionals working in two cities, Chennai (Tamil Nadu) and Ernakulum (Kerala) in South India. This included project managers, planning engineers, site engineers, and design engineers.

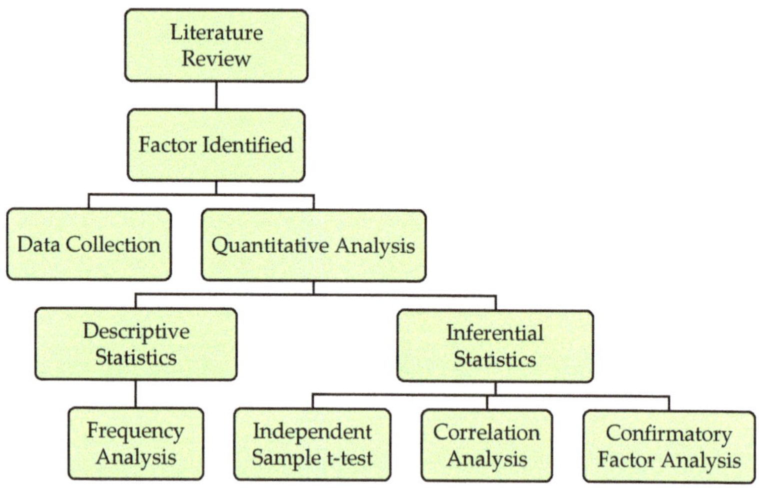

Figure 2. Methodology flow chart.

The questionnaire was designed based on the factors identified in the literature and expert interviews. Respondents were requested to rate the items on a five-point Likert scale ranging from 1 (strongly disagree) to 5 (strongly agree). Stress, personal satisfaction, work culture, and top-level acceptance were measured using a 25-item instrument. Respondents were requested to rate the items on a five-point Likert scale ranging from 1 (not at all) to 5 (almost all the time). The WLB was measured using a two-item instrument with a five-point Likert scale ranging from 1 (low) to 5 (very high). A total of 360 construction industry professionals, located in Chennai and Ernakulum in South India, were invited to participate in the survey. A total of 200 people completed the survey, 50% each from Chennai and Ernakulum. All the participants worked in reputable construction companies.

2.2. Data Analysis

Data analysis was performed using the SPSS version 23.0 software (Statistical Package for Social Sciences). Following the review of the literature, the study identified that seven factors (work environment, family, stress, personal satisfaction, work culture, top-level acceptance, and work–family balance) were essential for a detailed and practical study of the WFI among construction workers. The mean standard deviation and shape measures (skewness and kurtosis) were found. The result shows that personal satisfaction (3.55) was rated as the top factor for the WFI of employees in the construction industry, followed by work–family balance (3.15), stress (2.91), and work culture (2.83).

3. Results and Discussion

3.1. Frequency Analysis

The respondents were asked to provide their demographic details, such as sex, age, marital status, number of children, and years of experience. The total working hours per day, and whether they stayed with their family or needed to work night shifts, were also recorded. A total of 78% of the professionals stayed with their family, while 22% lived alone, having left their families in their hometowns. Table 1 shows the demographic characteristics.

Table 1. Demographic profile of the respondents.

Background Information	Category	Frequency	Percentage
Place	Chennai	100	50
	Ernakulum	100	50
Gender	Male	164	82
	Female	36	18
Marital status	Single	138	69
	Married	62	31
Type of family	Nuclear	158	79
	Joint	42	21
Staying with family	Yes	156	78
	No	44	22
Night shifts	Yes	20	10
	No	180	90
Age	22–32	98	49
	33–42	43	21.5
	43–52	37	18.5
	52 and above	22	11
Years of experience	3–5 years	10	5
	5–10 years	30	15
	11–15 years	20	10
	16–20 years	28	14
	21–25 years	32	16
	26–30 years	22	11
	31–35 years	30	15
	36–40 years	28	14

Notes: The background information of the respondents is collected in the form of categorical data. The data are analysed statistically to find the frequency and percentage distribution of the respondent's information.

The measure of data reliability was measured using the coefficient of reliability under Cronbach's alpha. From the analysis of all elements, the Cronbach's alpha scores were found to be 0.786, which were higher than the limit score of 0.7. Henceforth, it can be concluded that the design of the survey instrument for the current work is consistent.

3.2. Independent Sample t-Test

Three hypotheses, with their corresponding null hypotheses, were tested using independent sample *t*-tests:

H1. *There is no specified difference with respect to the perception of employees based on the location towards the WFI factors in the construction industry.*

H2. *There is no specified difference with respect to the employees' perception of gender towards the WFI factors in the construction industry.*

H3. *There is no specified difference with respect to the employees' perception of marital status towards the WFI factors in the construction industry.*

3.2.1. Hypothesis 1

From the results in Figure 3, it was made clear that the male and female respondents of the study differed significantly with respect to the mean rating for the factors work environment (t = 0.003, and p = 0.000), stress (t = 3.062, and p = 0.003), top-level acceptance (t = 3.035, and p = 0.003), and work–family balance (t = 1.981, and p = 0.049). However, the present study did not identify any significant differences in the mean score between employees working in Chennai and Ernakulum for factors, such as family, personal satisfaction, and work culture. Therefore, H1 is supported for the factors family, personal satisfaction, and work culture, but rejected for the factors work environment (1% level), stress (1% level), top-level acceptance (1% level), and work–family balance (5% level). Family, personal satisfaction, and work culture play a vital role in the WFI, even when tested based on two different locations. The analysis shows that personal satisfaction is an essential factor, which can be related to the theory of needs [36]. The highest level of needs according to this theory is self-esteem, which is equivalent to personal satisfaction. Next came work culture, which can be understood through the doctrines of organizational culture; every organization possesses its own culture, which should be aligned to its own people's culture. Another critical factor is the family, which is the basic element of any social set-up, and any behavioural shortcomings will have both direct and indirect effects on the family.

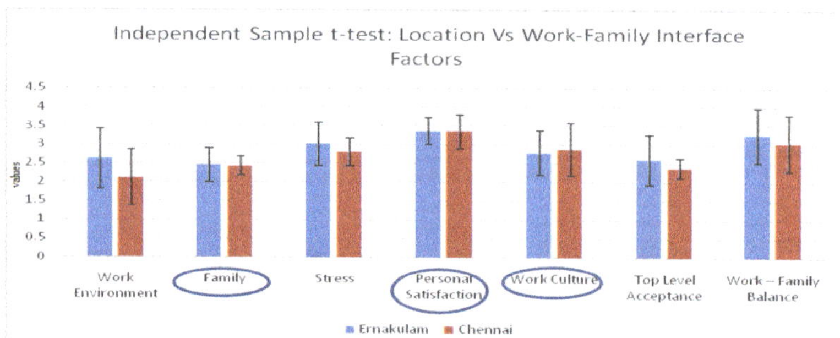

Figure 3. Independent sample t-test: location vs. work–family interface factors.

3.2.2. Hypothesis 2

H2 was tested using independent sample t-tests with gender as the independent variable and the WFI factors as the dependent variables. Figure 4 showed that male and female respondents differed significantly in the mean rating for the factors work environment (t = 2.0978, and p = 0.032), family (t = 2.6681, and p = 0.008), stress (t = 2.23, and p = 0.023), personal satisfaction (t = 2.654, and p = 0.008), and top-level acceptance (t = 2.671, and p = 0.007). However, the present work has not identified any significant differences in the mean rating of male and female employees for the factors work culture and work–family balance. Therefore, H2 is accepted for the factors work culture and work–family balance but rejected for the factors work environment (5% level), family (1% level), stress (5% level), personal satisfaction (1% level), and top-level acceptance (1% level). Work culture and WLB play a critical role in both genders. During the study, it is also seen that some organisational work culture helps employees to spend more time with family. However, at certain times, professional life is affected by family commitments and concerns.

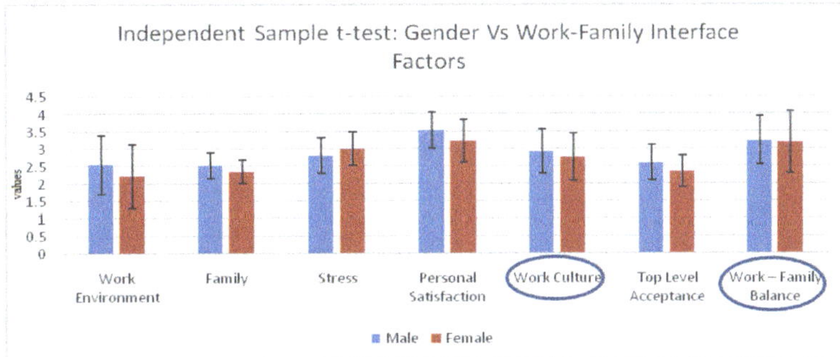

Figure 4. Independent sample *t*-test: gender vs. work–family interface factors.

3.2.3. Hypothesis 3

In Figure 5, it was clarified that the single and married respondents in the study differed significantly in the mean rating of factors, such as work environment (t = 2.635, and p = 0.010), family (t = 2.045, and p = 0.042), stress (t = 3.914, and p = 0.000), top-level acceptance (t = 2.7145, and p = 0.007), and balance (t = 3.281, and p = 0.001). However, the current study has failed to identify any significant change in the mean rating between single and married respondents for personal satisfaction and work culture. Therefore, H3 is accepted for the factors personal satisfaction and work culture but rejected for the factors work environment (5% level), family (5% level), stress (5% level), top-level acceptance (1% level), and work–family balance (1% level). Personal satisfaction does not play an important role in marital status, but the level of stress will be greater in the work culture of married people.

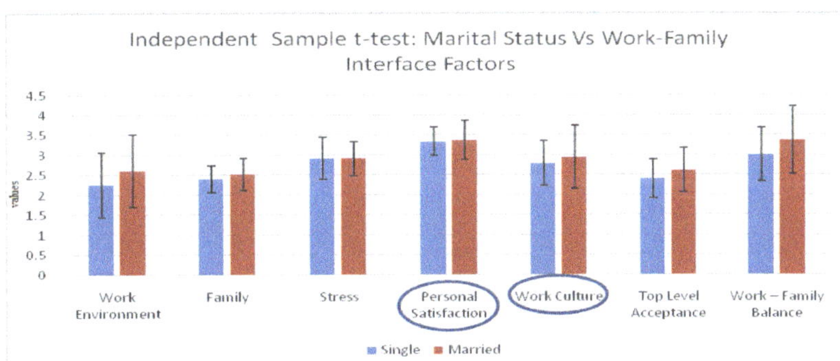

Figure 5. Independent sample *t*-test: marital status vs. work–family interface factors.

3.3. Correlation Analysis

A correlation analysis was conducted to find if there are any relationships between the factors governing the WFI of employees in the construction industry. In Table 2, stress was positively correlated with the work environment of construction management professionals (R = 0.143) at the 5% significance level. The relationship between top-level acceptance and work environment was also significant (R = 0.177) at the 5% significance level. Similarly, the work–family balance was positively correlated with the work environment (R = 0.708) at the 1% significance level. This implies that construction management professionals' productivity and performance could be negatively impacted if adequate WFI is not maintained. This finding has great implications for construction management leaders and chief executives, and they should formulate measures to make the working environment more conducive for their employees, including those working at the managerial and operational levels.

Table 2. Correlation coefficient matrix between the factors.

	Work Environment	Family	Stress	Personal Satisfaction	Work Culture	Top-Level Acceptance	Work–Family Balance
Work environment	1.000						
Family	0.127	1.000					
Stress	0.143 *	0.059	1.000				
Personal satisfaction	0.076	0.119	0	1.000			
Work culture	0.109	−0.015	0.128	−0.086	1.000		
Top-Level Acceptance	0.177 *	0.043	0.125	−0.019	0.087	1.000	
Work–family balance	0.708 **	0.127	0.093	0.121	0.181 *	0.062	1.000

Notes: * Correlation is significant at the 0.05 level (2-tailed). ** Correlation is significant at the 0.01 level (2-tailed).

Finally, the work–family balance was positively correlated with work culture (R = 0.181) at the 5% significance level. This study found that the work culture was not significantly correlated with the work environment, family, stress, personal satisfaction, or top-level acceptance.

3.4. Confirmatory Factor Analysis Using Structural Equation Modelling

Structural Equation Modelling (SEM) has been used based on the theoretical model developed in the present work. Figures 6 and 7 represent the unstandardised and standardised evaluations of the dimension model formulated in the present study. This study found that work environment and work–family balance had a significant impact on the WFI of construction professionals. The fit for goodness representing a statistical model defines the ways the model fits into the selected set of explanations. Table 3 predicts the results of the fit for goodness indices of the SPSS analysis of moment structures (AMOS) model developed for the study. The I-value was 0.172, which is higher than the suggested value of 0.05. The attained values of the goodness of fit index (GFI) and the adjusted Goodness of fit index (AGFI) were higher than the recommended value of 0.9 [37]. Similarly, the obtained value for the comparative fit index (CFI) was 0.969, above the threshold value of 0.9 [38]. The root mean square error of approximation (RMSEA) value was 0.042, which is relatively less than the suggested value of 0.09 [39]. Overall, the overall model fit values obtained in the current study are within an acceptable level. Therefore, it was determined that the measurement model was a fit. Table 4 shows the weights of SEM.

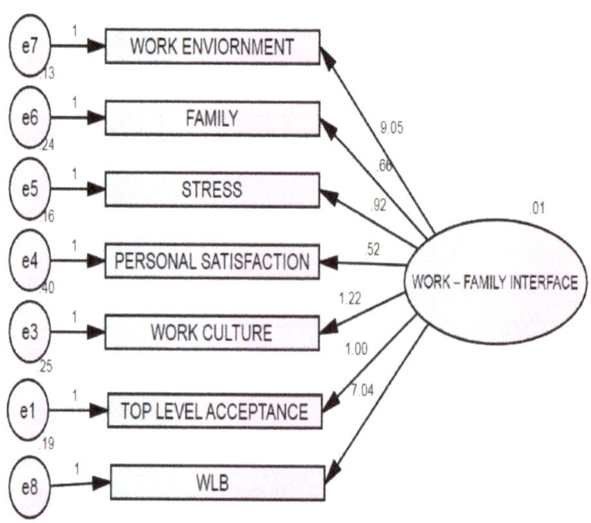

Figure 6. Unstandardised estimates.

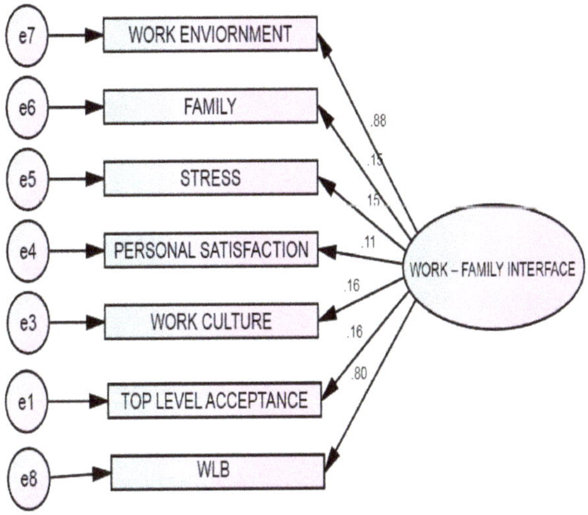

Figure 7. Standardised estimates.

Table 3. The goodness of fit: AMOS model.

Variable	Chi-Square Value	*p* Value	RMSEA	RMR	GFI	AGFI	CFI	TLI	RFI	NFI	Chi sq./df
Recommended value	-	>0.05	<0.09	<0.08	>0.90	>0.90	>0.90	>0.90	>0.90	>0.90	<5.0
Obtained Value	18.819	0.172	0.042	0.014	0.974	0.949	0.969	0.953	0.939	0.913	1.344

Table 4. Regression weights of the structural equation model.

Observed Variable			Unstandardised Estimate	Standardised Estimate	S.E.	C.R.	*p*
Work environment	<---	Work–family interface	9.051	0.885	4.379	2.067	0.039 *
Family	<---	Work–family interface	0.657	0.151	0.450	1.459	0.145
Stress	<---	Work–family interface	0.915	0.153	0.622	1.471	0.141
Personal satisfaction	<---	Work–family interface	0.516	0.106	0.444	1.163	0.245
Work culture	<---	Work–family interface	1.222	0.159	0.815	1.499	0.134
Top-Level Acceptance	<---	Work–family interface	1.000	0.164	0.713	1.426	0.112
Work–family balance	<---	Work–family interface	7.040	0.800	3.327	2.116	0.034 *

Notes: * Correlation is significant at the 0.05 level (2-tailed).

In particular, this study found that factors of work environment and work–family balance significantly impacted the WFI (see Figures 6 and 7).

4. Conclusions

This study found that family and personal satisfaction were the most prominent factors that play a vital role in the WFI with respect to location. At the same time, work culture and work–family balance were the dominant factors influencing the WFI in female and male construction professionals. However, this study also revealed that, for married respondents, personal satisfaction did not play a significant role for the WFI. Furthermore, it was revealed that work culture and a high level of stress were the main factors influencing the WFI among married construction professionals. In the future studies, information on their experience can be collected and the results can be compared. Using correlation analysis, this study found that the work environment, stress, work culture, personal satisfaction, and top-level acceptance played a key role. The model was a good fit, as established through confirmatory factor analysis and revealed that the work environment and work–family balance had a significant impact on the WFI. The SEM model developed emphasises the need to pay great attention to the work environment and the balance between work and family among construction professionals in India. In light of these findings, it can be concluded that construction organisations should consider the physical and overall mental health well-being of their employees and not focus only on the delivery of key project performance indicators, such as cost, time, quality, safety, and sustainability. Although this study was limited to South India, the findings could be applied to other parts of India and could direct future studies on WFI in other parts of the world. This study suggests that the prominent factors identified should be considered by management to design standard policies to improve WFI and to develop WLB strategies to create stability in the lives of construction professionals. In the future, longitudinal studies can be conducted as this study is considered to be cross-sectional. Additionally, with the help of a large sample, survey can be conducted and a qualitative method can be adopted for the in-depth exploration of the mindset of the construction professionals, since it differs for each country and context.

Author Contributions: Conceptualization, K.S.A. and M.G.S.P.; methodology, K.S.A.; software, E.I.D.; validation, K.S.A., K.P. and K.G.; formal analysis, M.G.S.P.; investigation, M.G.S.P.; resources, M.G.S.P. and D.S.; data curation, M.G.S.P. and D.S.; writing—original draft preparation, M.G.S.P.; writing—review and editing, K.S.A., M.S. and E.I.D.; visualization, M.S. and E.I.D.; supervision, K.S.A., K.P. and K.G.; project administration, K.S.A., M.S. and E.I.D. All authors have read and agreed to the published version of the manuscript.

Funding: This research received no external funding.

Data Availability Statement: The data presented in this study are available on request from the corresponding author. The data are not publicly available because some data are proprietary or confidential. Therefore, the data may only be provided with restrictions (e.g., anonymized data).

Conflicts of Interest: The authors declare no conflict of interest.

References

1. Indian Construction Industry: Indian Construction Industry Overview. Available online: https://www.indianconstructionindustry.com/overview.html (accessed on 11 October 2021).
2. Lingard, H.; Francis, V.; Turner, M. Work-life strategies in the Australian construction industry: Implementation issues in a dynamic project-based work environment. *Int. J. Proj. Manag.* **2012**, *30*, 282–295. [CrossRef]
3. Townsend, K.; Lingard, H.; Bradley, L.; Brown, K.; Lingard, H.; Brown, K. Working time alterations within the Australian construction industry. *Pers. Rev.* **2012**, *40*, 70–86. [CrossRef]
4. Lingard, H.; Francis, V. The decline of the 'traditional' family: Work-life benefits as a means of promoting a diverse workforce in the construction industry of Australia. *Constr. Manag. Econ.* **2005**, *23*, 1045–1057. [CrossRef]
5. Malone, E.K.; Issa, R.R.A. Work-life balance and organizational commitment of women in the U.S. construction industry. *J. Prof. Issues Eng. Educ. Pract.* **2013**, *139*, 87–98. [CrossRef]
6. Cattell, K.; Bowen, P.; Edwards, P. Stress among South African construction professionals: A job demand-control-support survey. *Constr. Manag. Econ.* **2016**, *34*, 700–723. [CrossRef]
7. Mustapha, F.H.; Naoum, S. Factors influencing the effectiveness of construction site managers. *Int. J. Proj. Manag.* **1998**, *16*, 1–8. [CrossRef]
8. Parakandi, M.; Behery, M. Sustainable human resources: Examining the status of organizational work-life balance practices in the United Arab Emirates. *Renew. Sustain. Energy Rev.* **2016**, *55*, 1370–1379. [CrossRef]
9. Jepson, J.M.; Kirytopoulos, K.; London, K. Exploring project managers' perception of stress when working in increasingly complex construction projects. *Constr. Econ. Build.* **2017**, *17*, 47–67. [CrossRef]
10. Anandh, K.S.; Gunasekaran, K.; Senthamizh Sankar, S. An envisage on emotional intelligence among superior-subordinate in construction sector of Chennai City, India. *AIP Conf. Proc.* **2020**, *2277*, 240012.
11. Sawicki, M.; Szóstak, M. Quantitative Assessment of the State of Threat of Working on Construction Scaffolding. *Int. J. Environ. Res. Public Health* **2020**, *17*, 5773. [CrossRef]
12. Sawicki, M.; Szóstak, M. Impact of Alcohol on Occupational Health and Safety in the Construction Industry at Workplaces with Scaffoldings. *Appl. Sci.* **2020**, *10*, 6690. [CrossRef]
13. Kiruthiga, K.; Anandh, K.S.; Gunasekaran, K. Assessment of influencing factors on improving effectiveness and productivity of construction engineers. *Int. J. Appl. Eng. Res.* **2015**, *10*, 13849–13854.

14. Youniss, A.; Ismail, A.B.; Khoiry, M.A. A conceptual model of delay factors affecting road construction projects in Liby. *J. Eng. Sci. Technol.* **2017**, *12*, 3286–3289.
15. Anandh, K.S.; Prasanna, K.; Priya, M.G.; Simon, S.M. An industrial study of just in time (JIT) management in precast construction projects. *AIP Conf. Proc.* **2020**, *2277*, 240011.
16. McCuen, R.H. Balancing corporate and personal values. *J. Manag. Eng.* **1998**, *14*, 40–44. [CrossRef]
17. Lingard, H.; Francis, V. *Managing Work-Life Balances in Construction*; Spon Press: London, UK; Routledge: London, UK, 2009.
18. Minarelli, J.; Baroudi, B. The Work-Life Balance of Construction Management Professionals. In Proceedings of the Construction, Building and Real Estate Research Conference of the Royal Institution of Chartered Surveyors, Toronto, ON, Canada, 19–22 September 2016; pp. 1–11.
19. Martínez-León, I.M.; Olmedo-Cifuentes, I.; Ramón-Llorens, M.C. Work, personal and cultural factors in engineers' management of their career satisfaction. *J. Eng. Technol. Manag.* **2018**, *47*, 22–36. [CrossRef]
20. Greenhaus, J.H.; Collins, K.M.; Shaw, J.D. The relation between work—Family balance and quality of life. *J. Vocat. Behav.* **2003**, *63*, 510–531. [CrossRef]
21. Sommerville, J.; Langford, V. Multivariate influences on the people side of projects: Stress and conflict. *Int. J. Proj. Manag.* **1994**, *12*, 234–243. [CrossRef]
22. Behson, S.J. The relative contribution of formal and informal organizational work-family support. *J. Vocat. Behav.* **2005**, *66*, 487–500. [CrossRef]
23. Cattell, K.S.; Bowen, P.A.; Cooper, C.L.; Edwards, P.J. *The State of Well-Being in the Construction Industry*; The Chartered Institute of Building: Bracknell, UK, 2017.
24. Hofstede, G. Dimensional zing Cultures: The Hofstede Model in Context. *Online Read. Psychol. Cult.* **2009**, *2*, 8.
25. Lingard, H.; Francis, V. The work-life experiences of office and site-based employees in the Australian construction industry. *Constr. Manag. Econ.* **2004**, *22*, 991–1002. [CrossRef]
26. Butler, A.B.; Grzywacz, J.G.; Bass, B.L.; Linney, K.D. Extending the demands-control model: A daily diary study of job characteristics, work-family conflict and work-family facilitation. *J. Occup. Organ. Psychol.* **2005**, *78*, 155–169. [CrossRef]
27. Francis, V.; Lingard, H.; Prosser, A.; Turner, M. Work-family and construction: Public and private sector differences. *J. Manag. Eng.* **2013**, *29*, 392–399. [CrossRef]
28. Krane, P.H.; Olsson, N.O.E. Uncertainty management of projects from the owners' perspective, with main focus on managing delivered functionality. *Int. J. Manag. Proj. Bus.* **2014**, *7*, 133–143. [CrossRef]
29. Lingard, H.; Francis, V. Does a supportive work environment moderate the relationship between work-family conflict and burnout among construction professionals? *Constr. Manag. Econ.* **2006**, *24*, 185–196. [CrossRef]
30. Anandh, K.S.; Gunasekaran, K.; Sunny, D. A review on the work-family interface of construction professionals in south India. *J. Adv. Res. Dyn. Control Syst.* **2018**, *10*, 405–420.
31. Anandh, K.S.; Gunasekaran, K. An investigation on stress among the professionals in the Indian construction industry. In Proceedings of the Construction Research Congress, ASCE, New Orleans, LA, USA, 2–4 April 2018; pp. 1–7.
32. Rani, H.A.; Radzi, A.R.; Alias, A.R.; Almutairi, S.; Rahman, R.A. Factors affecting workplace well-being: Building construction projects. *Buildings* **2022**, *12*, 910. [CrossRef]
33. Anandh, K.S.; Gunasekaran, K. Constructing a model to examine the influence of quality of work-life on work-life balance—Discernment of civil engineers from construction industry in Chennai. *Indian J. Sci. Technol.* **2016**, *9*, 1–12. [CrossRef]
34. Soundarya Priya, M.G.; Anandh, K.S.; Kamal, S.; Shanmuga Priya, S. Assessing Quality of Working Life (QWL) Among Construction Professionals in Private Sectors in Chennai. *Lect. Notes Civ. Eng.* **2022**, *284*, 635–647.
35. Zhang, R.P.; Bowen, P. Work-family conflict (WFC)—Examining a model of the work-family interface of construction professionals. *Saf. Sci.* **2021**, *144*, 105469. [CrossRef]
36. McLeod, S. Maslow's hierarchy of needs. *Simply Psychol.* **2007**, *1*, 1–18.
37. Hooper, D.; Coughlan, J.; Mullen, M. Structural equation modelling: Guidelines for determining model fit. *Electron. J. Bus. Res. Methods* **2008**, *6*, 53–60.
38. Hu, L.T.; Bentler, P.M. Cut off criteria for fit indexes in covariance structure analysis: Conventional criteria versus new alternatives. *Struct. Equ. Model.* **1999**, *6*, 1–55. [CrossRef]
39. Hair, J.F. *Multivariate Data Analysis*, 7th ed.; Prentice-Hall: Upper Saddle River, NJ, USA, 2010.

buildings

MDPI

Article

A Study of the Effects of Job Stress on the Psychosocial Safety Behavior of Construction Workers: The Mediating Role of Psychological Resilience

Chenning Lu [1], Deming Yu [1], Qingyue Luo [2] and Can Xu [1,*]

[1] School of Civil Engineering, Central South University, Changsha 410075, China
[2] China Machinery International Engineering Design & Research Institute Co., Ltd., Changsha 410007, China
* Correspondence: 174801040@csu.edu.cn; Tel.: +86-18973725958

Abstract: A large number of studies have been conducted to demonstrate that job stress negatively affects construction workers' physiological safety behaviors, but there is a lack of research on the impact of psychosocial safety behaviors on construction workers, which is an important component of overall workplace safety. This study modeled the effects between three job stressors (the job itself, job insecurity, and family-work conflict) and psychosocial safety behavior sub-dimensions (psychosocial safety compliance and psychosocial safety participation), using psychological resilience as a mediating variable. Data were obtained from 304 construction workers in China. The results showed that there were positive and negative effects between the three job stressors, psychosocial safety behavior sub-dimensions, and psychological resilience. Psychological resilience was mediated differently across the three job stressors and psychosocial safety behavior sub-dimensions. This study reveals the mechanisms by which job stress affects the psychosocial safety behaviors of construction workers and provides more empirical evidence to unravel the relationships between various job stressors and psychosocial safety behavior sub-dimensions. In addition, this paper discusses measures to improve psychosocial safety behaviors based on the perspectives of job stressors and psychological resilience.

Keywords: job stress; psychosocial safety behavior; psychological resilience; construction industry

Citation: Lu, C.; Yu, D.; Luo, Q.; Xu, C. A Study of the Effects of Job Stress on the Psychosocial Safety Behavior of Construction Workers: The Mediating Role of Psychological Resilience. *Buildings* **2023**, *13*, 1930. https://doi.org/10.3390/buildings13081930

Academic Editors: Mariusz Szóstak and Marek Sawicki

Received: 26 June 2023
Revised: 22 July 2023
Accepted: 27 July 2023
Published: 28 July 2023

1. Introduction

Construction workers are an important part of the construction industry, and their work is critical to the quality and safety of construction. Construction is a high-risk industry, and accidents occur from time to time. According to the domino model proposed by Heinrich [1], unsafe human behavior is the direct cause of safety accidents. Construction accident investigation studies have found that most accidents are caused by human factors [2–4]. Therefore, many studies have proposed the safety behaviors of construction workers as a prior indicator of the occurrence of safety accidents and combined psychological and behavioral sciences to explore the formation mechanism of construction workers' unsafe behaviors, aiming to improve the safety of construction workplaces and reduce the occurrence of safety accidents [5–11]. However, workplace safety in the construction industry includes not only physiological safety but also psychological safety. Compared with physiological safety and physical behaviors, psychosocial safety and psychological behaviors are often neglected [12]. Studies have found that work-related psychological injuries have a tendency to outweigh physical injuries related to musculoskeletal injuries [12].

Long-term exposure to job stress can cause workers to develop negative emotions such as anxiety, stress, and depression [13]. According to the Job Demands-Resources (JD-R) model proposed by Demerouti et al. [14], job stress can reinforce employees' feelings of exhaustion and negatively affect physical health, thus contributing to unsafe behaviors. There have been numerous studies that have confirmed the negative impact of stress on workers' safety behaviors [5,15–17]. However, these studies focused on physiological safety behaviors. Psychosocial hazards arise when job stress is not effectively managed [18,19]. Bronkhorst [20] extended the JD-R model by introducing the concept of psychosocial safety behaviors based on physiological safety behaviors. The introduction of psychosocial safety behaviors provides a new research direction for improving psychosocial safety outcomes [21]. Psychosocial safety behavior can be considered a prior indicator to assess the psychosocial safety of construction workers. In the construction industry, there is a lack of research on the relationship between the effects of workplace stress and the psychosocial safety behavior of workers.

According to JD-R theory, job resources buffer the process of stress that damages health [22]. Initially, job resources referred to resources provided by the organization, including job control, leadership support, and salary and benefits [23], and as research progressed, personal resources were found to play an important role in buffering the negative effects of stress. Thus, the role of personal resources needs to be considered when studying the impact of job stress on the psychosocial safety behaviors of construction workers. Psychological resilience, an important component of positive psychology, has been shown by research to be effective in combating psychological stress [24]. Thus, psychological resilience can be viewed as a personal resource that can reduce the negative effects of job stress. In this paper, the role of workers' psychological resilience will be considered when examining the effects of job stress on psychosocial safety behaviors.

Previous studies have focused on the safety behaviors of construction workers at the physiological level, ignoring the importance of psychosocial safety behaviors for overall safety in the workplace. Maintaining positive mental health is as important as maintaining physical health. Therefore, based on the above analysis, this paper focuses on the impact of construction workers' job stress on individuals' psychosocial safety behaviors and considers the role of individual psychological resilience from the perspective of personal resources. Examining the relationship between them can help implement different interventions to reduce the impact of work stress on psychosocial safety behaviors. Compared with the results of existing studies, it expands the theoretical study of construction workers' safety behaviors and provides a theoretical basis for future research on psychosocial safety behaviors in various industries.

2. Theoretical Basis and Research Hypothesis

2.1. Job Stress

Human stress arises from interactions between humans and their environment that strain or exceed their ability to adapt and threaten their health [25]. Job stress is a specific result of the work environment [17]. The construction industry is a highly competitive industry [26], and construction workers work in harsh, demanding, and dangerous environments. The work itself is characterized by heavy workloads, tight schedules, heavy responsibilities, and poor safety [5,27,28]. As a result, there are many factors that contribute to job stress among construction workers. According to previous studies, they include the job itself, role ambiguity, role conflict, role overload, job insecurity, job characteristics, employee competencies, interpersonal safety conflicts, safety constraints, and family-work conflicts [6,17]. These factors are known as job stressors and are also referred to as psychosocial risk factors [29]. Job stressors can have an impact on workers' mental health, causing a range of psychological, physical, and behavioral responses.

Different job stressors affect workers to different degrees. Wu et al. [17] found that the job itself, career development, and family-work conflict were the main job stressors. Construction workers often work at heights and are required to use a variety of dangerous machines and equipment. Therefore, the job itself is an important source of stress for construction workers. Construction workers are required to take on a lot of responsibilities and workload, and the stress and challenges brought by the job itself may have adverse effects on workers' mental health. Construction workers are always engaged in short-term tasks, and their jobs are unstable [5,15]. Especially in the post-epidemic era, when the COVID-19 Pandemic has triggered economic contraction and construction workers are vulnerable to unemployment and financial difficulties [30], the job instability of construction workers is highlighted [31]. Construction workers face challenges to their career development, leading to increased job insecurity, which can be detrimental to their physical health and cause adverse psychological, behavioral, and emotional feelings [32]. At the same time, construction workers are prone to family-work conflicts due to their long-term jobs and income instability. High levels of family-work conflicts can negatively affect work behaviors and affect employees' energy, commitment, and other resources, causing psychological distress. Therefore, this paper focuses on analyzing the effects of three job stressors—the job itself, job insecurity, and family-work conflict—on workers' psychosocial safety and behavioral aspects.

2.2. *Job Stress and Psychosocial Safety Behaviors*

Safety behaviors are safety-related behaviors performed by individuals in organizations [7]. Physiological and psychosocial safety behaviors are employee activities in the workplace to maintain physical and mental safety or to create an environment that supports physical and mental safety [20]. Griffin and Neal [21] proposed to distinguish between two types of physiological safety behaviors. Bronkhorst [20] built on this by dividing psychosocial safety behaviors into psychosocial safety compliance and psychosocial safety participation. Psychosocial safety compliance is the activity that an individual must perform to maintain psychosocial safety; psychosocial safety participation is the activity that an individual performs that does not directly enhance psychosocial safety but contributes to a positive psychosocial safety environment. Psychosocial behavioral research can address workplace mental health issues, mitigate psychosocial risks, foster good attitudes toward psychosocial safety behaviors, and reduce overall psychological harm rates [33].

Psychological demands such as job stress are usually associated with mental health outcomes such as emotional exhaustion [18]. A large body of prior research has demonstrated that job stress has a negative impact on physiological safety behavior. In the post-epidemic era, there has been a significant increase in psychological problems such as anxiety, stress, isolation, anxiety, stigma, and discrimination [34,35]. According to JD-R theory, job demands have a direct impact on occupational mental health, and job stress is an important component of job demands. In a study of job demands and psychosocial safety behaviors, Bronkhorst [20] found that workload was significantly and negatively related to psychosocial safety behaviors. Yaris [33] proposed the physical and psychosocial workplace safety (PPWS) model, suggesting that job stress negatively affects an individual's physical and psychological well-being. Zhang et al. [36] found that job role demands significantly and negatively affected psychosocial safety behaviors. Previous studies have proposed different mechanisms by which job stress affects physiological safety compliance and physiological safety participation [6,37]. For psychosocial safety behavior, construction workers may protect their own psychosocial safety as much as possible under job stress rather than devote more energy to maintaining the development of a supportive psychosocial safety environment. Therefore, this study explores the effects of three job stressors on two sub-dimensions of psychosocial safety behaviors and proposes the following hypotheses:

H1. *Job stress has a significant negative effect on psychosocial safety compliance.*

H1a. *The job itself has a significant negative effect on psychosocial safety compliance.*

H1b. *Job insecurity has a significant negative effect on psychosocial safety compliance.*

H1c. *Family-work conflict has a significant negative effect on psychosocial safety compliance.*

H2. *Job stress has a significant negative effect on psychosocial safety participation.*

H2a. *The job itself has a significant negative effect on psychosocial safety participation.*

H2b. *Job insecurity has a significant negative effect on psychosocial safety participation.*

H2c. *Family-work conflict has a significant negative effect on psychosocial safety participation.*

2.3. The Mediating Role of Psychological Resilience

The degree to which job stress affects individuals varies, and how individuals perceive job stress is more important than what they perceive it to be. According to Social Cognitive Theory (SCT), both environment and psychological perception affect individual behavior, and differences in individual characteristics affect the effect of psychological perception [38]. Psychological resilience is an important protective factor against stress [39–41]. Psychological resilience is activated when an individual encounters stress, prompting the individual to deal with it effectively [42]. Howard [43] defines psychological resilience as an individual's capability to adapt and adjust to negative situations like stress, difficulty, and danger. It supports individuals in coping with adversity and encourages them to adjust and develop positively in challenging environments [44].

Those who have more resources under the influence of job stress will be less vulnerable to the loss of resources [45]. Previous research has found that organizational resources have a positive impact on psychosocial safety behaviors [46,47]. Psychological resilience acts as an internal resource [48]. It mobilizes positive emotions to deal with stressful events and counteract the negative effects of stressors [49,50]. Bakker [51] suggested that the higher an employee's personal resources are, the more they are able to protect themselves from the demands of the job. The higher the psychological resilience, the higher the level of mental health, and the more stress-reducing the person is [52,53]. Gao et al. [54] found that psychological resilience mediates the relationship between COVID-19-related stressors and psychological well-being in expatriate employees in the international construction industry. Psychological resilience can have an impact on an individual's risky decision-making, and lower psychological resilience may trigger poor decision-making behavior [55]. It can be inferred that when construction workers face job stress, those with higher levels of psychological resilience are more likely to face the situation positively and avoid psychosocially insecure behaviors. Therefore, this study proposes the following hypothesis:

H3. *Job stress has a significant negative effect on psychological resilience.*

H3a. *The job itself has a significant negative effect on psychological resilience.*

H3b. *Job insecurity has a significant negative effect on psychological resilience.*

H3c. *Family-work conflict has a significant negative effect on psychological resilience.*

H4. *Psychological resilience has a significant positive effect on psychosocial safety compliance.*

H5. *Psychological resilience has a significant positive effect on psychosocial safety participation.*

H6. *Psychological resilience plays a mediating role in the relationship between job stress and psychosocial safety compliance.*

H6a. *Psychological resilience mediates the relationship between the job itself and psychosocial safety compliance.*

H6b. *Psychological resilience mediates the relationship between job insecurity and psychosocial safety compliance.*

H6c. *Psychological resilience mediates the relationship between family-work conflict and psychosocial safety compliance.*

H7. *Psychological resilience mediates the relationship between job stress and psychosocial sa fety participation.*

H7a. *Psychological resilience mediates the relationship between the job itself and psychosocial sa fety participation.*

H7b. *Psychological resilience mediates the relationship between job insecurity and psychosocial safety participation.*

H7c. *Psychological resilience mediates the relationship between family-work conflict and psychosocial safety participation.*

A conceptual model between the variables was developed based on the six hypotheses and 15 sub-hypotheses proposed above, as shown in Figure 1.

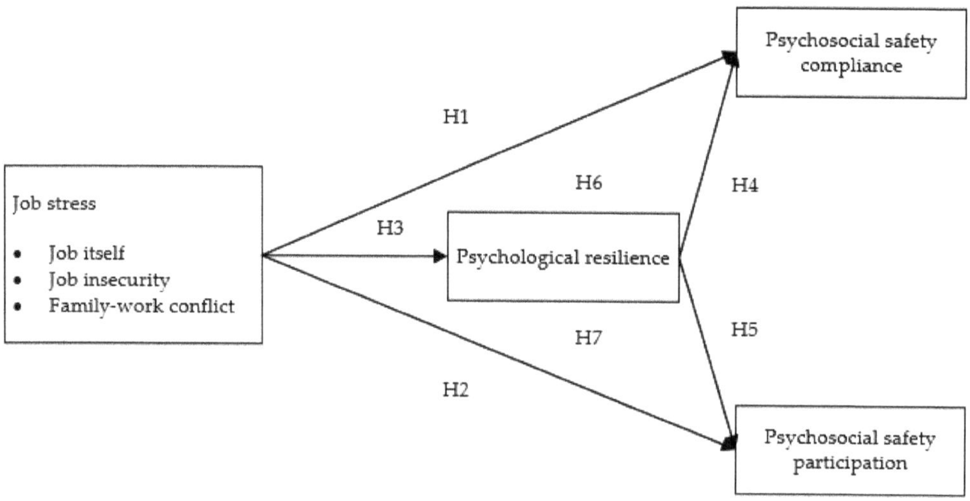

Figure 1. Conceptual model.

3. Materials and Methods

3.1. Participants and Processes

Prior to the formal survey, a preliminary questionnaire was sent to 20 workers at a construction site in Changsha, Hunan Province, China, in November 2022 for pretesting in this study. Based on the feedback from these 20 workers, the questionnaire questions were modified to make them easier to understand and to facilitate the formal survey.

The formal survey was conducted in March 2023 in the Chinese construction industry. The participants in this study were construction workers, and a combination of online and paper questionnaires was used to distribute 243 online questionnaires and 150 paper questionnaires asking them to assess their job stress, psychosocial safety behaviors, and psychological resilience. Ultimately, 243 web-based questionnaires and 131 paper-based questionnaires were returned in this study. We excluded 51 web-based questionnaires with a too-short response time and non-construction workers filling them out and 19 paper-based questionnaires with many missing values. The final sample consisted of 304 valid questionnaires, with a valid response rate of 77.35%. The questionnaires are shown in Appendix A.

The demographics of the workers included age, gender, education level, work experience, marital status, fertility, and type of work (Table 1). The age of workers was mostly distributed at 50 years and below (80.26%, n = 244), and a few (19.74%, n = 60) workers were above 50 years of age. The workers were mostly male (82.24%, n = 250), and the rest were female. In terms of educational attainment, the majority (77.63%, n = 236) had a low-level school diploma, and only 22.37% (n = 68) had a college diploma or higher. Furthermore, 76.97% of the workers had worked for more than 5 years (n = 234), while 23.03% had worked for less than 5 years. Moreover, 75.99% of the workers were married (n = 231), and the rest were unmarried. Other trades (odd jobs, small jobs, etc.), special equipment (elevators, tower cranes, lifts, etc.), and carpentry were the top three trades (36.51%, 18.75%, and 16.78%, respectively).

Table 1. Demographic characteristics of construction workers (N = 304).

Characteristics	Items	Frequency	Percentage (%)
Age	≤30	68	22.37
	31–40	96	31.58
	41–50	80	26.32
	≥51	60	19.74
Gender	Male	250	82.24
	Female	54	17.76
Educational level	Primary school and below	34	11.18
	Junior high school	135	44.41
	High school, junior college, technical school	67	22.04
	College and above	68	22.37
Work experience	≤5	70	23.03
	6–15	132	43.42
	16–25	53	17.43
	≥26	49	16.12
Marital status	Unmarried	73	24.01
	Married	231	75.99
Fertility	No children	75	24.67
	Only child	82	26.97
	Many children	147	48.36
Type of work	Carpenter	51	16.78
	Steelworker	44	14.47
	Bricklayer	27	8.88
	Erector	14	4.61
	Special equipment (elevator, tower crane, hoist, etc.)	57	18.75
	Other types of work (odd jobs, small jobs, etc.)	111	36.51

3.2. Questionnaire Development and Measurement

The questionnaire of this study contains three parts: the first is the basic information of the questionnaire, including the introduction of the purpose and use of the questionnaire so that the respondents have a preliminary understanding of the questionnaire, and the second is the demographic characteristics of the respondents, including age, gender, education level, work experience, marital status, fertility, and type of work. The third part was a questionnaire on three variables, job stress, psychosocial safety behaviors, and psychological resilience, developed by referring to relevant established scales; Table 2 lists the variables. All measures were scored on a five-point Likert scale.

Table 2. Variables, measures of the studied variables, number of items, and supporting literature.

Variables	Dimensions	Items	Supporting Literature
Job stress	Job itself	4	Wu et al. [17]
	Job insecurity	4	Vander et al. [56]
	Family-work conflict	3	Wu et al. [17]
Psychosocial safety behavior	Psychosocial safety compliance	3	Bronkhorst [20]
	Psychosocial safety participation	3	
Psychological resilience	Psychological resilience	10	Campbell-Sills and Stein [57]

3.2.1. Job Stress

This study analyzes three job stressors: the job itself, job insecurity, and family-work conflict. The measures of these stressors are referred to the scales developed by Wu et al. [17] and Vander et al. [56], respectively, and the scales are translated and modified. Among those, four items were used to measure the job itself, four items to assess job insecurity, and three items to assess family-work conflict. For example, "My job is simple and has a low workload." (job itself); "I could be without this job at any time." (job insecurity); and

"I have been away from my family for a long time and do not take care of them much." (family-work conflict).

3.2.2. Psychosocial Safety Behavior

In this paper, the measures of psychosocial safety behaviors are referred to the scales developed by Bronkhorst [20], and the scale is translated and modified. The scale includes two dimensions: psychosocial safety compliance and psychosocial safety participation. Each dimension is measured by three items. For example, "I use measures to prevent or minimize psychological strain in my job." (psychosocial safety compliance) and "I promote the psychological safety program within the organization." (psychosocial safety participation).

3.2.3. Psychological Resilience

In this paper, we refer to Campbell-Sills and Stein's [57] modified Connor-Davidson Resilience Scale to measure the psychological resilience of construction workers with 10 questions. For example, "I can adapt to change".

3.3. Analysis Strategy

To test the hypotheses proposed in this paper, the data were analyzed using SPSS 26.0 and AMOS 24.0 software. First, descriptive analysis, reliability tests, and correlation analyses were performed on the variables using SPSS 26.0. Second, Confirmatory Factor Analysis (CFA) was conducted on the sample data using AMOS 24.0 to test the practicability and veracity of the construct validity of the questionnaire. Finally, path analysis and mediating effects tests were conducted on the sample data using AMOS 24.0 to analyze the direct effects between the variables and the mediating effects of psychological resilience, and the theoretical hypotheses were tested.

4. Results

4.1. Descriptive Analysis, Reliability Test, and Correlation Analysis

In Table 3, the descriptive statistics, Cronbach's alphas, and Pearson correlation coefficients between the variables of the sample data are presented. Cronbach's alpha is a standard measure commonly used for reliability testing, and it must be higher than 0.7 in order to obtain good reliability [58]. The closer it is to 1, the better the reliability and stability of the questionnaire [59]. The Cronbach's alphas for the six variables of job itself, job insecurity, family-work conflict, psychosocial safety compliance, psychosocial safety participation, and psychological resilience were all above 0.7, showing good reliability. In addition, the Cronbach's alpha coefficients of each second-order variable were also above the critical values, and the Cronbach's alpha values of job stress and psychosocial safety behavior were 0.865 and 0.879, respectively. Based on this, all variables had good reliability.

Table 3. Descriptive statistics, Cronbach's alpha values, and Pearson's correlation coefficients among variables.

Variables	Mean	SD	1	2	3	4	5	6
1. The job itself	3.220	0.866	(0.812)					
2. Job insecurity	3.036	0.821	0.425 **	(0.835)				
3. Family-work conflict	3.249	0.923	0.432 **	0.488 **	(0.808)			
4. Psychosocial safety Compliance	3.235	0.862	−0.534 **	−0.428 **	−0.296 **	(0.826)		
5. Psychosocial safety participation	3.070	0.826	−0.535 **	−0.481 **	−0.369 **	0.711 **	(0.801)	
6. Psychological resilience	3.336	0.808	−0.506 **	−0.533 **	−0.403 **	0.632 **	0.637 **	(0.956)

Notes: ** $p < 0.01$; SD = standard deviation; Cronbach's alpha values are reported in the brackets.

According to Pearson's correlation coefficient, all three job stressors (the job itself, job insecurity, and family-work conflict) were negatively correlated with psychosocial safety compliance and negatively correlated with psychosocial safety participation. Psychological resilience was negatively related to all three job stressors and positively related to psychosocial safety compliance and psychosocial safety participation.

4.2. Validity Testing

In this study, using AMOS 24.0, CFA analysis was conducted on the job stress measurement model, psychosocial safety behavior measurement model, and psychological resilience measurement model. The model fit was verified against professional indicators, and the general better fit criteria were: chi-square/degrees of freedom (χ^2/df) ≤ 5 [60], root-mean-square error of approximation (RMSEA) < 0.1, incremental fit index (IFI) > 0.9, Tucker-Lewis index (TLI) > 0.9 [61], comparative fit index (CFI) ≥ 0.9, standardized root mean square residual (SRMR) ≤ 0.08 [62]. According to the test results, all three measurement models have a good fit.

Hair et al. [63] proposed to verify convergent validity using standardized factor loadings (SFL), construct reliability (CR), and average variance extracted (AVE). When these three metrics satisfy SFL > 0.5 [63], CR > 0.6, and AVE > 0.5 [64], it indicates good convergent validity. Table 4 shows the results of the convergent validity of the variables. For the job itself (psychosocial safety compliance, etc.), the SFLs range (0.533–0.899) is higher than the critical value of 0.5, The CRs range (0.802–0.956) is higher than the critical value of 0.6, and the AVEs range (0.522–0.685) is higher than the critical value of 0.5. Therefore, the measures all met the convergent validity requirement.

Table 4. Results of convergent validity for variables.

Variables	Item	SFL	CR	AVE
The job itself	1	0.698	0.813	0.522
	2	0.741		
	3	0.740		
	4	0.709		
Job insecurity	1	0.788	0.839	0.572
	2	0.533		
	3	0.809		
	4	0.853		
Family-work conflict	1	0.899	0.815	0.599
	2	0.670		
	3	0.734		
Psychosocial safety compliance	1	0.746	0.828	0.616
	2	0.819		
	3	0.788		
Psychosocial safety participation	1	0.774	0.802	0.575
	2	0.725		
	3	0.774		
Psychological resilience	1	0.804	0.956	0.685
	2	0.811		
	3	0.802		
	4	0.773		
	5	0.846		
	6	0.844		
	7	0.842		
	8	0.825		
	9	0.850		
	10	0.872		

Notes: SFL = standard deviation; CR = construct reliability; AVE = average variance extracted.

If the square root of the AVE of a variable is greater than its correlation coefficient, it indicates good discriminant validity [48]. In Table 5, the results of discriminant validity calculations for the six variables are shown. For example, the square root of AVE for the job itself is 0.723, which is higher than the correlation coefficients associated with it, indicating good discriminant validity. The other five variables had the same results, indicating good discriminant validity for all.

Table 5. Results of discriminant validity.

Variables	1	2	3	4	5	6
1. The job itself	(0.723)					
2. Job insecurity	0.425	(0.756)				
3. Family-work conflict	0.432	0.488	(0.775)			
4. Psychosocial Safety Compliance	−0.534	−0.428	−0.296	(0.785)		
5. Psychosocial Safety participation	−0.535	−0.481	−0.369	0.711	(0.758)	
6. Psychological resilience	−0.506	−0.533	−0.403	0.632	0.637	(0.828)

Notes: The square roots of AVEs are reported in brackets.

4.3. *Hypotheses Testing*

The Structural Equation Model (SEM) technique was used to test the fit of the sample data to the hypothetical measurement model, and the fit indices of the hypothetical model were: χ^2/df = 1.766 (p = 0.000), RMSEA = 0.050, IFI = 0.956, TLI = 0.949, CFI = 0.955, SRMR = 0.042. The fit indices were all within the acceptable range.

The hypothesis was supported by judging the p-value to be less than 0.05 and the estimate sign to be positive or negative in line with the hypothesized positive or negative relationship. For hypotheses with multiple sub-hypotheses (H1, H2, H3, H6, and H7), the hypothesis was considered partially supported if some sub-hypotheses were supported [7]. Figure 2 shows the results of the path analysis, including the structural model and the standard path coefficients labeled on each path. In addition, links with significant hypothesis test results are indicated by solid lines, and links with insignificant hypothesis tests are indicated by dashed lines. Table 6 shows the path coefficients of the structural model.

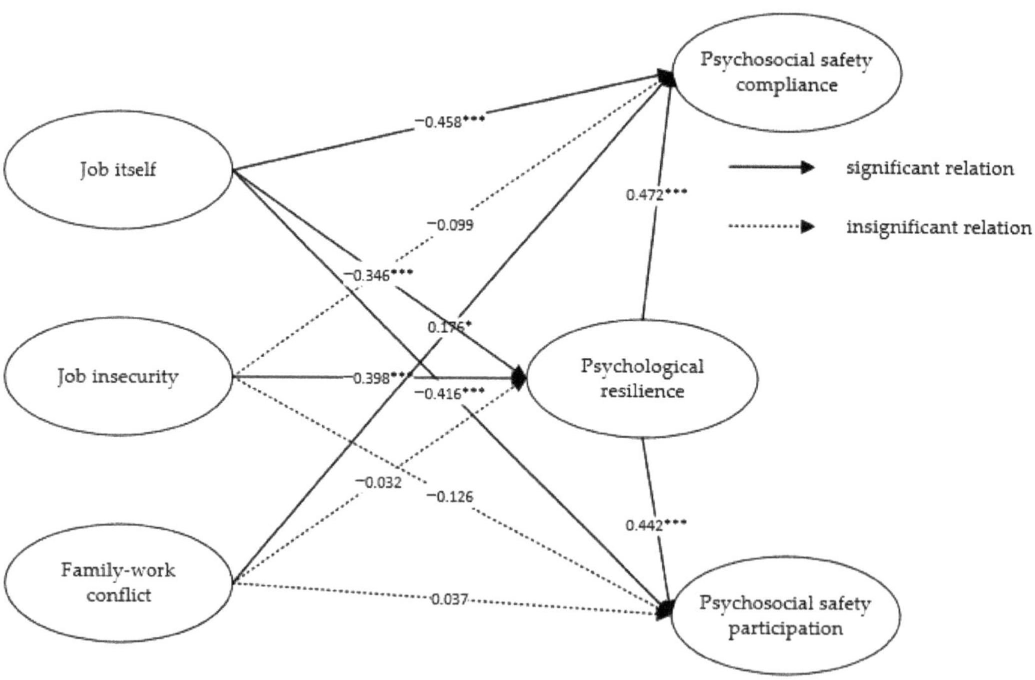

Figure 2. Structural model with standard path coefficients. Note: * $p < 0.05$; *** $p < 0.001$.

Table 6. Path coefficients of the model.

Path	Estimate	S.E.	C.R.	P
The job itself → Psychosocial safety compliance	−0.458	0.083	−5.596	***
Job insecurity → Psychosocial safety compliance	−0.099	0.065	−1.344	0.179
Family-work conflict → Psychosocial safety compliance	0.176	0.066	2.502	*
Job itself → Psychosocial safety participation	−0.416	0.082	−5.353	***
Job insecurity → Psychosocial safety participation	−0.126	0.066	−1.735	0.083
Family-work conflict → Psychosocial safety participation	0.037	0.066	0.550	0.582
The job itself → Psychological resilience	−0.346	0.072	−4.679	***
Job insecurity → Psychological resilience	−0.398	0.061	−5.527	***
Family-work conflict → Psychological resilience	−0.032	0.063	−0.457	0.648
Psychological resilience → Psychosocial safety compliance	0.472	0.077	6.423	***
Psychological resilience → Psychosocial safety participation	0.442	0.076	6.300	***

Notes: * $p < 0.05$, *** $p < 0.001$; Estimate = standardized regression coefficients; S.E. = standardized error; C.R. = critical ratio.

The results of the hypothesis tests showed that H1a was supported, H1b was rejected, and H1c, although also rejected, yielded the opposite conclusion of the hypothesis, that family-work conflict has a significant positive effect on psychosocial safety compliance. Therefore, H1 was partially supported. H2a was supported, H2b was rejected, and H2c was rejected. Therefore, H2 was partially supported. H3a and H3b were supported, and H3c was rejected, where job insecurity had a stronger effect on psychological resilience than the job itself. Therefore, H3 was partially supported. H4 and H5 were both supported, and the effect of psychological resilience on psychosocial safety compliance was slightly higher than its effect on psychosocial safety engagement.

This study used SEM to assess the mediating mediation effects, which allows for the inclusion of measurement error compared to multiple regression models [65,66]. In this paper, we use Bootstrap for mediating effects testing, with 5000 draws on the original sample size (N = 304). The results are shown in Table 7. H6a and H6b were partially supported, and psychological resilience partially mediated the effect of the job itself on construction workers' psychosocial safety compliance and psychosocial safety participation, with mediating effects of 26.232% and 26.923%, respectively. H6b and H7b were supported. Psychological resilience fully mediates the effect of job insecurity on psychosocial safety compliance and participation. H6c and H7c were rejected.

Table 7. Results of the mediation effect test.

Path	Direct Effects	Indirect Effects	Total Effects
The job itself → Psychological resilience → Psychosocial safety compliance	−0.464 ***	−0.165 **	−0.629 ***
Job insecurity → Psychological resilience → Psychosocial safety compliance	−0.087	−0.166 **	−0.253 *
Family-work conflict → Psychological resilience → Psychosocial safety compliance	0.166 *	−0.014	0.151
The job itself → Psychological resilience → Psychosocial safety participation	−0.437 ***	−0.161 **	−0.598 ***
Job insecurity → Psychological resilience → Psychosocial safety participation	−0.115	−0.161 **	−0.276 *
Family-work conflict → Psychological resilience → Psychosocial safety participation	0.036	−0.014	0.023

Notes: * $p < 0.05$, ** $p < 0.01$, *** $p < 0.001$.

5. Discussion

This study investigated the influence of job stressors (job itself, job insecurity, and family-work conflict) on construction workers' psychosocial safety behaviors in terms of psychosocial safety compliance and psychosocial safety participation and considered the mediating role of psychological resilience. The results of the study showed that the three job stressors had different effects on psychosocial safety compliance and psychosocial safety participation, and it was found that: (1) The job itself had a negative effect on all psychosocial safety behavior sub-dimensions; job insecurity had no effect on all psychosocial safety behavior sub-dimensions; family-work conflict had a positive effect on psychosocial safety compliance; and there was no effect on psychosocial safety participation. (2) Psychological resilience had a positive effect on all psychosocial safety behavior sub-dimensions. (3) Psychological resilience partially mediates the effect of both work itself and psychosocial safety behavior sub-dimensions; all play a fully mediating role in the effect of job insecurity on psychosocial safety behavior sub-dimensions; and none in the effect of family-work conflict on psychosocially safe behavior sub-dimensions.

5.1. Theoretical Implications

First, this study verifies the extended JD-R theoretical model proposed by Bronkhorst, and the results show that the application of the theoretical model in the construction industry has certain feasibility and validity. At present, there are many research results on physiological safety and less research on psychosocial safety. Psychosocial safety behavior provides a good framework for studying psychosocial safety [21]. Just as physiological safety behavior is a prior indicator for measuring physiological safety accidents, psychosocial safety behavior is a prior indicator for measuring psychosocial safety hazards. The results of this study suggest that the extended JD-R theoretical model provides an effective theoretical basis for the study of psychosocial safety issues in the construction industry and can help companies develop intervention strategies that can reduce the psychosocial risks arising from construction workers at work. Oil, gas, and mining are also high-risk industries, and their employees are also prone to psychosocial safety problems [36,67], and this study can provide a theoretical basis and methodology for the study of psychosocial safety behaviors in these industries. Therefore, this study can provide strong support for research on psychosocial safety behaviors in the construction industry and even other industries.

Second, this study found that different job stressors have different effects on psychosocial safety compliance and psychosocial safety participation. The job itself can negatively affect construction workers' psychosocial safety compliance and psychosocial safety participation, which echoes Bronkhorst's [20] study. The demands of high workload, poor security, and high responsibility for construction workers impose psychological costs that subsequently affect workers' psychosocial safety behaviors. Job insecurity has no direct effect on psychosocial safety compliance and participation and differs from Bronkhorst's [20] findings in that it can negatively affect psychosocial safety compliance and participation through psychological resilience. For many, continued employment is an essential source of income [68]. Job insecurity responds to the expectation that employees will lose their jobs [69], and this expectation can threaten psychological resources [70]. Construction workers' psychological resilience is a psychological resource and is significantly and positively associated with psychosocial safety compliance and psychosocial safety participation. Interestingly, this study found a positive effect of family-work conflict on psychosocial safety compliance, which is the exact opposite of the results of the study on the effect of physical safety behaviors [17]. Zhu et al. [71] also obtained fat conclusions contrary to the hypothesis and found that psychosocial stress symptoms had a positive and significant effect on the sub-dimension of psychosocial safety behaviors among Ghanaian miners. In the course of research on the JD-R model, some scholars have suggested that stressors can be classified as hindering or challenging based on the nature of their effects [72,73]. Hindering stressors are detrimental to goal accomplishment, while challenging stressors promote goal accomplishment and personal development. A high level of workload may bring growth and benefits to individuals [23]. In China, construction workers are the primary source of income for their families, have significant family responsibilities, and play an important role in their families. In this context, appropriate family-work conflict stress is beneficial for construction workers to proactively maintain psychological safety and thus better manage family-work conflicts. By demonstrating the different effects of different job stressors on psychosocial safety compliance and psychosocial safety participation. This reveals the intrinsic mechanism by which construction workers' job stress affects psychosocial safety behaviors and provides help for effective interventions to mitigate the effects of job stress on workers' psychosocial safety. These findings are informative for further research on the factors influencing psychosocial safety behaviors.

Finally, this study contributes to psychosocial safety behavior research by considering the role of individual resources in a hypothetical model. Prior research has identified that job stress can have an impact on psychosocial safety behaviors; however, the role of considering individual resources in the impact mechanism is lacking. High levels of job stress among employees are largely objective and often uncontrollable [14,74]. The findings of this study provide an individual corrective perspective in which we demonstrate the positive effects of psychological resilience on psychosocial safety compliance and psychosocial safety participation, as well as the ability of psychological resilience to reduce the negative effects of job itself and job insecurity on psychosocial safety compliance and psychosocial safety participation. These findings can broaden our understanding of the mechanisms influencing psychosocial safety behaviors, which can lead to the better development and implementation of relevant measures to ensure the psychosocial safety of construction workers.

5.2. Practical Implications

First, this study is important for enhancing overall workplace safety. The psychosocial safety behaviors of workers in high-risk industries should be addressed. Construction projects can incorporate psychosocial safety behaviors into safety performance assessments. This will promote workers' awareness of the importance of psychosocial safety, which in turn will lead to a greater focus on their own psychological health and safety at work, thus improving the safety level and performance of the entire project. At the same time, construction projects should also strengthen training and promotion for workers to improve their knowledge and understanding of psychosocial safety and promote full participation to improve overall safety in the workplace.

Secondly, this study found that three job stressors positively and negatively affect psychosocial safety compliance and participation. Therefore, only by clarifying the mechanisms of action between them can we better improve psychosocial safety. According to the research results, managers need to focus on the negative effects of the job itself and job insecurity. Managers need to reasonably arrange the workload and working hours of construction workers to avoid overload. It is also essential to ensure the safety of construction workers and to ensure that safety equipment is in good condition [75]. Construction managers need to promote the continued employment of construction workers. Construction workers have a low level of education and a single channel to obtain employment information. In this case, realistic job information can be provided to broaden their access to job information [69]. For family-work conflicts, managers can take steps to reward construction workers for good psychosocial safety behaviors. For example, monetary rewards, since money is the main reason why workers go to construction sites in China [76], are a way to create a positive cycle.

Finally, the results of this study suggest that interventions can be made at the individual level that can mitigate the negative effects of job stress on construction workers' psychosocial safety behaviors. Project management teams can take steps to enhance the psychological resilience of construction workers. To improve workers' stress tolerance and resilience, managers can consider implementing "psychological resilience education", such as sharing lessons learned from bad experiences [6]. Building the psychological resilience of construction workers at the organizational level, such as by providing training in psychological resilience-related knowledge and skills, is also an effective way to enhance their psychological resilience.

5.3. Limitations and Future Research

First, the present study is cross-sectional and cannot detect longitudinal relationships among job stress, psychological resilience, and psychosocial safety behaviors. Future research could examine potential relationships among them over time. Second, there are a variety of job demands and resources, and the present study considered only three job stressors and one personal resource. Therefore, future research could introduce more job demands and job resources, such as interpersonal conflict and psychosocial safety climate. Furthermore, this study found that different job stressors have different effects on psychosocial safety behaviors, and future research can further investigate the mechanisms of action of challenging and hindering stressors. Finally, this study explored only the mechanisms that influence psychosocial safety behaviors, and the relationship between the impact of psychosocial safety and physiological safety was not clear. According to the PPWS model proposed by Yaris, psychosocial safety and physiological safety will interact with each other, and future studies can further explore the relationship between psychosocial safety behaviors and physiological safety behaviors.

6. Conclusions

Psychosocial safety is an important component of overall safety in the construction workplace. In this study, a hypothesized relationship between job stress, psychosocial safety behaviors, and mental toughness was proposed. The data comes from 304 construction workers in China, and the hypotheses were tested through SEM analysis. This study found that three job stressors have different effects on two dimensions of psychosocial safety behaviors through different pathways. The job itself had a direct negative impact on construction workers' psychosocial safety compliance and psychosocial safety participation, with psychological resilience playing a partially mediating role. Job insecurity had a negative effect on construction workers' psychosocial safety compliance and psychosocial safety participation through psychological resilience. Family-work conflict had a direct positive effect on construction workers' psychosocial safety compliance but no direct effect on psychosocial safety participation. The study also found a direct positive effect of psychological resilience on psychosocial safety compliance and psychosocial safety participation among construction workers. The findings of this study are discussed to increase understanding of the interaction between job stress and psychosocial safety behaviors among construction workers. By revealing the mechanisms by which job stress triggers psychosocial safety behaviors, this study helps broaden the body of research on construction workers' safety behaviors and has implications for the study of psychosocial safety behaviors in other high-risk occupations.

Author Contributions: Conceptualization, C.L. and D.Y.; Methodology, C.L. and D.Y.; Software, C.L. and C.X.; Validation, C.L., D.Y. and C.X.; Formal analysis, C.L.; Investigation, C.L. and Q.L.; Resources, C.L., D.Y., Q.L. and C.X.; Data curation, C.L. and Q.L.; Writing—original draft, C.L.; Writing—review & editing, D.Y., Q.L. and C.X.; Visualization, D.Y.; Supervision, D.Y.; Project administration, D.Y., Q.L. and C.X.; Funding acquisition, D.Y. All authors have read and agreed to the published version of the manuscript.

Funding: This research received no external funding.

Data Availability Statement: Research data for this article is available through the corresponding author.

Conflicts of Interest: The authors declare no conflict of interest.

Appendix A

Construction Workers' Job Stress, Psychosocial Safety Behavior, and Psychological Resilience Questionnaire

Dear Respondents:

Hello! I am a master's student at Central South University. Thank you for taking the time out of your busy schedule to participate in this questionnaire. The purpose of this questionnaire is to understand construction workers' job stress, psychosocial safety behaviors, and psychological resilience so as to propose intervention measures for the impact of construction workers' job stress on mental health. This questionnaire is completely anonymous, and your answers will be treated with strict confidentiality, so please do not have any concerns. We assure you that the data obtained from this questionnaire will be used purely for academic research.

This questionnaire is not a test; there are no standard answers, no "right", "wrong", "good", or "bad" answers to any question; they are all single-choice. What we want to know are your true state and feelings. Please fill out the actual situation according to your own truthfulness. Thank you for your support!

Part I: Personal Information (Please tick the options that match your situation.)

1. Your age: A. ≤30 years old; B. 31–40 years old; C. 41–50 years old; D. ≥51 years old
2. Your gender: A. Male; B. Female
3. Your education level: A. Primary school and below; B. Junior high school; C. High school, junior college, or technical school; D. College and above
4. Your work experience: A. ≤5 years; B. 6–15 years; C. 16–25 years; D. ≥26 years
5. Your marital status: A. Unmarried; B. Married
6. Your fertility: A. No children; B. Only child; C. Many children
7. Your type of work: A. Carpenter; B. Steelworker; C. Bricklayer; D. Erector; E. Special equipment (elevator, tower crane, hoist, etc.); F. Other types of work (odd jobs, small jobs, etc.)

Part II: Job Stress, Psychosocial Safety Behavior, and Psychological Resilience Measurement (Please answer the following questions based on your real feelings and put a tick on the number that best matches. "1" = strongly disagree, "2" = disagree, "3" = not sure, "4" = agree, "5" = strongly agree.)

The job itself:

8. My job is simple and has a low workload.
9. At work, I never worry about personal safety.
10. I never work overtime.
11. I am not afraid of being accountable at work.

Job insecurity:

12. I could be without this job at any time.
13. I believe I will be able to keep this job.
14. I am uneasy about my future job income and opportunities.
15. I may lose this job in the near future.

Family-work conflict:

16. I have been away from my family for a long time and do not take much care of them.
17. My family does not understand and support my work enough.
18. My financial income is not high, and I have a heavy family burden.

Psychosocial safety compliance:

19. I use measures to prevent or minimize psychological strain in my job.
20. I use the correct regulations and protocols for psychological safety when carrying out my job.
21. I ensure the highest levels of psychological safety when I carry out my job.

Psychosocial safety participation:

22. I promote the psychological safety program within the organization.
23. I put in extra effort to improve the psychological safety of the workplace.
24. I voluntarily carry out tasks or activities that help improve workplace psychological safety.
25. I can adapt to change.
26. No matter what happens, I can handle it.
27. I can see the humorous side of things.
28. Coping with stress makes me feel empowered.
29. I will recover quickly from illness and hardship.
30. I can overcome obstacles to achieve my goals.
31. Under pressure, I can focus and think clearly.
32. I do not get discouraged by failure.
33. I feel that I am a strong person.
34. I can handle my negative emotions.

References

1. Heinrich, H.W. *Industrial Accident Prevention: A Scientific Approach*; McGraw-Hill: New York, NY, USA, 1931.
2. Rupasinghe, N.K.A.H.; Panuwatwanich, K. Extraction and analysis of construction safety hazard factors from open data. *IOP Conf. Ser. Mater. Sci. Eng.* **2020**, *849*, 012008. [CrossRef]
3. Hong Kong Special Administrative Region, L.D. Occupational Safety and Health Statistics 2011. Available online: http://www.labour.gov.hk/tc/osh/pdf/archive/statistics/OSH_Statistics_2011.pdf (accessed on 8 January 2023).
4. Li, H.; Lu, M.; Hsu, S.-C.; Gray, M.; Huang, T. Proactive behavior-based safety management for construction safety improvement. *Saf. Sci.* **2015**, *75*, 107–117. [CrossRef]
5. Leung, M.-Y.; Liang, Q.; Olomolaiye, P. Impact of Job Stressors and Stress on the Safety Behavior and Accidents of Construction Workers. *J. Manag. Eng.* **2016**, *32*, 04015019. [CrossRef]
6. Wang, D.; Wang, X.; Xia, N. How safety-related stress affects workers' safety behavior: The moderating role of psychological capital. *Saf. Sci.* **2018**, *103*, 247–259. [CrossRef]
7. He, C.; Jia, G.; McCabe, B.; Chen, Y.; Sun, J. Impact of psychological capital on construction worker safety behavior: Communication competence as a mediator. *J. Saf. Res.* **2019**, *71*, 231–241. [CrossRef] [PubMed]
8. Shin, M.; Lee, H.-S.; Park, M.; Moon, M.; Han, S. A system dynamics approach for modeling construction workers' safety attitudes and behaviors. *Accid. Anal. Prev.* **2014**, *68*, 95–105. [CrossRef]
9. Fang, D.; Huang, Y.; Guo, H.; Lim, H.W. LCB approach for construction safety. *Saf. Sci.* **2020**, *128*, 104761. [CrossRef]
10. Newaz, M.T.; Davis, P.; Jefferies, M.; Pillay, M. The psychological contract: A missing link between safety climate and safety behaviour on construction sites. *Saf. Sci.* **2019**, *112*, 9–17. [CrossRef]
11. Zhang, Z.; Guo, H.; Gao, P.; Wang, Y.; Fang, Y. Impact of owners? safety management behavior on construction workers? unsafe behavior. *Saf. Sci.* **2023**, *158*, 105944. [CrossRef]
12. Carlos Aldasoro, J.; Luisa Cantonnet, M. The management of the new and emerging musculoskeletal and psychosocial risks by EU-28 enterprises. *J. Saf. Res.* **2021**, *77*, 277–287. [CrossRef]
13. Onubi, H.O.; Yusof, N.A.; Hassan, A.S. Perceived COVID-19 Safety Risk and Safety Behavior on Construction Sites: Role of Safety Climate and Firm Size. *J. Constr. Eng. Manag.* **2021**, *147*, 04021153. [CrossRef]
14. Demerouti, E.; Bakker, A.B.; Nachreiner, F.; Schaufeli, W.B. The job demands-resources model of burnout. *J. Appl. Psychol.* **2001**, *86*, 499–512. [CrossRef] [PubMed]
15. Choudhry, R.M.; Fang, D. Why operatives engage in unsafe work behavior: Investigating factors on construction sites. *Saf. Sci.* **2008**, *46*, 566–584. [CrossRef]
16. Liang, Q.; Zhou, Z.; Ye, G.; Shen, L. Unveiling the mechanism of construction workers' unsafe behaviors from an occupational stress perspective: A qualitative and quantitative examination of a stress-cognition-safety model. *Saf. Sci.* **2022**, *145*, 105486. [CrossRef]
17. Wu, X.; Li, Y.; Yao, Y.; Luo, X.; He, X.; Yin, W. Development of Construction Workers Job Stress Scale to Study and the Relationship between Job Stress and Safety Behavior: An Empirical Study in Beijing. *Int. J. Environ. Res. Public Health* **2018**, *15*, 2409. [CrossRef]
18. Bailey, T.S.; Dollard, M.F.; McLinton, S.S.; Richards, P.A.M. Psychosocial safety climate, psychosocial and physical factors in the aetiology of musculoskeletal disorder symptoms and workplace injury compensation claims. *Work Stress* **2015**, *29*, 190–211. [CrossRef]
19. Derdowski, L.A.; Mathisen, G.E. Psychosocial factors and safety in high-risk industries: A systematic literature review. *Saf. Sci.* **2023**, *157*, 105948. [CrossRef]
20. Bronkhorst, B. Behaving safely under pressure: The effects of job demands, resources, and safety climate on employee physical and psychosocial safety behavior. *J. Saf. Res.* **2015**, *55*, 63–72. [CrossRef]
21. Weaver, B.; Kirk-Brown, A.; Goodwin, D.; Oxley, J. Psychosocial safety behavior: A scoping review of behavior-based approaches to workplace psychosocial safety. *J. Saf. Res.* **2023**, *84*, 33–40. [CrossRef]
22. Bakker, A.B.; Demerouti, E. The Job Demands-Resources model: State of the art. *J. Manag. Psychol.* **2007**, *22*, 309–328. [CrossRef]
23. QI, Y.-J.; Wu, X.-C. Job Demands-Resources Model: The Development of Theoretical and Empirical Research. *J. Beijing Norm. Univ. (Soc. Sci. Ed.)* **2018**, *270*, 28–36.
24. Zhao, Q.; Huangfu, C.; Li, J.; Liu, H.; Tang, N. Psychological Resilience as the Mediating Factor Between Stigma and Social Avoidance and Distress of Infertility Patients in China: A Structural Equation Modeling Analysis. *Psychol. Res. Behav. Manag.* **2022**, *15*, 391–403. [CrossRef] [PubMed]
25. Kanji, G.K.; Chopra, P.K. Psychosocial system for work well-being: On measuring work stress by causal pathway. *Total Qual. Manag. Bus. Excell.* **2009**, *20*, 563–580. [CrossRef]
26. Naoum, S.G.; Herrero, C.; Egbu, C.; Fong, D. Integrated model for the stressors, stress, stress-coping behaviour of construction project managers in the UK. *Int. J. Manag. Proj. Bus.* **2018**, *11*, 761–782. [CrossRef]
27. Lingard, H.; Sublet, A. The impact of job and organizational demands on marital or relationship satisfaction and conflict among Australian civil engineers. *Constr. Manag. Econ.* **2002**, *20*, 507–521. [CrossRef]
28. Lv, X.; Wu, X.; Ci, H.; Liu, Q.; Yao, Y. Empirical research on the influencing factors of the occupational stress for construction workers. *IOP Conf. Ser. Earth Environ. Sci.* **2017**, *61*, 012102. [CrossRef]
29. Manapragada, A.; Bruk-Lee, V.; Thompson, A.H.; Heron, L.M. When safety climate is not enough: Examining the moderating effects of psychosocial hazards on nurse safety performance. *J. Adv. Nurs.* **2019**, *75*, 1207–1218. [CrossRef]

30. Tong, R.; Wang, L.; Cao, L.; Zhang, B.; Yang, X. Psychosocial factors for safety performance of construction workers: Taking stock and looking forward. *Eng. Constr. Archit. Manag.* **2023**, *30*, 944–962. [CrossRef]
31. King, T.L.; Lamontagne, A.D. COVID-19 and suicide risk in the construction sector: Preparing for a perfect storm. *Scand. J. Public Health* **2021**, *49*, 774–778. [CrossRef]
32. Liang, H.; Liu, T.; Yang, W.; Xia, F. Impact of COVID-19 Pandemic Perception on Job Stress of Construction Workers. *Int. J. Environ. Res. Public Health* **2022**, *19*, 169. [CrossRef]
33. Yaris, C.; Ditchburn, G.; Curtis, G.J.; Brook, L. Combining physical and psychosocial safety: A comprehensive workplace safety model. *Saf. Sci.* **2020**, *132*, 104949. [CrossRef]
34. Rossi, R.; Socci, V.; Talevi, D.; Mensi, S.; Niolu, C.; Pacitti, F.; Di Marco, A.; Rossi, A.; Siracusano, A.; Di Lorenzo, G. COVID-19 Pandemic and Lockdown Measures Impact on Mental Health Among the General Population in Italy. *Front. Psychiatry* **2020**, *11*, 790. [CrossRef] [PubMed]
35. Wise, T.; Zbozinek, T.D.; Michelini, G.; Hagan, C.C.; Mobbs, D. Changes in risk perception and self-reported protective behaviour during the first week of the COVID-19 pandemic in the United States. *R. Soc. Open Sci.* **2020**, *7*, 200742. [CrossRef] [PubMed]
36. Zhang, Y.; Chen, X.; Gong, S.; Tong, R. Dual hierarchical modelling for the influence of job role demands on psychosocial safety behavior: Evidence from coal industry. *Int. J. Ind. Ergon.* **2022**, *89*, 103291. [CrossRef]
37. Sampson, J.M.; DeArmond, S.; Chen, P.Y. Role of safety stressors and social support on safety performance. *Saf. Sci.* **2014**, *64*, 137–145. [CrossRef]
38. Bandura, A. Social Cognitive Theory: An Agentic Perspective. *Asian J. Soc. Psychol.* **1999**, *2*, 21–41. [CrossRef]
39. Liu, X.; Tang, W.; Xiao, R.; Zhou, W.; Huang, H.; Lin, R.; Tan, C.; Teng, X. Reducing Psychological Stress of Chinese Adolescents by Mindfulness-Based Intervention: The Mediating Role of Resilience. *Child Youth Care Forum* **2023**, *52*, 351–370. [CrossRef]
40. Voon, S.P.; Lau, P.L.; Leong, K.E.; Jaafar, J.L.S. Self-Compassion and Psychological Well-Being Among Malaysian Counselors: The Mediating Role of Resilience. *Asia-Pac. Educ. Res.* **2022**, *31*, 475–488. [CrossRef]
41. Kocjan, G.Z.; Kavcic, T.; Avsec, A. Resilience matters: Explaining the association between personality and psychological functioning during the COVID-19 pandemic. *Int. J. Clin. Health Psychol.* **2021**, *21*, 100198. [CrossRef]
42. Ma, W.; Sang, B.; Hong, S. A review of research on psychological resilience and its mechanisms of action. *J. East China Norm. Univ. (Educ. Sci. Ed.)* **2008**, *26*, 89–96. [CrossRef]
43. Howard, D.E. Searching for resilience among African-American youth exposed to community violence: Theoretical issues. *J. Adolesc. Health* **1996**, *18*, 254–262. [CrossRef]
44. Rutter, M. Implications of Resilience Concepts for Scientific Understanding. *Ann. N. Y. Acad. Sci.* **2006**, *1094*, 1–12. [CrossRef] [PubMed]
45. Shelef, L.; Schiff, M.; Pat-Horenczyk, R.; Dekel, R. COVID-19 vs. terrorism: Contribution of the COR theory to the process of coping with invisible threats. *J. Psychiatr. Res.* **2022**, *147*, 176–182. [CrossRef] [PubMed]
46. Ahmed, F.; Zhao, F.; Faraz, N.A.; Qin, Y.J. How inclusive leadership paves way for psychological well-being of employees during trauma and crisis: A three-wave longitudinal mediation study. *J. Adv. Nurs.* **2021**, *77*, 819–831. [CrossRef] [PubMed]
47. Lecours, A.; St-Hilaire, F.; Daneau, P. Fostering mental health at work: The butterfly effect of management behaviors. *Int. J. Hum. Resour. Manag.* **2022**, *33*, 2744–2766. [CrossRef]
48. Noh, D.; Park, S. Mediating Effects of Social Support and Resilience on the Association between COVID-19-Related Stress and Mental Health in Korean Young Adults. *Int. J. Environ. Res. Public Health* **2022**, *19*, 6935. [CrossRef]
49. Tugade, M.M.; Fredrickson, B.L.; Barrett, L.F. Psychological resilience and positive emotional granularity: Examining the benefits of positive emotions on coping and health. *J. Personal.* **2004**, *72*, 1161–1190. [CrossRef]
50. Su, P.; Yi, J.; Chen, X.; Xiao, Y. Visual Analysis of Psychological Resilience Research Based on Web of Science Database. *Psychol. Res. Behav. Manag.* **2023**, *16*, 465–481. [CrossRef]
51. Bakker, A.B.; Demerouti, E. Job Demands-Resources Theory: Taking Stock and Looking Forward. *J. Occup. Health Psychol.* **2017**, *22*, 273–285. [CrossRef]
52. Millear, P.; Liossis, P.; Shochet, I.M.; Biggs, H.; Donald, M. Being on PAR: Outcomes of a Pilot Trial to Improve Mental Health and Wellbeing in the Workplace with the Promoting Adult Resilience (PAR) Program. *Behav. Chang.* **2008**, *25*, 215–228. [CrossRef]
53. Kermott, C.A.; Johnson, R.E.; Sood, R.; Jenkins, S.M.; Sood, A. Is higher resilience predictive of lower stress and better mental health among corporate executives? *PLoS ONE* **2019**, *14*, e0218092. [CrossRef] [PubMed]
54. Gao, L.; Deng, X.; Yang, W.; Fang, J. COVID-19 related stressors and mental health outcomes of expatriates in international construction. *Front. Public Health* **2022**, *10*, 961726. [CrossRef] [PubMed]
55. Yang, Y.; Zhang, Z.; Li, X.; Zhao, D.; Liu, S.; Zhang, S. The Relationship between Mental Toughness and Risky Decision Making Behaviors of Volleyball Players. *Chin. J. Sports Med.* **2015**, *34*, 681–687. [CrossRef]
56. Vander Elst, T.; De Witte, H.; De Cuyper, N. The Job Insecurity Scale: A psychometric evaluation across five European countries. *Eur. J. Work Organ. Psychol.* **2014**, *23*, 364–380. [CrossRef]
57. Campbell-Sills, L.; Stein, M.B. Psychometric analysis and refinement of the Connor-Davidson Resilience Scale (CD-RISC): Validation of a 10-item measure of resilience. *J. Trauma. Stress* **2007**, *20*, 1019–1028. [CrossRef] [PubMed]
58. Nunnally, J.; Bernstein, I. *Psychometric Theory*, 3rd ed.; McGraw-Hill: New York, NY, USA, 1994.
59. Li, S.; Fan, M.; Wu, X. Effect of Social Capital between Construction Supervisors and Workers on Workers' Safety Behavior. *J. Constr. Eng. Manag.* **2018**, *144*, 04018014. [CrossRef]

60. Han, Y.; Li, J.; Cao, X.; Jin, R. Structural Equation Modeling Approach to Studying the Relationships among Safety Investment, Construction Employees' Safety Cognition, and Behavioral Performance. *J. Constr. Eng. Manag.* **2020**, *146*, 04020065. [CrossRef]
61. Bentler, P.M. Comparative fit indexes in structural models. *Psychol. Bull.* **1990**, *107*, 238. [CrossRef]
62. Hu, L.T.; Bentler, P.M. Cutoff criteria for fit indexes in covariance structure analysis: Conventional criteria versus new alternatives. *Struct. Equ. Model. A Multidiscip. J.* **1999**, *6*, 1–55. [CrossRef]
63. Hair, J.F.; Black, W.C.; Babin, B.J.; Anderson, R.E.; Tatham, R. *Multivariate Data Analysis*, 6th ed.; Pearson Prentice Hall: New Jersey, NJ, USA, 2006.
64. Fornell, C.; Larcker, D.F. Structural equation models with unobservable variables and measurement error: Algebra and statistics. *J. Mark. Res.* **1981**, *18*, 382–388. [CrossRef]
65. Bagozzi, R.P. Evaluating structural equation models with unobservable variables and measurement error: A comment. *J. Mark. Res.* **1981**, *18*, 375–381. [CrossRef]
66. Brown, R.L. Assessing specific mediational effects in complex theoretical models. *Struct. Equ. Model. A Multidiscip. J.* **1997**, *4*, 142–156. [CrossRef]
67. Bergh, L.I.V.; Hinna, S.; Leka, S.; Jain, A. Developing a performance indicator for psychosocial risk in the oil and gas industry. *Saf. Sci.* **2014**, *62*, 98–106. [CrossRef]
68. Sverke, M.; Hellgren, J.; Näswall, K. No security: A meta-analysis and review of job insecurity and its consequences. *J. Occup. Health Psychol.* **2002**, *7*, 242–264. [CrossRef]
69. Chih, Y.-Y.; Kiazad, K.; Li, M.; Capezio, A.; Zhou, L.; Restubog, S.L.D. Broken promises: Implications for the job insecurity and job performance of Chinese construction workers. *J. Constr. Eng. Manag.* **2017**, *143*, 04016114. [CrossRef]
70. Hobfoll, S.E. The influence of culture, community, and the nested-self in the stress process: Advancing Conservation of Resources theory. *Appl. Psychol.* **2001**, *50*, 337–370. [CrossRef]
71. Zhu, Y.; Quansah, P.E.; Obeng, A.F.; Cobbinah, E. Investigating the Effects of Role Demands, Psychosocial Stress Symptoms and Safety Leadership on Mineworkers' Safety Performance. *Psychol. Res. Behav. Manag.* **2020**, *13*, 419–436. [CrossRef]
72. Podsakoff, N.P.; LePine, J.A.; LePine, M.A. Differential challenge stressor-hindrance stressor relationships with job attitudes, turnover intentions, turnover, and withdrawal behavior: A meta-analysis. *J. Appl. Psychol.* **2007**, *92*, 438–454. [CrossRef]
73. Van den Broeck, A.; De Cuyper, N.; De Witte, H.; Vansteenkiste, M. Not all job demands are equal: Differentiating job hindrances and job challenges in the Job Demands-Resources model. *Eur. J. Work Organ. Psychol.* **2010**, *19*, 735–759. [CrossRef]
74. Hoffmeister, K.; Gibbons, A.M.; Johnson, S.K.; Cigularoy, K.P.; Chen, P.Y.; Rosecrance, J.C. The differential effects of transformational leadership facets on employee safety. *Saf. Sci.* **2014**, *62*, 68–78. [CrossRef]
75. Leung, M.-y.; Chan, Y.-S.; Yuen, K.-W. Impacts of Stressors and Stress on the Injury Incidents of Construction Workers in Hong Kong. *J. Constr. Eng. Manag.* **2010**, *136*, 1093–1103. [CrossRef]
76. Kazaz, A.; Ulubeyli, S. Drivers of productivity among construction workers: A study in a developing country. *Build. Environ.* **2007**, *42*, 2132–2140. [CrossRef]

buildings

MDPI

Article

Identifying Critical Factors and Trends Leading to Fatal Accidents in Small-Scale Construction Sites in Korea

Jong-Moon Hwang [1], Jeong-Hun Won [2,3,4], Hyeon-Ji Jeong [3] and Seung-Hyeon Shin [4,*]

1 Occupational Safety and Health Research Institute, Korea Occupational Safety and Health Agency, Ulsan 44429, Republic of Korea; bm0722@kosha.or.kr
2 Department of Safety Engineering, Chungbuk National University, Cheongju 28644, Republic of Korea; jhwon@chungbuk.ac.kr
3 Department of Disaster Prevention Engineering, Chungbuk National University, Cheongju 28644, Republic of Korea; gus1895@naver.com
4 Department of Big Data, Chungbuk National University, Cheongju 28644, Republic of Korea
* Correspondence: shshin0317@chungbuk.ac.kr; Tel.: +82-10-5813-5959

Abstract: Small-scale construction sites in South Korea account for about 91.5% of all construction workplaces and contribute to 72.3% of the total accidents and fatalities. Safety measures at these sites are often underestimated, and proper safety education is lacking. In particular, the fatality rate is about 4.43 times higher compared to medium-/large-scale construction sites. In this study, a systematic analysis was conducted to examine the causes and trends of industrial accidents in small-scale construction sites to address these issues. This study analyzed industrial accidents in small-scale construction sites using statistical analysis, LDA topic modeling, and network analysis based on data from the Korea Occupational Safety and Health Agency (KOSHA) from 2018 to 2022. The analysis revealed that the most critical cause of accidents in small-scale construction sites is 'Scaffolding and working platforms', with accidents primarily involving 'Fall'. Furthermore, various risk factors and accident trends were identified in apartment construction, new building projects, and mobile scaffolding usage. This study systematically analyzed the causes and trends of industrial accidents at small-scale construction sites, providing important evidence to enhance safety management and preventive measures. The results are expected to play a crucial role in establishing a safety culture at construction sites and ensuring the wellbeing of construction workers.

Keywords: small-scale construction sites; safety management; accident causes; accident types; accident trends

Citation: Hwang, J.-M.; Won, J.-H.; Jeong, H.-J.; Shin, S.-H. Identifying Critical Factors and Trends Leading to Fatal Accidents in Small-Scale Construction Sites in Korea. *Buildings* **2023**, *13*, 2472. https://doi.org/10.3390/buildings13102472

Academic Editors: Mariusz Szóstak and Marek Sawicki

Received: 5 September 2023
Revised: 25 September 2023
Accepted: 27 September 2023
Published: 28 September 2023

1. Introduction

Infrastructure facilities such as roads, bridges, and buildings are essential for improving the quality of life for citizens and driving national economic development. Consequently, the construction industry is a core driving force behind a country's progress. The fact that the construction industry is both a crucial sector and entails high risks cannot be ignored [1]. Globally, construction workers constitute only about 7% of the total workforce across industries, yet they account for a significantly larger proportion of fatalities [2,3]. While global efforts by governments and researchers have led to decreased construction-related industrial accidents, aiming to improve the industry's hazardous image, the construction industry still experiences more fatal accidents than other industries [4,5]. In Korea, as depicted in Figure 1, from the years 2013 to 2022, the accident fatality rate in the construction industry was consistently higher compared to the manufacturing sector. On average, the fatality rate in the construction sector was 3.09 times higher than in the manufacturing sector. This disparity was even more pronounced in 2020, where the difference in fatality rates between the two sectors reached a staggering fourfold. Such a stark contrast underscores the urgent need for addressing safety concerns in the construction sector and offers a baseline for understanding the safety advancements and best practices in the manufacturing sector that could potentially be adopted in construction. When comparing accident fatality rates across industries, the construction industry consistently exhibits higher fatality rates than the manufacturing sector. In 2020, the fatality rate in construction (2.00) was about four times higher than that of manufacturing (0.50). Analyzing the trend in fatality rates, while the manufacturing and overall industries have experienced continuous decreases, the fatality rate of the construction industry has been on the rise again since 2016. In 2013, for the first time since then, the fatality rate in construction surpassed 2.00.

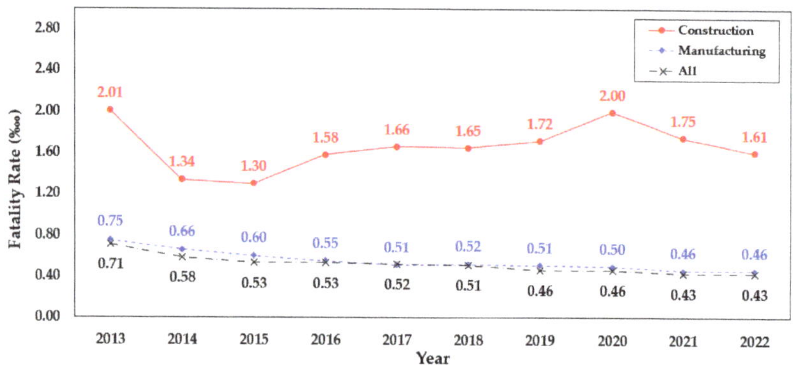

Figure 1. Work-related accident fatality rate across construction, manufacturing, and all industries in Korea.

The work-related accident fatality rate in the construction industry exhibits significant variations based on factors such as the number of on-site workers and the project costs. Workers at small-scale construction sites are exposed to higher accident risks than those at large-scale construction sites, a trend observed not only in developing countries but also in advanced countries [6–8]. In the case of South Korea, in 2020, there were 301,271 construction workplaces with construction costs of less than KRW 5 billion, constituting about 91.5% of all construction workplaces (329,279), and they accounted for about 72.3%, with 331 out of 458 fatalities. These data indicate that construction industrial accidents are concentrated in these smaller-scale areas (USD 1 $\approx$ KRW 1340). In particular, the fatality rate at small-scale construction sites is about 4.43 times higher than that of construction sites with project costs not less than KRW 12 billion. Thus, to effectively reduce industrial accidents in the construction industry, it is necessary to concentrate efforts on managing them in small-scale construction sites.

The various factors influencing the severity of fatal accidents occurring at small-scale construction sites can vary based on the safety and health regulations of the country, the culture, and the construction stakeholders. For instance, in the USA, the Occupational Safety and Health Administration (OSHA) sets and enforces standards to assure safe working conditions, particularly emphasizing training, outreach, and education. The UK's Health and Safety Executive (HSE) is focused on reducing work-related death and serious injury across all sectors, with a specific framework for construction that mandates specific safety roles and responsibilities for projects. In Australia, the Safe Work Australia body develops national policy relating to WHS (work health and safety) and workers' compensation, emphasizing consultation, cooperation, and coordination among various stakeholders in construction. However, according to multiple studies, small-scale construction projects are found to be facing common safety and health issues [5–8]; the significance of workplace safety measures is underestimated, and there is a lack of sufficient safety education for both workers and managers. In particular, construction clients often perceive industrial accidents as unpredictable events, leading to a tendency not to employ managers or experts with specialized knowledge in safety and health.

One of the key strategies to reduce industrial accidents at construction sites and enhance workplace safety is to conduct in-depth analyses of past incidents, identifying high-risk factors and implementing effective preventive measures to eliminate these risks [7]. Several countries, including South Korea, the United Kingdom, the United States, Japan, and Singapore, apply this methodology to decrease industrial accidents at small-scale construction sites. They periodically disclose industrial accident statistics and formulate safety and health management policies based on these data. Furthermore, various researchers analyze industrial accident cases to identify significant safety and health issues that the government may not recognize, presenting specific solutions to address them [7–9].

Most studies on improving safety and health at small-scale construction sites have recently focused on identifying the causes of accidents, their occurrences, and the populations of high-risk groups [7–11]. While these studies have significantly contributed to easily understanding the primary causes of accidents, they have overlooked the complex interactions between the complex causes and outcomes of accidents. In large-scale construction sites, the support of dedicated safety and health managers or experts enables the interpretation of and response to how complex accident causes interact and lead to accidents in a site-specific manner. However, in small-scale construction sites lacking specialized safety and health management support, simply providing accident causes is insufficient for understanding how the causes and outcomes interact to lead to accidents. Therefore, relying solely on statistical analysis may not be enough to prevent or respond to these incidents adequately. Therefore, in small-scale construction sites, providing specific accident scenarios explaining how the causes interact to trigger accidents can be more effective than simply listing the causes.

Furthermore, the existing research focused on small-scale construction sites in South Korea mostly relies on historical data, failing to capture the recent changes in work environments and trends. Therefore, the primary objective of this study is to systematically analyze the causes and trends of industrial accidents occurring in recent small-scale construction sites within South Korea. This aims to provide essential scientific evidence for establishing effective safety and health management strategies. In particular, this study identifies the critical causes and accident types that have the most significant impact on small-scale construction sites and presents how these causes and accident types actually lead to specific accidents.

The structure of this paper is as follows. Section 2, titled 'Background', delves into a comprehensive literature review of the existing research related to accidents in small-scale construction sites and explains the accident status in the domestic construction industry. In Section 3, the 'Materials and Methods' section is covered, including the data collection process for this study and the plan and methodology of this study. Next, in Section 4, the 'Results' are presented. A comparative analysis focusing on the causes and accident types was conducted between industrial accidents at small-scale construction sites with a construction cost of less than KRW 5 billion and medium-/large-scale construction sites. Additionally, the critical causes and accident types at small-scale construction sites were analyzed. For accidents related to the most critical cause and accident types at small-scale construction sites, LDA topic modeling and network analysis were conducted in this study, and the main results regarding accident trends were identified. Moving forward to Section 5, the 'Discussion' was carried out regarding the analysis results. Finally, in Section 6, the 'Conclusion' discusses the study's outcomes and outlines future research directions.

2. Background

2.1. Literature Review

2.1.1. Accident Characteristics in Small-Scale Construction Sites

The recent literature primarily emphasizes the identification of accident causes, their frequency, and the groups most susceptible to these risks in small-scale construction sites. Cheng et al. [6] analyzed the characteristics of industrial accidents occurring at small-scale construction sites in Taiwan. Dumrak et al. [7] found that 57.5% of all fatal accidents in south Australia happened at small-scale construction sites with less than 20 workers. Similarly, Camino López et al. [8] revealed that sites with fewer than 25 workers accounted for a staggering 58.1% of the total fatalities. Bang et al. [9] analyzed fatality rates based on construction cost and facility type using accidents from 2013 to 2019. The study revealed that serious accidents frequently occurred at sites of less than 0.008 billion USD in South Korea. Collectively, these studies, including contributions by Daba et al. [10] and Berhe et al. [11], consistently show that workers in smaller construction sites are at a higher risk compared to those in larger settings.

2.1.2. Contemporary Trends and Data Limitations in Construction Safety Research

While significant studies have been undertaken concerning accident characteristics in small-scale construction sites, there remains a gap in the literature concerning the specific differences between accidents in small-scale settings versus larger ones. Studies such as those by Lim et al. [5], Choi et al. [12], and Kang and Ryu [13] have predominantly focused on historical data, potentially overlooking contemporary shifts in work environments and trends. The reliance on pre-2019 data, given the escalating severity of recent industrial accidents in smaller construction sites, underscores the need for updated study methodologies and data sources.

2.1.3. The Role and Impact of Safety Training

Emerging studies have highlighted the pivotal role of safety training in enhancing hazard recognition and risk perception among construction workers. Fu et al. [14] underscored the potential of integrating visual cues in safety education to enhance hazard recognition abilities, especially among novice workers. Such innovative training approaches were further supported by Namian et al. [15], who emphasized the significance of high-engagement training methods in improving both hazard recognition and safety risk perception. Their findings revealed that high-engagement training led to superior hazard recognition and an elevated perception of safety risk. Furthermore, Perlman et al. [16] highlighted the discrepancies between how construction superintendents assess risk levels and the ratings provided by formal safety risk assessment methods. Their study accentuated the importance of training and education in bridging this gap. Moreover, Uddin et al. [17] identified specific hazard categories that construction workers are more proficient at recognizing, emphasizing the need for targeted training interventions.

In conclusion, while there has been considerable study on the characteristics and causes of accidents in small-scale construction sites, a profound need exists for innovative training methodologies that are rooted in a deep understanding of workers' hazard recognition patterns and challenges. The integration of technology, personalized learning interventions, and high-engagement training methods promises a more holistic and effective approach to construction safety education.

2.2. Accidents in Small-Scale Construction Sites in Korea

The safety and health management systems of construction sites in South Korea are based on the "Occupational Safety and Health Act", and different regulations apply depending on the scale of the construction project. The scale of construction is determined by the total construction cost, and generally, sites with construction costs less than KRW 5 billion are defined as small-scale construction sites. Currently, small-scale construction sites are mostly in blind spots regarding safety and health management regulations. Construction sites with construction costs of not less than KRW 5 billion and less than KRW 12 billion are considered medium-scale construction sites. They are subject to most safety and health management regulations, and safety managers must be appointed for on-site safety management. However, safety managers for medium-scale construction sites are allowed to have dual roles. On the other hand, construction sites with construction costs not less than KRW 12 billion are considered large-scale construction sites. Similar to medium-scale sites, safety managers must be appointed, but safety managers for large-scale construction sites are dedicated safety managers and are not allowed to take on other duties outside of safety management works.

In the past 10 years, an analysis of the accident fatality rate due to accidents in the construction industry based on the construction cost revealed significant differences in South Korean construction sites (Figure 2). For construction sites with construction costs of less than KRW 12 billion, where there is no requirement for a dedicated safety manager, the fatality rate due to accidents was analyzed to be at least 2.5 times higher (in 2014) and up to 4.43 times higher (in 2020) compared to construction sites with construction costs of KRW 12 billion or more. The fatality rate in large-scale construction sites gradually increased from 2015 to 2019, which was followed by a decrease after 2020. Medium-scale construction sites exhibited a significant increase in 2016 followed by a continuous decrease until 2018, but the trend shifted to an increase in 2019. On the other hand, the fatality rate in small-scale construction sites showed repeated fluctuations until 2019, which was followed by a sharp increase in 2020 and a slight decrease in 2021.

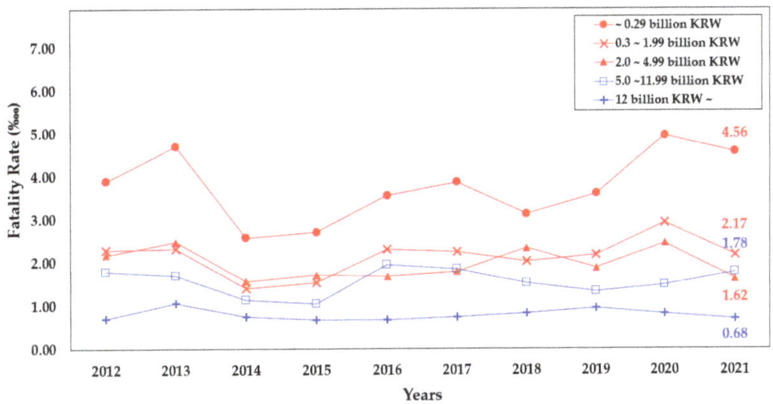

Figure 2. Work-related accident fatality rate according to construction costs in Korea.

3. Materials and Methods

3.1. Data Collection

In this study, accidents at small-scale construction sites were analyzed using the industrial accident status statistics data (referred to as "industrial accident data") reported to the Korea Occupational Safety and Health Agency (KOSHA) from 2018 to 2022. The industrial accident data include various fields such as industry category, construction scale, victim's name, diagnosis, cause, accident type, disability grade, country, occupation, date of death, age, gender, duration of employment, date of the accident, time of the accident, accident overview, employment status, and worker's position. Each field was prepared according to KOSHA's classification criteria.

In response to the importance of diverse data sourcing, it is worth mentioning that while our primary dataset originated from official sources, the inclusion of data from broader platforms, such as social media enterprises, could potentially enhance the comprehensiveness of the analysis. Such datasets, although rich in capturing real-time sentiments, can pose challenges related to data validity, consistency, and potential biases [18].

In this study, the analysis focused primarily on the cause and the accident type directly associated with industrial accidents. The classification criteria and definitions for the cause and the accident type are presented in Appendix A. In this study, any items or details within the causes and accident types that fell under other fields or with insufficient information were excluded from the analysis. Ultimately, 1511 cases of work-related fatalities due to accidents were analyzed in this study.

3.2. Methods

This study aimed to identify critical causes and accident types from industrial accident data in small-scale construction sites and to present trends in accidents associated with the most critical causes and accident types. This study involved several key steps, including data collection, statistical analysis, topic modeling, and network analysis. The research flowchart detailing these steps is presented in Figure 3.

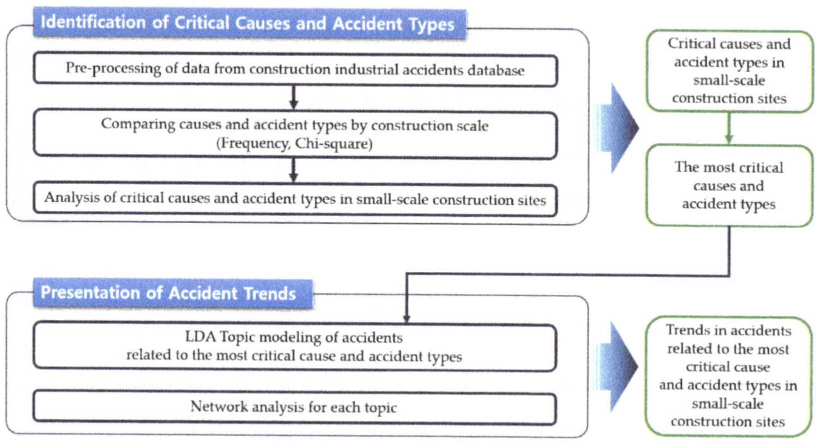

Figure 3. Research flowchart.

A chi-squared test was conducted to investigate the relationship of industrial accidents between small-scale construction sites and other construction sites in terms of causes and accident types. In this case, the chi-squared test was performed using the SPSS v26 software. If the chi-squared test indicated a difference in industrial accidents based on causes and accident types between small-scale construction sites and other construction sites, the critical causes and accident types for small-scale construction sites were identified using absolute and relative frequencies. In this case, the absolute frequency represents the frequency of accidents occurring in small-scale construction sites based on the causes or accident types. The relative frequency indicates the proportion of accident occurrences in small-scale construction sites compared to the total accident occurrences across all construction sites. The relative frequency is calculated according to Formula (1):

$$RF = (\text{Frequency at small-scale construction}/\text{Frequency at all construction}) \quad (1)$$

RF represents the relative frequency, where Frequency at small-scale construction refers to the frequency of causes or accident types in small-scale construction sites, and Frequency at all construction refers to the frequency of causes or accident types across entire construction sites. Critical causes and accident types were presented in graph form by comprehensively considering absolute and relative frequencies. Through the analysis of the critical causes and accident types, it could be determined whether certain causes or accident types caused more accidents in small-scale construction sites relative to the total construction sites or if they were frequent in the entire construction industry. In the graph, causes or accident types located in the upper right corner frequently occurred in small-scale construction sites and were particularly hazardous, indicating that they pose a significant risk compared to the entire construction industry.

Next, to analyze specific trends in industrial accidents related to the critical causes and accident types identified in small-scale construction sites, latent Dirichlet allocation (LDA) topic modeling and network analysis were conducted. Topic modeling is a methodology that involves utilizing statistical and optimization algorithms to extract latent topics from large text collections. It identifies topics that constitute the themes within an entire text collection, categorizing the text into relevant topics [19,20]. In topic modeling, the latent Dirichlet allocation (LDA) model is commonly used. The primary goal of LDA is to effectively identify latent topic information by analyzing the co-occurrence patterns of words within documents [21]. Despite the success and popularity of LDA, it has limitations in terms of capturing relationships between keywords derived from LDA in the way the text is represented [22]. Therefore, the study of Gerlach et al. [22] suggested that integrating topic modeling with network analysis could improve this aspect. Network analysis is a methodology that models complex systems as graphs composed of nodes and edges to analyze their structure and behavior. It is commonly used in data science to analyze complex relationships and patterns between nodes [23]. Therefore, this study analyzed the overview of industrial accidents caused by the most critical causes and accident types using LDA topic modeling to classify them into several accident types. In this case, the LDA topic modeling and network analysis were conducted using Python 3.9. Before conducting the topic modeling, the industrial accident overview data in the Korean natural language were tokenized as a preprocessing step. Initially, special characters, numbers, and characters other than Korean were removed. Tokenization was carried out using the Kkma morphological analyzer from the Konlpy package. Among the tokens obtained through tokenization, common Korean stopwords such as postpositional particles and conjunctions, along with proper nouns, terms related to hospital transfers, and other terms like confirmation, fact, and incident, which do not significantly impact accident analysis, were mostly removed as stopwords. After this preprocessing, the documents were transformed into a bag-of-words format to create a corpus. Next, the LDA topic modeling was performed using various packages from the Python gensim package, including LdaModel, Dictionary, MmCorpus, copora, models, TfidfModel, etc. The analysis to identify an appropriate number of topics was based on a comprehensive consideration of perplexity and coherence scores. Perplexity and coherence scores are quantifiable metrics that measure the predictability of a given set of texts. Generally, lower perplexity and higher coherence scores indicate a better prediction quality [24,25]. Subsequently, the LDA topic modeling was conducted based on the optimal number of topics. The network analysis using the network package was performed for each topic to analyze the complex interactions of accidents and extract accident trends. The network visualization was created using word pairs with the top 50 frequencies. The size of nodes and the thickness of edges were increased proportionally to the frequency of nodes and the connectivity of edges. Additionally, it's important to note that due to the proximity of certain keywords in the network visualization, there might be an overlap, causing certain keywords to be obscured. This visual overlap is a consequence of closely related keywords having a shorter distance between them in the network representation.

4. Results

4.1. Identification of Critical Causes and Accident Types

In this section, the critical causes and accident types leading to work-related fatal accidents in small-scale construction sites are presented. A comparative analysis was conducted on the causes and accident types of fatal accidents occurring in small-scale construction sites and other construction sites. Subsequently, the absolute frequency (AF) and relative frequency (RF) of accidents that occurred in small-scale construction sites were calculated to identify the critical causes and accident types contributing to fatal accidents.

4.1.1. Comparing Causes and Accident Types by Construction Scale

The analysis examined the differences in industrial accidents based on causes and types of fatal accidents between small-scale construction sites and other construction sites. The results are summarized in Table 1. Among the 1511 work-related fatal accidents in the construction industry that were analyzed, there were 1013 accidents reported on small-scale construction sites with construction costs of less than KRW 5 billion, while 498 accidents occurred on medium-/large-scale construction sites with construction costs of not less than KRW 5 billion. Work-related fatalities in small-scale construction sites accounted for about 67%.

Table 1. Chi-squared test and frequency analysis of accident causes and types in small-scale and other construction sites.

Category	<KRW 50 Billion (*n*)	≥KRW 50 Billion (*n*)	Total (*n*)
Accident Causes (χ^2= 76.482, p < 0.001)			
Scaffolding and working platforms	190	72	262
Stepped structure and opening	128	61	189
Means of land transportation	138	50	188
Transport and lifting equipment/machinery	99	81	180
Construction/mining machinery	81	66	147
Stairs and ladders	96	22	118
General manufacturing and processing equipment/machinery	73	28	101
Floors, surfaces, etc.	39	22	61
Molds and supporting post	24	31	55
Electrical equipment, parts	31	17	48
Equipment/machinery, parts, and accessories	19	15	34
Materials	23	7	30
Components and accessories of buildings/structures	15	10	25
Humans, animals/plants	15	2	17
Portable power tools	12	2	14
Chemical products	12	1	13
Non-metallic mineral products	8	1	9
Containers, packaging and devices	5	3	8
Work environment, natural phenomena such as atmospheric conditions, etc.	2	3	5
Manual mechanical equipment	2	2	4
Hand tools	1	0	1
Fragments, debris, waste	0	1	1
Means of air, water transportation	0	1	1
Total	1013	498	1511
Accident Types (χ^2= 39.582, p < 0.001)			
Fall	547	223	770
Collision	109	67	176
Struck by object	59	60	119
Pressed under/overturned	73	30	103
Caught in between	45	30	75
Collapse	42	32	74
Fire	33	18	51
Electrocution	35	14	49
Explosion/rupture	24	5	29
Chemical leakage	14	4	18
Tripping	12	5	17
Fallen in/Drown	12	5	17
Cutoff/cut/stab	4	2	6
Animal injury	3	1	4
Workplace traffic accident	1	0	1
Oxygen deficiency	0	1	1
Abnormal temperature contact	0	1	1
Total	1013	498	1511

A chi-squared test was conducted to determine the statistical significance of the proportional differences in the causes of industrial accidents based on the construction costs. The chi-squared test results indicated that the p-value was less than 0.001, indicating a significant difference in the proportion of industrial accident causes based on the construction costs. Among the twenty-three causes, except for four causes, including 'Molds and supporting post', 'Work environment, natural phenomena such as atmospheric conditions, etc.', 'Fragments, debris, waste', and 'Means of air, water transportation', it was analyzed that accidents caused by all other causes were more frequent on construction sites with construction costs of less than KRW 5 billion. Next, a chi-squared test was performed to verify the frequency differences of accident types based on construction costs. The analysis revealed that the p-value was less than 0.001, indicating a significant difference in the proportion of industrial accident types based on the construction cost. Among the seventeen accident types, except for three types (struck by object, oxygen deficiency, and abnormal temperature contact), it was analyzed that all the other types had a higher frequency on construction sites with construction costs of less than KRW 5 billion.

4.1.2. Analysis of Critical Causes and Accident Types in Small-Scale Construction Sites

Considering the stark disparity in the frequency of industrial accidents, based on their causes and accident types between small-scale and medium-/large-scale construction sites, we undertook a comprehensive analysis. Figure 4 delineates the principal causes of fatal accidents in small-scale construction sites. The absolute frequency (AF) and relative frequency (RF) for scaffolding and working platforms emerged as the highest, with AF = 190 and RF = 0.725. This underscores that scaffolding and working platforms were the predominant factors leading to fatal accidents in small-scale construction sites. Subsequent to this, the major causes in small-scale construction sites were identified in the following sequence: means of land transportation (AF = 138, RF = 0.734), stepped structure and opening (AF = 128, RF = 0.677), and stairs and ladders (AF = 96, RF = 0.813).

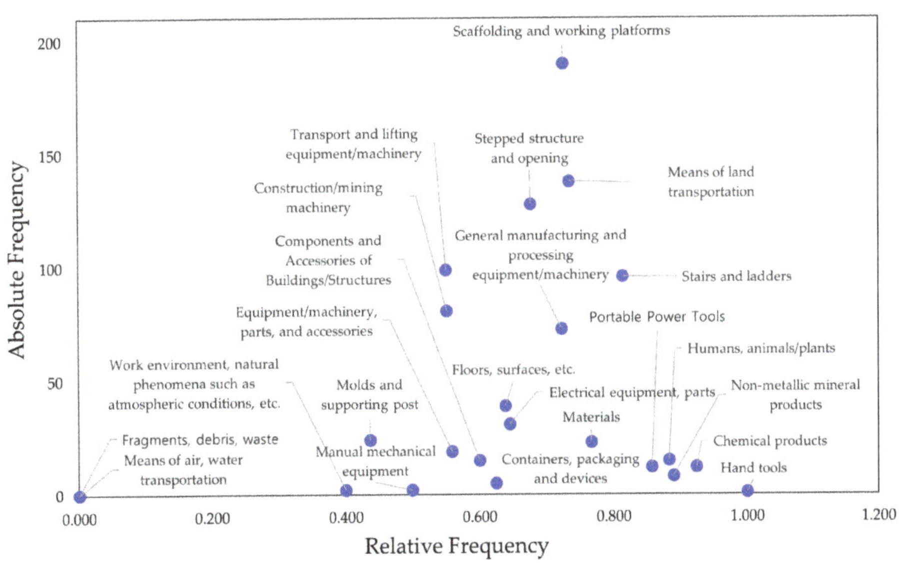

Figure 4. Critical causes in small-scale construction sites.

The results of the analysis of critical accident types in fatal accidents at small-scale construction sites are indicated in Figure 5. The analysis revealed that fatal accidents caused by falls had an overwhelmingly higher absolute frequency and relative frequency than other accident types (AF = 547, RF = 0.710). Furthermore, fatal accidents resulting from collisions were also relatively high compared to other fatal accidents (AF = 109, RF = 0.619). Following collisions, accidents involving being pressed under or overturned (AF = 73, RF = 0.709), being struck by an object (AF = 59, RF = 0.496), and being caught in between (AF = 45, RF = 0.600) were identified as critical factors contributing to fatal accidents.

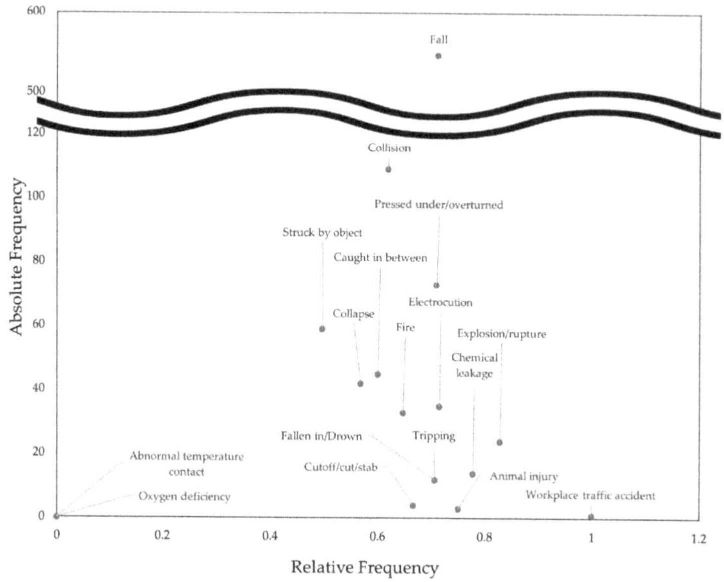

Figure 5. Critical accident types in small-scale construction sites.

The analysis of the most critical causes and accident types in fatal accidents at small-scale construction sites revealed that the most critical cause was scaffolding and working platforms, while the most critical accident type was falls.

4.2. *Presentation of Accident Trends in Small-Scale Construction Sites*

An analysis of the accident trends was conducted regarding fatal accidents occurring at small-scale construction sites, specifically those caused by the critical factors of scaffolding and working platforms as well as falls. The scope of the analysis encompassed industrial accident cases reported to the Occupational Safety and Health Agency from 2018 to 2022. In small-scale construction sites, there were 190 work-related fatal accidents due to scaffolding and work platform accidents and 547 work-related fatal accidents due to falls. The trends in fatal accidents were identified through LDA topic modeling and network analysis applied to the overview of the industrial accident data.

It was recognized that upon an initial review, the network analysis figures for each topic might have appeared to have overlapping characteristics. This perception was believed to stem from the broader categorization of accident types. However, upon closer examination, distinct differences in keyword interrelations across these networks were observed. These subtleties, while nuanced, were deemed instrumental in understanding the underlying patterns and causative factors of accidents in unique contexts. The decision to present these topics in detailed individual figures, instead of consolidating them, was based on the commitment to ensure that these critical distinctions were clearly communicated. By delving into the granular details of each topic, a comprehensive overview was intended to be provided, ensuring that the intricate relationships and patterns intrinsic to each accident type were distinctly captured and interpreted. A holistic representation of the findings was considered paramount to convey the multifaceted nature of accidents in small-scale construction sites.

4.2.1. Topic Modeling and Network Analysis for the Most Critical Cause

Accident trends related to work-related fatal accidents caused by scaffolding and working platforms in small-scale construction sites were analyzed using LDA topic modeling and network analysis. Perplexity and coherence scores were calculated on the pre-tokenized data to determine the optimal number of topics for fatal accidents related to scaffolding and working platforms (Figure 6). Based on the calculated perplexity and coherence score, the optimal number of topics was determined to be five.

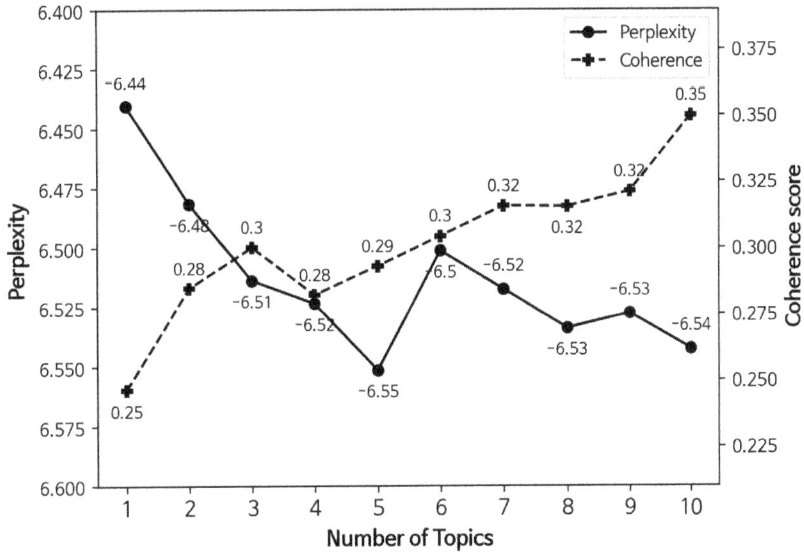

Figure 6. Perplexity and coherence score in accidents caused by scaffolding and working platforms.

An inter-topic distance map (IDM) of the topic modeling results is presented in Figure 7. The IDM visualizes the relationships and distances between different topics derived from the LDA topic modeling. In the IDM shown in Figure 7, numbers 1 through 5 inside the circles represent the topic numbers. The size of each circle reflects the prevalence of the corresponding topic: larger circles indicate topics that are more prevalent in the dataset. The spatial distance between circles on the map illustrates the relevance between topics: topics that are closer to each other are more closely related in terms of content, while those farther apart are less related. The IDM represents the entire set of topics from the learned topic model on a two-dimensional scale [26]. Topics are depicted as circles, where larger circles indicate a higher prevalence. Closer distances between topics indicate a higher relevance, while greater distances imply a lower relevance [27]. The analysis of the IDM revealed that there was no overlapping area between topics, and the distribution of data for each topic was appropriately balanced: Topic 1—48 incidents, topic 2—31 incidents, topic 3—29 incidents, topic 4—41 incidents, topic 5—41 incidents.

Through LDA topic modeling, it was analyzed that accidents in small-scale construction sites caused by scaffolding and working platforms could be broadly classified into five distinct types. Next, network analysis was performed to understand the accident trends considering the interactions between factors within each topic. The results of the network analysis for topic 1 are indicated in Figure 8. The analysis of topic 1 revealed that 20 nodes and 50 edges were confirmed. The average degree of connectivity was 5.0, and the network density was about 0.26. The important nodes identified included 'Scaffolding', 'Floor', 'Apartment', 'Rope', 'Outer wall', etc. In particular, the high frequency of 'Scaffolding' and 'Floor' indicates that they were likely to be the main causes of fatal accidents when scaffolding work was performed with unstable platforms or inadequate floor conditions. Furthermore, the connections with 'Apartment', 'Rope', and 'Outer wall' point toward potential risk factors during rope or outer wall operations at apartment construction sites.

Figure 7. Inter-topic distance map of accidents caused by scaffolding and working platforms.

The results for topic 2 regarding accidents caused by scaffolding and working platforms in small-scale construction sites are indicated in Figure 9. In topic 2, the main factors related to fatal accidents involving scaffolding and working platforms in small-scale construction sites were analyzed. For the network presented, 21 nodes and 50 edges were confirmed, exhibiting an average degree of connectivity of 4.76 and a network density of approximately 0.24. The important nodes identified included 'Installation', 'Floor', 'New construction', 'Footing', and 'Scaffolding'. In particular, the frequency between 'Installation' and 'Floor', as well as 'New construction', was notably high. This indicates that the condition of the floor during the initial stages of scaffolding or platform installation in new construction buildings can have a significant impact. The analysis of accident trends revealed that fatal accidents were more likely to occur during the early stages of working platform installation on new construction buildings with inadequate floor conditions. These results emphasize the importance of enhancing safety inspections during the installation of platforms and scaffolding at construction sites.

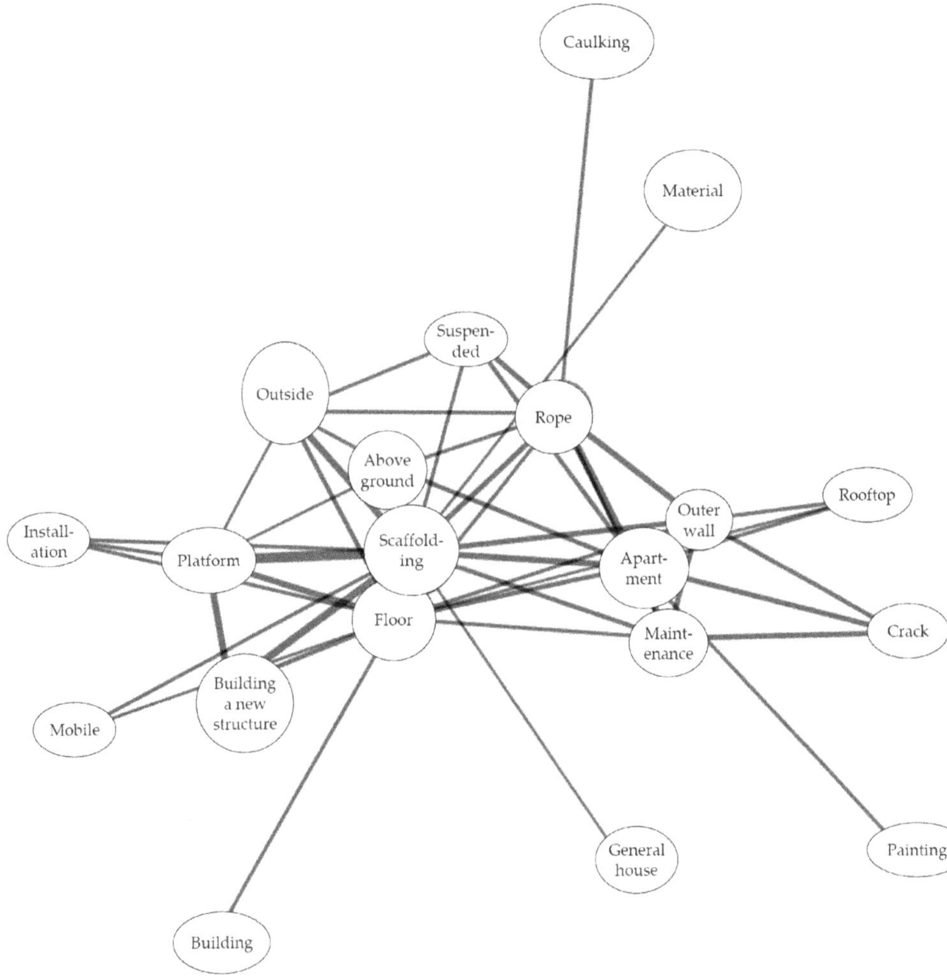

Figure 8. Network analysis results for topic 1 in scaffolding and working platforms accidents.

Additionally, it is worth mentioning that Figure 9 also illustrates a secondary network consisting of the nodes 'Apartment' and 'Rooftop', which appear separate from the primary network. This secondary network emerged from our analysis but was deemed less directly related to the central topic of our study. While we did not provide an extensive interpretation for this secondary network in the main content, its inclusion in the figure serves to ensure transparency and provide a complete view of our analytical results.

The network analysis of topic 3 identified complex causes and interactions of fatal accidents related to scaffolding and working platforms in small-scale construction sites (Figure 10). In the analysis, 19 nodes and 50 edges were identified, displaying an average degree of connectivity of about 5.26 and a network density of approximately 0.29. The important nodes identified included 'Mobile', 'Scaffolding', 'New construction', 'Floor', 'Platform', and 'Concrete'. In particular, the high connectivity frequency between 'Scaffolding' and 'New construction' indicates the risks of fatal accidents occurring during scaffolding work. These results were generally similar to topic 2. However, considering the strong connections involving 'Mobile', 'Scaffolding', and 'Platform' in topic 3, it can be inferred that accidents related to mobile scaffolding and working platforms were emphasized. It was analyzed that the nodes 'Apartment' and 'Rooftop' were connected only to each other and not to other nodes. This could have two main interpretations: first, the connection between 'Apartment' and 'Rooftop' may signify accidents occurring under specific circumstances or conditions. In this case, these two nodes can be interpreted as related only under specific conditions and insignificant in more general situations. The second possibility is data incompleteness. The lack of connections between 'Apartment', 'Rooftop', and other nodes could imply that there might not have been enough data collected for these nodes. The analysis of accident trends associated with topic 3 revealed that in the case of a newly constructed building with mobile scaffolding installed and the concrete floor being unfinished, the instability of the working platform follows, resulting in an increased risk of fatal accidents. This analysis emphasizes the importance of safety checks for mobile scaffolding and working platforms, particularly re-evaluation after concrete work and floor conditions are addressed.

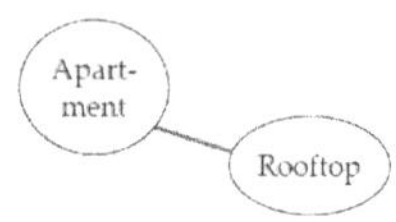

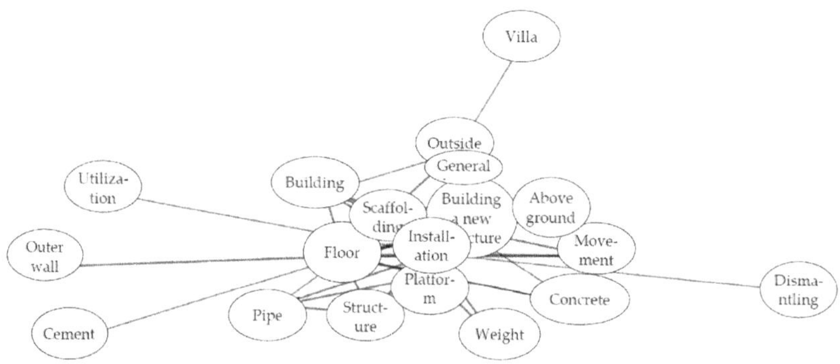

Figure 9. Network analysis results for topic 2 in scaffolding and working platforms accidents.

Figure 11 shows the analysis of various causes of fatal accidents related to scaffolding and working platforms in small-scale construction sites corresponding to topic 4. In the network containing 26 nodes and 50 edges, the average degree of connectivity was about 3.85, and the network density was about 0.154. The important nodes identified included 'Scaffolding', 'Floor', 'Platform', 'Outside', and 'Handrail'. Among these, 'Scaffolding', 'Floor', and 'Platform' overlapped with other topics, while 'Outside' and 'Handrail' were found to play a unique role only in topic 4. The analysis of accident trends revealed that external environmental factors (e.g., strong winds) and the absence or instability of handrails could increase the instability of working platforms, thus elevating the risk of fatal accidents. These results emphasize the importance of safety checks for external environmental factors and proper handrail installation. Furthermore, a node labeled 'Subcontracting' was identified, which was unlike the other topics. This result indicates that fatal accidents related to scaffolding and working platforms are also frequent among subcontracted workers.

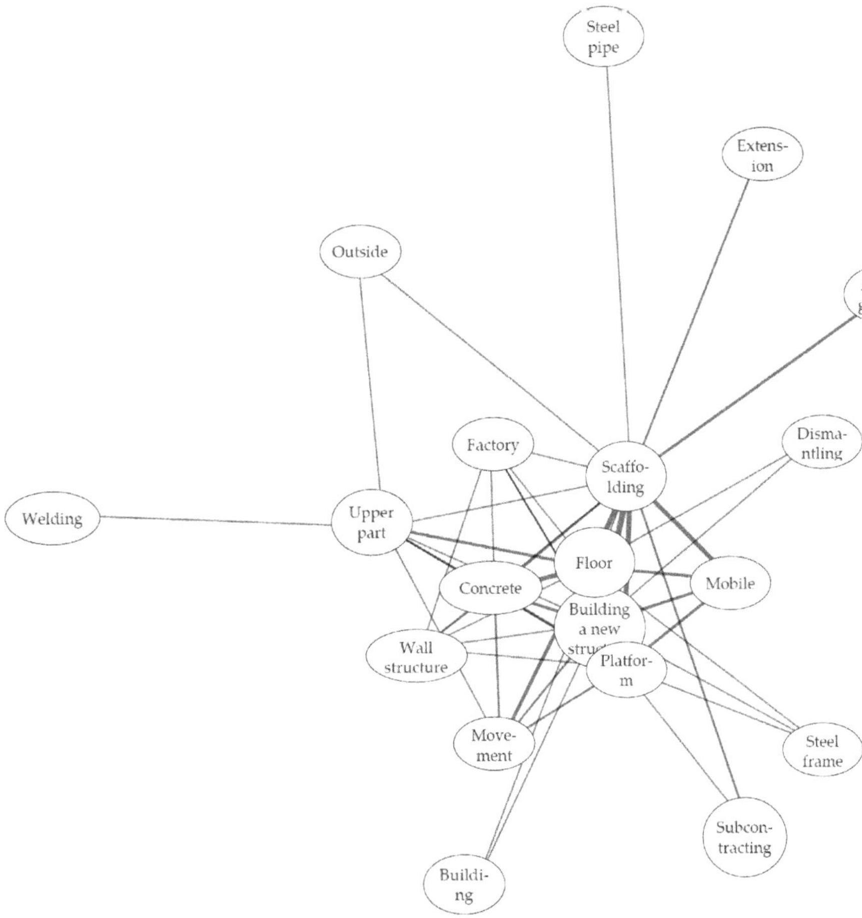

Figure 10. Network analysis results for topic 3 in scaffolding and working platforms accidents.

The results of the network analysis for topic 5 are presented in Figure 12. This network consisted of 20 nodes and 50 edges, with an average degree of connectivity of 5.0 and a network density of about 0.26. The important nodes identified included 'Outer wall' which refers to the external facade or surface of a building or structure, 'Floor', 'Scaffolding', 'Rope', and 'Installation'. In particular, the high centrality of 'Outer wall' and 'Rope' elevated the risk of accidents related to work on the outer wall and the use of ropes. The analysis of the accident trends revealed that during work on the outer wall using scaffolding and rope, instability in the rope or installation errors could lead to an increased instability of working platforms, thereby increasing the risk of fatal accidents. These results emphasize the importance of thorough safety checks for ropes when using scaffolding for operations on the outer wall.

Figure 11. Network analysis results for topic 4 in scaffolding and working platforms accidents.

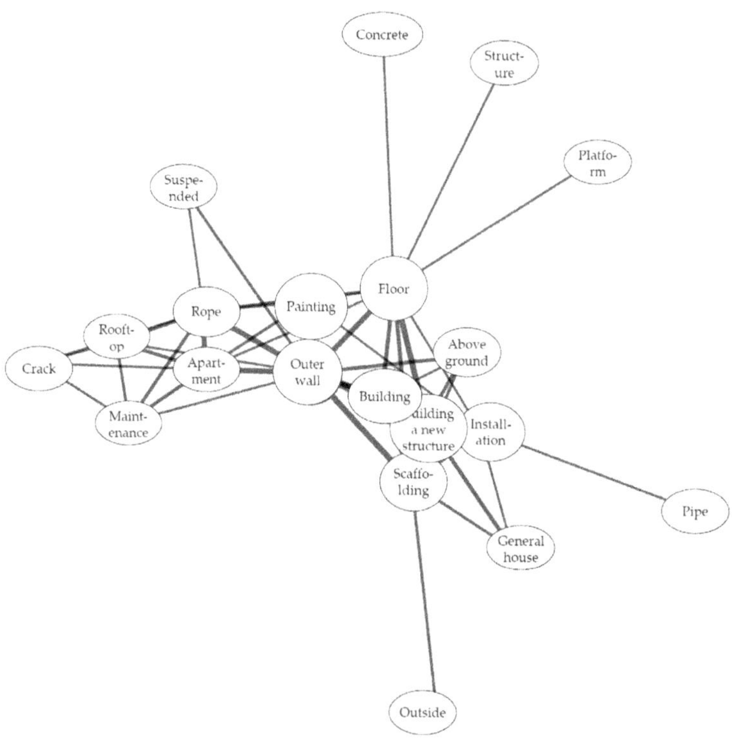

Figure 12. Network analysis results for topic 5 in scaffolding and working platforms accidents.

4.2.2. Topic Modeling and Network Analysis for the Most Critical Accident Type

The trends of accidents caused by falls in work-related fatal accidents at small-scale construction sites were analyzed using LDA topic modeling and network analysis. Perplexity and coherence scores were calculated on the pre-tokenized data to determine the optimal number of topics for fatal accidents related to falls (Figure 13). Based on the calculated perplexity and coherence score, the optimal number of topics was seven.

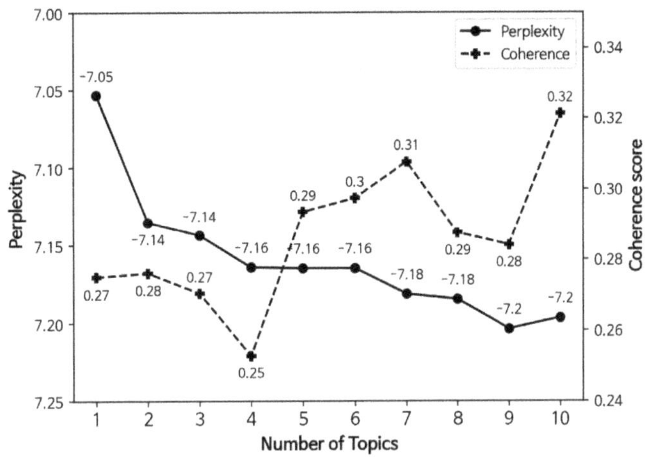

Figure 13. Perplexity and coherence score of fall accidents.

The IDM for accidents caused by falls is presented in Figure 14. In the IDM shown in Figure 14, numbers 1 through 7 inside the circles represent the topic numbers. The analysis of the IDM revealed that there was no overlapping area between topics except for topics 2 and 7. While there was some overlapping area between topics 2 and 7, the extent of the overlap was relatively limited. In terms of the amount of data for each topic, topics 6 and 2 had relatively more data compared to the other topics, and topic 3 appeared to have less data: topic 1—86 incidents, topic 2—110 incidents, topic 3—49 incidents, topic 4—63 incidents, topic 5—56 incidents, topic 6—122 incidents, topic 7—61 incidents.

Through LDA topic modeling, it was determined that accidents in small-scale construction sites caused by falls could be broadly classified into seven distinct types. Next, network analysis was performed to understand accident trends considering seven interactions between the factors within each topic.

The results of the network analysis for topic 1 are presented in Figure 15. A network analysis focused on topic 1, which pertained to accidents caused by falls at small-scale construction sites, was conducted. Through this analysis, the interaction between key connecting factors and their importance was identified. The highest frequency of connections was observed between 'Floor' and 'New construction'. In other words, this suggests that accidents involving falls from heights to the ground were frequent at construction sites, particularly in new construction sites. Specifically, accidents falling under topic 1 involved incidents where individuals fell from 'Scaffolding', 'Working platforms', 'Molds', and 'Openings' during the construction of facilities such as 'Neighborhood living facilities', 'Detached houses', and 'Apartments'. Operations frequently associated with fall accidents falling under topic 1 included brick masonry, working while moving, and operations involving the use of ladders. Topic 1 was considered to be the most common type of falling accident, and to prevent accidents related to topic 1, thorough safety checks during floor work in newly constructed buildings are crucial. In particular, ensuring stability before starting work is a crucial measure for preventing "Fall" accidents, which can be concluded from the analysis. The network of topic 1 consisted of 29 nodes and 50 edges, with an average degree of connectivity of about 3.45 and a network density of about 0.123.

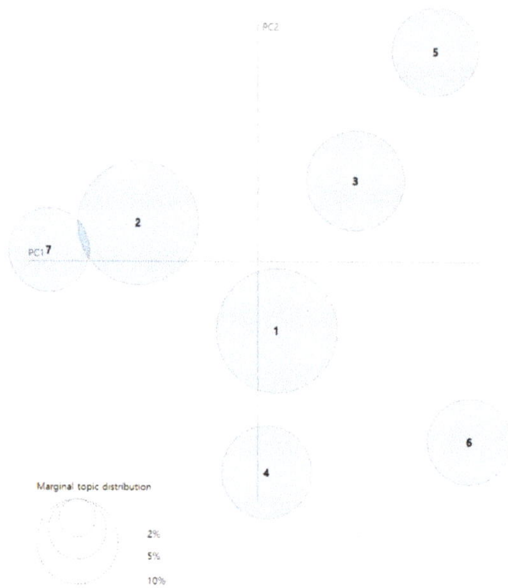

Figure 14. Inter-topic distance map of fall accidents.

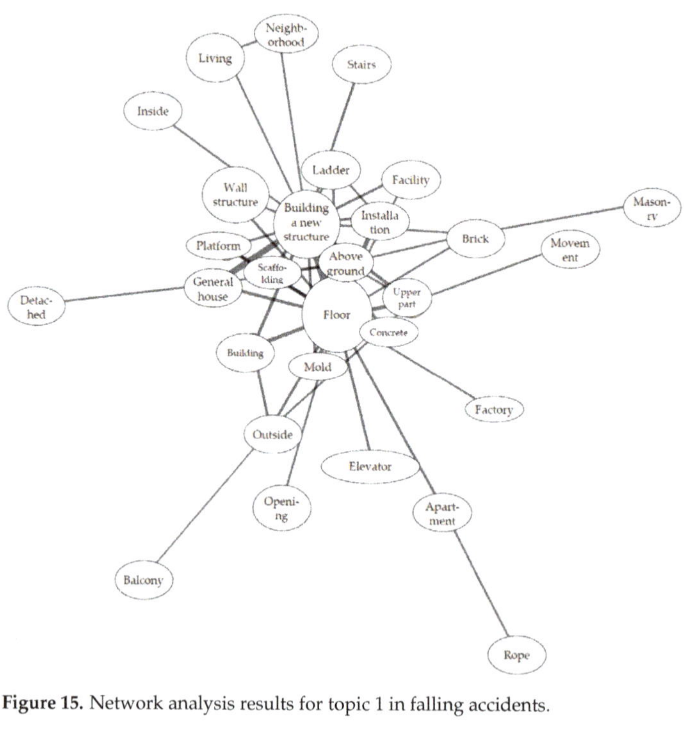

Figure 15. Network analysis results for topic 1 in falling accidents.

The network for topic 2 consisted of 20 nodes and 50 edges, with an average degree of connectivity of 5.0 and a network density of about 0.263 (Figure 16). According to the results of the network analysis, fatal accidents involving falls from neighborhood living facility construction sites or general house construction sites, similar to topic 1, formed the central focus. However, in topic 2, a higher frequency of connections was observed between 'Scaffolding', 'Working platforms', 'Outside', and 'Installation'. This indicates that accidents involving falls during the construction process of new buildings where scaffolding and working platforms were used or during the installation process of this equipment were frequent. These accident trends highlight the risks associated with the installation and usage of scaffolding, indicating an increased risk of falling accidents if the installation status and the safety of workers using scaffolding and working platforms are not adequately considered. Therefore, to prevent falling accidents related to topic 2, rigorous safety inspections and management are necessary during the installation and usage of scaffolding and working platforms.

Figure 16. Network analysis results for topic 2 in falling accidents.

The results of the network analysis for topic 3 are presented in Figure 17. The network for topic 3 consisted of 25 nodes and 50 edges, with an average degree of connectivity of 4.0 and a network density of about 0.167. For topic 3, it was determined that accidents involving falls during the installation of ladders, scaffolding, working platforms, and equipment were found to be predominant.

The network analysis results for topic 4 are indicated in Figure 18. The network of topic 4 consisted of 22 nodes and 50 edges, with an average degree of connectivity of about 4.55 and a network density of about 0.216. Unlike the other topics, topic 4 indicated that accidents involving falls from ladders during work were a major topic, in addition to accidents occurring at construction sites of new buildings. It was determined that accidents involving falling from ladders frequently occurred during the maintenance and inspection of infrastructure such as roads and bridges. In general, during maintenance or inspection of bridges, there are often multiple instances where site managers are absent and a small number of workers are involved. Due to these factors, safety and health management may be lacking compared to regular construction sites, creating an environment that is more vulnerable to falling accidents than regular construction sites. However, considering these accidents were identified as the major topics of falling accidents occurring at small-scale construction sites, it is necessary to enhance occupational health and safety management at such sites.

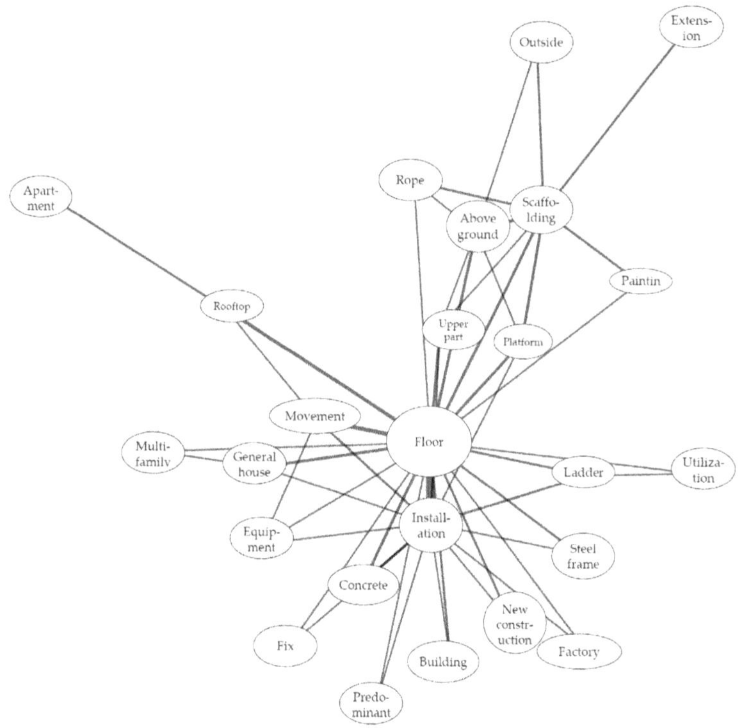

Figure 17. Network analysis results for topic 3 in falling accidents.

The results of the network analysis for topic 5 are presented in Figure 19. This network consisted of 22 nodes and 50 edges, with an average degree of connectivity of about 4.55 and a network density of about 0.216. The core keywords of topic 5 were 'Aerial', 'Boarding', 'Installation', 'Vehicle', 'Work platform', 'Loading', 'Painting', and 'Outer wall'. Analyzing the network structure of these keywords revealed that accidents related to vehicles and other equipment were a major type of accident during aerial operations. In particular, the connection structure between 'Boarding' and 'Vehicle', as well as 'Loading', indicates high risks associated with boarding and loading processes on vehicles or other mobile equipment. The connection between 'installation' and 'Outer wall' points to falling accidents during outer wall installation work. Furthermore, the connection structure between 'Painting' and 'Suspended scaffolding' signifies falling accidents while performing painting works using suspended scaffoldings. Through this connectivity structure, the risks associated with operations such as aerial operation, particularly involving the boarding and loading of vehicles and equipment and painting operations, were recognized on construction sites. Through this analysis, it can be concluded that accidents related to vehicles and equipment during aerial operation in construction sites were a major accident type in topic 5. Therefore, the safe usage of vehicles and equipment, providing worker education and training, and conducting thorough safety checks during operations are necessary to prevent such accidents.

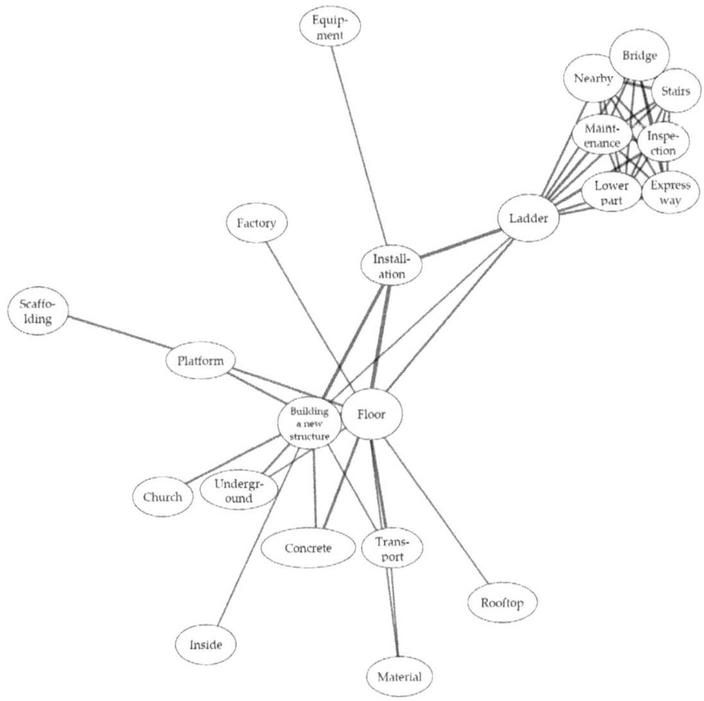

Figure 18. Network analysis results for topic 4 in falling accidents.

The network for topic 6 consisted of 23 nodes and 50 edges, with an average degree of connectivity of about 4.35 and a network density of about 0.198 (Figure 20). According to the results of the network analysis, the core accident patterns in topic 6 were related to accidents involving scaffolding and working platforms. In particular, accidents related to falls from scaffolding installed on the outer walls of apartments, accidents occurring during the installation of scaffolding and working platforms, and accidents during the demolition and dismantling processes were identified to fall under topic 6. In comparison with the other topics, a characteristic feature of topic 6 was the strong network formation of falling accidents from scaffolding during work on the outer wall of apartments and during demolition and dismantling operations.

According to the results of the network analysis for topic 7, similar to topic 6, accidents involving falls from scaffolding during the installation process on apartment and building outer walls appeared as a major topic (Figure 21). However, for topic 7, in contrast to the other topics, accidents involving falls while using scaffolding during maintenance work were found to be predominant in this analysis. The network for topic 7 consisted of 24 nodes and 50 edges. The average degree of connectivity in the network was about 4.17, and the network density was about 0.181.

The analysis results of topics 6 and 7 highlight the importance of safety and installation processes for scaffolding and working platforms used in work on the outer wall of apartments and in the general construction of buildings. Moreover, these results emphasize the importance of safety during maintenance, dismantling, and demolition work, emphasizing the necessity for proper safety education and thorough safety inspections for workers involved in such operations.

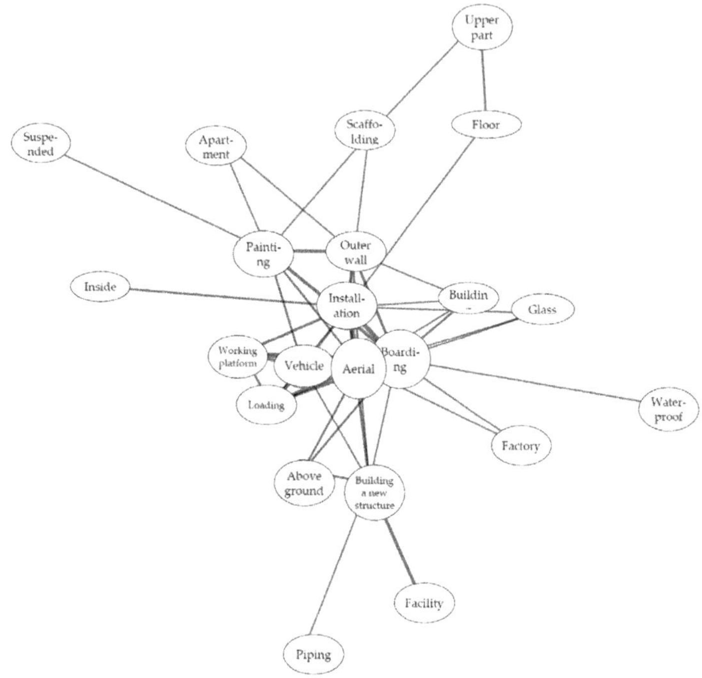

Figure 19. Network analysis results for topic 5 in falling accidents.

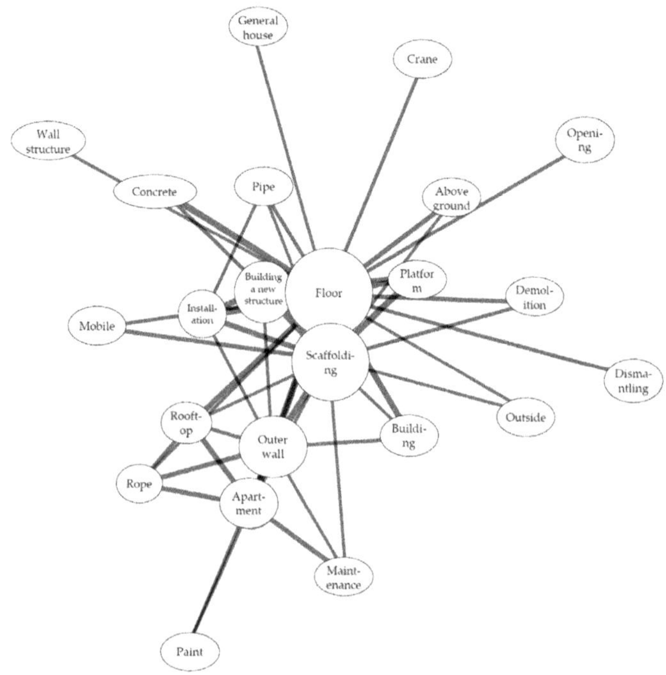

Figure 20. Network analysis results for topic 6 in falling accidents.

Figure 21. Network analysis results for topic 7 in falling accidents.

5. Discussion

This study systematically analyzed the causes and trends of industrial accidents in small-scale construction sites in Korea. In particular, the primary objective was to provide detailed accident scenarios for fatal accident causes and how these causes lead to actual accidents. To achieve this objective, various analytical methods were applied in this study. The results aimed to deeply understand the accident causes in small-scale construction sites and provide the scientific basis for establishing effective safety and health management strategies.

The results of this study have revealed several important findings compared to previous studies [5,9]. First, the proportion of fatal accidents in small-scale construction sites was found to be higher than that in medium-/large-scale construction sites. This indicates that workers in small-scale construction sites are exposed to higher accident risks. Second, it was confirmed that the causes and types of fatal accidents differ between small-scale and medium-/large-scale construction sites. This disparity is directly related to the scale of the construction site, allowing for the identification of specific critical accident causes and types that require particular attention in construction sites of a certain scale.

Through this study, the most critical accident cause on small-scale construction sites was identified as 'Scaffolding and working platforms', and the most critical accident type was identified as 'Fall'. For accidents involving scaffolding and working platforms and falling accidents on small-scale construction sites, topic modeling and network analysis were conducted to analyze how the various factors interact to cause fatal accidents and to systematically analyze accident trends. In the analysis of accidents caused by scaffolding and working platforms, various topics were utilized to understand the main causes and patterns of accidents occurring in various situations, such as apartment construction sites, new buildings, and situations involving mobile scaffolding and working platforms. In particular, the analysis focused on aspects such as the connectivity between apartments and rooftops, the risks during the initial stages of installing working platforms in newly constructed buildings, and the risks associated with mobile scaffolding and working platforms.

These results are expected to significantly assist construction site safety managers and workers in recognizing risk factors in specific situations and taking appropriate preventive measures. Furthermore, in the analysis of accidents caused by falling, a detailed analysis was conducted on the main causes and patterns of falling accidents in various situations, including falls in new construction sites, falls during the installation and use of scaffolding and working platforms, and falls during demolition and dismantling work.

Recognizing the importance of delineating the theoretical contribution, managerial implications, limitations, and future research directions in academic research, we are grateful for drawing our attention to this aspect. Our study offers a novel perspective on the safety measures and accident trends in small-scale construction sites, particularly in the Korean context. It bridges the gap in the existing literature by providing a comprehensive analysis using statistical tools, LDA topic modeling, and network analysis. This methodological approach in itself is a significant contribution, as it offers a multidimensional understanding of the data. Our findings can guide construction site managers, policy makers, and safety trainers in designing more effective safety protocols. Furthermore, while our study provides valuable insights into small-scale construction sites in Korea, the generalizability of our findings to larger construction sites or to different cultural contexts might be limited. Additionally, our analysis was based on data from 2018 to 2022, which, although recent, might not capture the very latest trends in the industry. For future research directions, subsequent research can expand on our findings by incorporating qualitative insights, perhaps through interviews with construction site workers and managers. This can provide a richer understanding of the on-the-ground realities and the nuances behind the data. A comparative study with medium- or large-scale construction sites or sites from different countries can further broaden our understanding of construction site safety.

In conclusion, this study systematically analyzed the main causes and patterns of various accidents occurring on small-scale construction sites. While previous studies have generally covered the causes of industrial accidents on construction sites from a general perspective, this study provided a more specific analysis of the causes of industrial accidents on small-scale construction sites. Through this, it offers a deeper understanding of the causes of industrial accidents in small-scale construction sites and their corresponding solutions. These analytical results are expected to greatly contribute to enhancing safety management and preventive measures on construction sites. Furthermore, these study results are expected to contribute to establishing a stronger safety culture within construction sites and to play a crucial role in ensuring the safety of construction workers.

Lastly, this study presents future research directions. First, since this study mainly focused on small-scale construction sites in Korea, analyzing the causes and trends of industrial accidents in different countries or on sites of different scales is necessary. Second, while this study analyzed the causes and trends of industrial accidents, additional time-series analysis is required to understand how these causes and trends change over time. Such future research could complement and expand upon the results of this study.

6. Conclusions

This study conducted a systematic analysis on 1511 records from 2018 to 2022, identifying prevalent accident types and causes in small-scale construction sites in Korea using LDA topic modeling and network analysis. In particular, the primary causes of falling accidents on construction sites and the associated risk factors were clearly identified, providing important evidence for establishing prevention and management strategies for falling accidents. These analytical results provide important evidence for enhancing safety management and preventive measures in construction sites.

The results of this study have greatly contributed to a comprehensive understanding of the causes and trends of accidents in small-scale construction sites. Furthermore, these study results are expected to contribute to establishing a stronger safety culture within the construction sites. In particular, through the identified major causes and risk factors from this study, it will be possible to formulate and implement more systematic safety management strategies in construction sites.

While this study has extensively analyzed the critical factors leading to fatal accidents in small-scale construction sites, focusing specifically on 'scaffolding and working platforms' and 'falling' accidents using LDA topic modeling and network analysis, it's essential to note that the scope was limited to specific scenarios. As a result, validation for the applicability of these results in diverse construction environments becomes crucial [14–17].

In future studies, based on the results identified in this study, it is necessary to analyze the causes and trends of accidents in construction sites of various scales and types. Furthermore, based on the major causes and risk factors identified in this study, the focus should be on developing effective safety management strategies and preventive measures.

It is expected that such follow-up studies guided in this direction will significantly contribute to enhancing overall safety in the construction industry.

Author Contributions: Conceptualization, J.-M.H., J.-H.W. and S.-H.S.; methodology, J.-M.H. and S.-H.S.; software, J.-M.H. and H.-J.J.; validation, J.-H.W. and S.-H.S.; formal analysis, J.-M.H. and S.-H.S.; investigation, J.-M.H. and S.-H.S.; resources, J.-M.H.; data curation, J.-H.W. and H.-J.J.; writing—original draft preparation, J.-M.H.; writing—review and editing, J.-H.W. and S.-H.S.; visualization, H.-J.J. and S.-H.S.; supervision, J.-H.W. and S.-H.S.; project administration, J.-M.H., J.-H.W., and S.-H.S.; funding acquisition, J.-M.H. and S.-H.S. All authors have read and agreed to the published version of the manuscript.

Funding: This work was supported by the Occupational Safety and Health Research Institute of Korea (2021-OSHRI-520), Chungbuk National University BK21 program (2021) and the Human Resources Development of the Korea Insitute of Energy Technology Evaluation and Planning (KETEP) grant funded by the Korea government (No. 20224000000070).

Data Availability Statement: Not applicable.

Acknowledgments: Not applicable.

Conflicts of Interest: The authors declare no conflict of interest.

Appendix A

Table A1. Classification criteria and definitions for accident cause.

Main Category Code	Subcategory Code	Cause	Classification Criteria and Definitions
0		Equipment/Machinery	This category includes equipment powered by energy and machinery mechanisms and devices used in specific processes. It includes parts or accessories attached to machinery, forming part of the machinery structure. This classification also includes cases where such equipment and machinery were operating for their intended work purposes and caused an accident during operation. Appliances like computers, audio and video devices, and heating/washing/cooling machines are included. Special-purpose vehicles designed for specific operations rather than transportation purposes are classified under the same code. However, vehicles designed for transporting people, goods, etc., are classified as means of transportation (6). Portable power machinery is classified as portable machinery-power (11). If components were attached to the machine at the time of the accident or it is suspected to have been so, the entire machine is classified as the cause. However, if parts or accessories have been separated from the machine or the entire machine is irrelevant to the accident, parts/accessories (2) are classified as the cause.

Table A1. *Cont.*

Main Category Code	Subcategory Code	Cause	Classification Criteria and Definitions
	00	Equipment/Machinery with Insufficient Information	
	01	General Manufacturing and Processing Equipment/Machinery	This category classifies machines that process and handle materials such as metal, wood, rubber, plastic, non-metallic minerals, etc., through cutoff, shaping, crushing, etc., to create secondary products. Equipment/machinery in this category refers to industrial equipment/machinery used in various industries, including manufacturing and other sectors. Equipment with general-purpose applications across multiple industries falls under this category. However, specialized equipment used for specific purposes in certain industries is classified separately as specialized process equipment/machinery (02), agriculture, forestry, and fishery equipment/machinery (04), and construction and mining equipment/machinery (05) categories. Inclusions: Bending, Rolling, Shaping Machines; Boring, Drilling, Planing, Milling Machines; Extrusion, Injection, Molding, Casting Machines; Grinding, Polishing Machines; Lathes; Presses (excluding printing); Sawing Machines; Screw Thread and Female Screw Cutting Machines; Laser Cutting Machines, Fluid Pressure Cutting Machines, Spot Welding Machines, etc. Exclusions: Agricultural/Horticultural Equipment (041); Logging/Woodworking Machines (042); Conveying/Handling Equipment (03); Construction/Mining Equipment (05); Food Cutting Machines (02101); Meat Grinders (02102); Paper Machines (022); Textile, Clothing, Leather Production Machines (024); Non-Power Hand Tools (13); Portable Power Saws (11102); Portable Power Surface Finishing Tools (11104)
	02	Specialized Process Equipment/Machinery	This category classifies equipment and machinery not classified under general manufacturing and processing equipment/machinery. It includes equipment/machinery exclusively used for the production of specific products.
	03	Transport and Lifting Equipment/Machinery	Machines used for the transportation and handling of specific materials. If it is known whether the related parts were attached to the entire machine during the accident, the entire machine is classified as the cause. However, if the parts are separated from the machine or if the machine is not related to the accident, only the specific parts are classified as the cause. Conveying and handling equipment/machinery are often composed of numerous small parts. For example, hoists, cranes, lifts, and elevators operate using pulleys and wheels. Such parts are classified under codes 22 (Machine Components) and 21 (Electrical Parts). Inclusions: Power Conveyors; Cranes; Hoists; Lifts; Elevators; Jacks Exclusions: Agricultural and Horticultural Equipment (041); Construction and Mining Equipment (05); Logging and Woodworking Equipment (042); Woodworking Machinery (014); Crane Accessories (22399); Electrical Parts (21); Means of Transport (6)
	04	Agriculture, Forestry, and Fishery Equipment/Machinery	

Table A1. *Cont.*

Main Category Code	Subcategory Code	Cause	Classification Criteria and Definitions
	05	Construction/Mining Machinery	
	09	Other Equipment/Machinery	
1		Portable and Manual Mechanical Equipment	This category includes portable hand tools (both non-powered and powered) and manually operated mechanical equipment. If parts causing the accident were attached to a tool, the entire tool is classified as the cause. However, if the parts were separated from the tool or if the primary purpose and function of the tool were unrelated to the accident, the parts are classified separately. For portable hand tools manufactured with both powered and non-powered capabilities, if it is unclear whether they were powered or non-powered at the time of the accident, they are classified as portable tools with no clear power status (12). Inclusions: Hand Tools; Portable Power Tools; Portable Tools-Unclear Power Status; Ladders; Medical Instruments; Unspecified Sewing Equipment Exclusions: Containers, Utensils, Furniture, and Equipment (4); Tool Storage Boxes (43103); Equipment/Machinery (0); Machine Jacks (03902); Parts, Accessories and Materials (2); Crane Accessories (223); Electrical Parts (21); Waterproof Sheets (24902), Drill Bits (22104), Saw Blades (22106)
	10	Portable and Manual Mechanical Equipment with Insufficient Information	
	11	Portable Power Tools	Handheld tools that require an energy source (electricity, gasoline, diesel, coal, air, steam, etc.) for operation and are held in the hand while being used. Portable tools are classified based on their general functions. Inclusions: Nail Guns; Portable Spray Equipment; Stapling Tools Exclusions: Equipment/Machinery (0); Agricultural and Horticultural Equipment (041); Power Lawnmowers (04111); Hair and Hand Dryers (09401); Vacuum Cleaners (09401); Hydraulic/Compressed Air Jacks (03902); Metalworking Machinery (013), Woodworking Machinery (014); Stationary Drills (01203); Stationary Woodworking Circular Saws (01401); Drill Bits (22104), Saw Blades (22106); Hand Tools (13)
	12	Portable Tools (Unclear Power Status)	This category is used for classifying portable tools designed to be powered and unpowered. It is specifically used when determining whether a portable tool was powered or unpowered at the time of the accident is difficult. Exclusions: Hand Tools; Portable Power Tools
	13	Hand Tools	This classification encompasses various hand tools operated manually without relying on power sources such as electricity, fuel (gasoline, coal), air, steam, fluids, explosives, etc. Inclusions: Hand Tools for Drilling; Hand Tools for Cutting; Hand Tools for Mining; Hand Tools for Fastening; Hand Tools for Measuring; Hand Tools for Striking; Hand Tools for Surface Treatment; Hand Tools for Cleaning; Crowbars; Hammers; Pitchfork; Rakes; Stapling Tools. Exclusions: Worktables (43107); Crane Accessories (223); Fasteners (Nails, Nuts, Bolts, etc.) (225); Drill Bits (22104); Saw Blades (22106); Portable Power Tools (11); Carts and Wheelbarrows (14102)

Table A1. *Cont.*

Main Category Code	Subcategory Code	Cause	Classification Criteria and Definitions
	14	Manual Mechanical Equipment	
	19	Other Portable and Manual Mechanical Equipment	
2		Parts, Accessories, and Materials	This classification categorizes components of equipment/machinery, parts and accessories of automobiles, subcomponents of buildings/structures, and materials such as metals, non-metallic minerals, wood, plastics, etc. When parts and accessories were involved in accidents independently from the entire machinery/equipment, means of transportation, buildings/structures, and other objects, they are classified as causes. If parts and accessories were attached to machinery, equipment, or means of transportation, the machinery/equipment or means of transportation are classified as causes. Even when parts, accessories, or equipment/machinery were attached, if the primary function of the equipment/machinery was not related to the accident, specific components are classified as causes. Inclusions: Rope; Crane Accessories; Electrical Parts; Metallic and Non-Metallic Mineral Materials Exclusions: Chemical Substances and Chemical Products (5); Containers (4); Furniture (4); Machinery (0); Components of Buildings/Structures (33); Means of Transportation (6)
	20	Parts, Accessories, and Materials with Insufficient Information	
	21	Electrical Equipment, Parts	Inclusions: Electrical Parts/Accessories for Machinery, Equipment, and Tools Exclusions: Machinery/Equipment (0); Hand tools (13); Means of Transportation (6)
	22	Equipment/Machinery, Parts, and Accessories	This classification includes non-attached parts necessary for operating and connecting equipment/machinery. It categorizes products used directly in assembly and other processes without any alteration in form. If it is known whether the related parts were attached to the entire machine during the accident, the entire machine is classified as the cause. However, if the parts were separated from the machine or if the machine was not related to the accident, only the specific parts are classified as the cause. Power transmission devices used in industrial machinery are classified under 221 (Parts and Accessories for Machinery/Equipment), while transmission devices for vehicles are classified under 224 (Parts and Accessories for Means of Transportation). Electrical equipment used in internal combustion engines or automobiles (excluding batteries) is classified under 211 (Electrical Equipment and Parts). Inclusions: Dies, Molds; Chains, Leather, Fabric, V-belt Power Transmission Devices; Drums, Pulleys, Sheaves, Cables, Winches; Engines, Turbines; Clutches; Gears; Rollers.
	23	Non-metallic Mineral Products	This classification categorizes non-metallic mineral products used directly in the detailed elements of buildings and structures (such as bricks, tiles, etc.) without any alteration in form. It excludes components and accessories of buildings (e.g., bathtubs, toilets).

Table A1. *Cont.*

Main Category Code	Subcategory Code	Cause	Classification Criteria and Definitions
	24	Materials	This classification categorizes products or materials that are not limited to specific equipment/machinery or buildings/structures but can be cut or altered in various forms for general and versatile use. It is used for cases not classified in other codes, where the product exists in its initial state, as raw materials, or as unfinished products. It also includes cases where the product exists as components that have been installed and then dismantled.
	25	Fragments, Debris, and Waste	
	29	Other Parts, Accessories, and Materials	
3		Buildings/Structures and Surfaces	This classification categorizes components of completed or under-construction buildings, structures (bridges, tunnels, towers, dams, etc.), provisional structures installed for construction purposes, and other components of structures, as well as surfaces like the ground and bedrock. If an entire structure was affected by a disaster in an independent state for fabrication or maintenance (where the material became a cause), it is classified under code 2: Parts, Accessories, and Materials.
	30	Buildings/Structures and Surfaces with Insufficient Information	
	31	Scaffolding and Working Platforms	
	32	Molds and Supporting Post	
	33	Stepped Structure and Opening	
	34	Stairs and Ladders	
	35	Floors, Surfaces, etc.	
	36	Other Provisional Structures	
	37	Other Buildings/Structures	
	38	Components and Accessories of Buildings/Structures	
	39	Other Buildings/Structures and Surfaces	
4		Containers, Utensils, Furniture, and Equipment	This category classifies non-industrial items and equipment, such as containers, packaging, and furniture used for transporting/handling goods. It includes protective gear, leisure/sports and corrective equipment, and wheelchairs.
	40	Containers, Utensils, Furniture, and Equipment with Insufficient Information	

Table A1. *Cont.*

Main Category Code	Subcategory Code	Cause	Classification Criteria and Definitions
	41	Containers, Packaging, and Devices	
	42	Household Utensils and Equipment	
	43	Furniture and Office Equipment	
	44	Clothing/Protective Gear and Accessories	
	49	Other Containers, Utensils, Furniture, and Equipment	
5		Chemical Substances and Chemical Products	This category classifies various chemical substances and chemical products in different states, such as liquids, gases, fumes, vapors, and solids. Generally, when the specific name of a chemical substance or its state is known, it is classified under the Chemical Substance code (Subcategories 51–56). When only the state of the product is known, it is classified under the code corresponding to that product (Subcategory 57). Emissions from vehicles, furnace gases, and gases generated from kilns are classified under carbon monoxide (55401), while combustion gases and smoke resulting from fires are classified under flame/fire smoke (81505). Inclusions: Acids; Alkalis; Aromatic and Aliphatic Hydrocarbons; Halogens and their Compounds; Metal Dust and Fumes; Pesticides and Insecticides; Coal, Natural Gas, Petroleum Fuels and their Products; Other Chemical Substances and Chemical Products Exclusions: Metallic Materials (241); Non-Metallic Mineral Materials excluding Fuels (242); Fragments, Splinters, Debris (25)
	50	Chemical Substances and Chemical Products with Insufficient Information	
	51	Acids	This category classifies various forms of acids. Inclusions: Acid Gases—Halogens; Inorganic Acids—Halogens; Inorganic Acids—Others; Organic Acids Exclusions: Benzoic Acid and Phenylacetic Acid (Herbicides, etc.) (57104); LSD (Lysergic Acid Diethylamide; Hallucinogen) (57204)
	52	Alkalis	This category classifies chemical substances known as alkalis, bases, and corrosive agents. For corrosive substances with insufficient information, they are classified under 520. For cement mixtures, mortars, and lime (excluding chlorinated lime), these are classified under 521 with calcium hydroxide and calcium oxide. For ash solutions and their products (wastewater and oven cleaners including ash solution), they are classified under 524 with sodium hydroxide, potassium hydroxide, and potassium carbonate. Chlorinated lime is classified under 542 as chlorine and chlorine compounds. Inclusions: Calcium Hydroxide, Calcium Oxide, Calcium Carbonate, Sodium Carbonate, Cement, Lime, Lithium Hydroxide, Sodium Hydroxide, Potassium Hydroxide, Sodium Carbonate Exclusions: Chlorinated Lime (542); Non-alkaline Oven Cleaners

Table A1. *Cont.*

Main Category Code	Subcategory Code	Cause	Classification Criteria and Definitions
	53	Aromatic and Aliphatic Hydrocarbons	This category classifies non-halogenated substances such as alcohols, aldehydes, amines, aromatic compounds, ethers, ketones, and peroxides, with chlorine, fluorine, bromine, iodine, and astatine. Halogenated compounds are classified under Subcategory 54: Halogens and Halogen Compounds. Inclusions: Alcohols, Antifreezes, Aldehydes, Aliphatic Amines, Aromatic Compounds, Ethers, Ketones, Peroxides.
	54	Halogens and Halogen Compounds	This category classifies halogens such as bromine, chlorine, fluorine, iodine, astatine, and their compounds. Compounds containing both fluorine and chlorine are classified under '541 Fluorine and Fluorine Compounds'. Vinyl chloride and polyvinyl chloride are classified under the '57209 Plastics and Resins' code, but molded or extruded plastic products are classified under '245'. Chlorinated hydrocarbons used as pesticides are classified under 57105. Halogen-containing acids are classified under the '51 Acids' subcategory. Inclusions: Bromine and Bromine Compounds, Chlorine and Chlorine Compounds, Fluorine and Fluorine Compounds, Iodine and Iodine Compounds, Carbon Tetrachloride. Exclusions: Halogenated Acids (51); Pesticides (57105), Non-chlorine Bleaching Agents (57203), Vinyl Chloride, Polyvinyl Chloride (57209)
	55	Other Chemical Substances	This category classifies ammonia and its compounds, cryogenic gases, cyanide compounds, oxygen and specific oxides, sewage and mine gases, methane, sulfur and sulfur compounds, and other chemical substances not classified elsewhere. Inclusions: Ammonia and Ammonia Compounds, Carbon Monoxide, Carbon Dioxide, Cryogenic Gases, Cyanide and Cyanide Compounds, Dry Ice, Methane, Mine Gases, Oxygen and Oxygen Compounds, Sewage Gas, Sulfur and Sulfur Compounds, Sulfur Dioxide
	56	Metal Particles, Trace Elements, Dust, and Fumes	This category classifies metal dust, particles, and mists, excluding dissolved metals. It also classifies fumes generated during heating/melting/welding processes. Accidents caused by metal radiation are classified under radioactive minerals (24301) or ionizing radiation (81301) codes based on the exposure/contact method. Finished metal products are classified under appropriate codes based on their functionality. Inclusions: Arsenic, Arsenic Compounds, Beryllium, Beryllium Compounds, Cadmium, Cadmium Compounds, Lead, Lead Compounds, Mercury, Mercury Compounds, Aluminum, Aluminum Compounds, Antimony, Antimony Compounds, Iron, Iron Compounds, Magnesium, Magnesium Compounds, Manganese, Nickel, Nickel Compounds, Zinc and Zinc Compounds, Fumes Generated During Welding or Joining Exclusions: Finished Metal Products; Dissolved or Solid-State Metals (241); Radioactive Metals (24301); Coal Dust (57301); Grain Dust (71405); Non-metallic Dust (24201), Ionizing Radiation (81301)
	57	Chemical Products	This category is used for classification when the substance cannot be categorized under other specific items or when the specific chemical substance is unknown.

Table A1. *Cont.*

Main Category Code	Subcategory Code	Cause	Classification Criteria and Definitions
	59	Other Chemical Substances and Chemical Products	
6		Means of Transportation	This category classifies means of transportation that move on land, water, or air and are used primarily for transporting people or goods (e.g., cars, passenger trains) or leisure (e.g., canoes, bicycles, jet skis). Vehicles and machinery used directly for agriculture, construction, logging, mining, manufacturing, etc., are classified under 0 (Facilities and Machinery). When accidents involve means of transportation, if a part of the means of transportation causing the accident was attached to the entire means of transportation, the entire means of transportation is the cause. If parts of the means of transportation were detached or separated, or if the means of transportation was not related to the accident, only specific parts become the cause. In other words, unattached vehicle parts and accessories are classified under 224 (Parts and Components of Means of Transportation) and unattached trailers under 22403. Unattached windshields and windows of means of transportation are classified under 22499, and if the cause was the floor surface of the means of transportation, it is classified under 31307. Inclusions: Means of Air, Water, and Land Transportation; Non-Industrial Transport Vehicles Except for Roads; Means of Railway Transportation Exclusions: Machinery (0), Agricultural/Horticultural Equipment (041), Construction/Mining Equipment (05), Logging Machinery (042), Transport and Lifting Machinery (03), Street Sweepers (09999), Vehicle Parts and Accessories (224), Floor Surface of Means of Transportation (31307)
	60	Means of Transportation with Insufficient Information	
	61	Means of Land Transportation	
	62	Means of Air, Water Transportation	
	69	Other Means of Transportation	This category classifies means of transportation used outside roads or not powered by internal combustion or other internal engines.
7		Humans, Animals/Plants	This category classifies living organisms (including infectious and parasitic organisms) and products made from them. HIV related to work is classified under clause 71603 (Virus). Inclusions: Animals and Animal Products, Raw or Processed Food, Infectious and Parasitic Organisms, Humans-Victims, Humans-Humans Other than Victims, Unprocessed Plants, Trees, Vegetables Exclusions: Chemical Substances (5); Wood (244)
	71	Humans, Animals/Plants	

Table A1. *Cont.*

Main Category Code	Subcategory Code	Cause	Classification Criteria and Definitions
8		Work Environment, Natural Phenomena such as Atmospheric Conditions, etc.	This category classifies the work environment and conditions. Natural phenomena such as atmospheric pressure, temperature, and other atmospheric conditions are classified only in limited cases when they were the only identifiable cause, such as climate, atmospheric conditions, and geographic events (floods, earthquakes, avalanches) Inclusions: Atmospheric Pressure; Avalanches, Landslides; Earthquakes; Fire, Smoke; Floods; High/Low-Temperature Environments; Climate and Atmospheric Conditions; Noise
	81	Work Environment, Natural Phenomena such as Atmospheric Conditions, etc.	
9		Other Causes	

Table A2. Classification criteria and definitions for accident type.

Classification Code	Type of Occurrence	Classification Criteria and Definitions
01	Fall (person falls from a height)	Incidents where a person falls from an elevated location such as a building, structure, provisional structure, tree, ladder, etc.
02	Tripping (person slips or trips)	Incidents where a person slips or falls on a flat surface, sloping surface, stairs, etc.
03	Pressed Under/ Overturned (object falls or overturns)	Incidents where leaning or standing objects are knocked down and pressed underneath, and construction machinery such as forklifts or other equipment overturns or tips over during operation or movement.
04	Collision (contact with the object)	Incidents where a person's movement or action results in contact or collision with an object (cause). It also includes situations where objects are set in motion (regularly or irregularly) and collide or make contact due to movement while not detaching from their fixed position.
05	Struck by object (struck by falling or flying object)	Incidents where a person is impacted by an object that becomes dislodged from a fixed position due to forces like gravity, centrifugal force, inertia, etc., or when material is expelled from equipment or other sources, causing harm to the person.
06	Collapse (building or piled material collapse)	Incidents where soil, piled material, structures, buildings, provisional structures, etc., collapse entirely or when significant parts break, causing the collapse.
07	Caught In Between (caught or entangled in machinery)	Incidents where a person gets caught or entangled in machinery due to movement between two objects. It includes situations when caught between objects moving linearly, between rotating parts and fixed components, caught in rotating parts such as rollers, or entangled in rotating parts or protrusions.
08	Cutoff/ Cut/ Stab	Incidents involving direct contact with sharp objects, such as knives or blades, that result in a part of a person's body being severed or cut. It also includes cases where the body comes into contact with rotating blade parts of saws/cutting tools.

Table A2. *Cont.*

Classification Code	Type of Occurrence	Classification Criteria and Definitions
09	Electrocution	When a person's body comes into direct contact with parts of electrical equipment or is exposed to induced current, resulting in effects such as muscle contraction, difficulty breathing, ventricular fibrillation, etc. It also includes cases where a person comes into contact with special high-voltage sources or is exposed to arcs due to flash contact, short circuit/mixed contact, etc.
10	Explosion/Rupture	Explosion refers to a rapid process where a substance undergoes chemical or physical changes, accompanied by the sudden release of heat, sound, and pressure. It can occur within buildings, containers, or the atmosphere, including intentional and unintentional events. Rupture, on the other hand, involves the tearing or bursting of pipes, containers, etc., due to physical pressure without a blast pressure.
11	Fire	Incidents where the unintentional ignition of combustible materials is due to an ignition source.
12	Imbalance and Excessive Action	Incidents where sudden and rapid bodily movements/actions, without proper handling of objects, result in an accident. This can also involve situations where excessive muscular force is exerted while handling objects, leading to an accident.
13	Abnormal Temperature Contact	Incidents where a person is exposed to high- or low-temperature environments or objects.
14	Chemical Leakage/Contact	Incidents where a person is exposed to hazardous or dangerous substances through leakage, contact, or inhalation.
15	Oxygen Deficiency	Incidents where a person is exposed to an environment with insufficient oxygen, regardless of hazardous substances, leading to inadequate respiratory function.
16	Fallen In/Drown	Incidents where a person falls into the water and drowns.
31	Workplace Traffic Accident	Accidents that occur on roads within the workplace.
32	Off-site Traffic Accident	Accidents that occur on roads outside the workplace.
33	Maritime or Aviation Traffic Accident	Accidents related to maritime/aviation activities.
41	Accidents in Sports Events, etc.	Accidents occurring in work-related recreational or sports events/workshops, gatherings, etc.
42	Intentional Violence	Intentional or unclearly intentional risky behavior (influenced by drugs, mental disorders, etc.) that causes harm to others through physical violence or assault. This category may also include threats, verbal abuse, and sexual violence.
43	Animal Injury	Incidents where a worker is injured by an animal (dog/cow/horse, etc.), including cases where a worker is bitten or kicked by an animal.
49	Others	
Z	Unclassifiable	

References

1. Fung, I.W.H.; Tam, V.W.Y.; Lo, T.Y.; Lu, L.L.H. Developing a risk assessment model for construction safety. *Int. J. Proj. Manag.* **2009**, *28*, 593–600. [CrossRef]
2. Pinto, A.; Nunes, I.L.; Ribeiro, R.A. Occupational risk assessment in construction industry—Overview and reflection. *Saf. Sci.* **2011**, *49*, 616–624. [CrossRef]
3. Sunindijo, R.Y.; Zou, P.X.W. Political skill for developing construction safety climate. *J. Constr. Eng. Manag.* **2012**, *138*, 605–612. [CrossRef]

4. Hallowell, M.R. Safety-Knowledge Management in American Construction Organizations. *J. Manag. Eng.* **2012**, *28*, 203–211. [CrossRef]
5. Lim, S.; Oh, A.; Won, J.; Chon, J. Improvement of Inspection System for Reduction of Small-Scale Construction Site Accident in Korea. *Ind. Health* **2018**, *56*, 466–474. [CrossRef] [PubMed]
6. Cheng, C.W.; Leu, S.S.; Fan, C. Characteristic analysis of occupational accidents at small construction enterprises. *Saf. Sci.* **2010**, *48*, 698–707. [CrossRef]
7. Dumrak, J.; Mostafa, S.; Kamardeen, I.; Rameezdeen, R. Factors associated with the severity of construction accidents: The case of South Australia. *AJCEB* **2013**, *13*, 32–49. [CrossRef]
8. Camino López, M.A.; Ritzel, D.O.; Fontaneda, I.; González Alcantara, O.J. Construction industry accidents in Spain. *J. Saf. Res.* **2008**, *39*, 497–507. [CrossRef]
9. Bang, S.; Jeong, J.; Lee, J.; Jeong, J.; Soh, J. Evaluation of Accident Risk Level Based on Construction Cost, Size and Facility Type. *Sustainability* **2023**, *151*, 1565. [CrossRef]
10. Daba, C.; Atamo, A.; Gebrehiwot, M. Small Scale Enterprise Workers Require Attention: What Predicts the Level of Occupational Injuries? *Environ. Health Insights* **2022**, *16*, 11786302221104949. [CrossRef]
11. Berhe, A.; Yemane, D.; Gebresilassie, A.; Terefe, W.; Ingale, L. Magnitude of Occupational Injuries and Associated Factors among Small- Scale Industry Workers in Mekelle City, Northern Ethiopia. *Occup. Med. Health Aff.* **2015**, *3*, 1000197.
12. Choi, S.D.; Guo, L.; Kim, J.; Xiong, S. Comparison of fatal occupational injuries in construction industry in the United States, South Korea, and China. *Int. J. Ind. Ergon.* **2019**, *71*, 64–74. [CrossRef]
13. Kang, K.; Ryu, H. Predicting types of occupational accidents at construction sites in Korea using random forest model. *Saf. Sci.* **2019**, *120*, 226–236. [CrossRef]
14. Fu, H.; Xia, Z.; Tan, Y.; Guo, X. Influence of Cues on the Safety Hazard Recognition of Construction Workers during Safety Training: Evidence from an Eye-Tracking Experiment. *J. Civil Eng. Educ.* **2023**, *150*, 04023009. [CrossRef]
15. Namian, M.; Albert, A.; Zuluaga, C.M.; Behm, M. Role of Safety Training: Impact on Hazard Recognition and Safety Risk Perception. *J. Constr. Eng. Manag.* **2016**, *142*, 04016073. [CrossRef]
16. Perlman, A.; Sacks, R.; Barak, R. Hazard recognition and risk perception in construction. *Saf. Sci.* **2014**, *64*, 22–31. [CrossRef]
17. Uddin, S.M.J.; Albert, A.; Alsharef, A.; Pandit, B.; Patil, Y.; Nnaji, C. Hazard Recognition Patterns Demonstrated by Construction Workers. *Int. J. Environ. Res. Public Health* **2020**, *17*, 7788. [CrossRef]
18. Zhou, L.; Wang, J.; Ashuri, B.; Chen, J. Discovering the Research Topics on Construction Safety and Health Using Semi-Supervised Topic Modeling. *Buildings* **2023**, *13*, 1169. [CrossRef]
19. Orusa, T.; Viani, A.; Moyo, B.; Cammareri, D.; Borgogno-Mondino, E. Risk Assessment of Rising Temperatures Using Landsat 4–9 LST Time Series and Meta® Population Dataset: An Application in Aosta Valley, NW Italy. *Remote Sens.* **2023**, *15*, 2348. [CrossRef]
20. Jiang, H.; Qiang, M.; Lin, P. A topic modeling based bibliometric exploration of hydropower research. *Renew. Sustain. Energy Rev.* **2016**, *57*, 226–237. [CrossRef]
21. Blei, D.M.; Ng, A.; Jordan, M.I. Latent Dirichlet Allocation. *J. Mach. Learn. Res.* **2001**, *3*, 993–1022.
22. Gerlach, M.; Peixoto, T.P.; Altmann, E.G. A network approach to topic models. *Sci. Adv.* **2018**, *4*, eaaq1360. [CrossRef] [PubMed]
23. Camacho, D.; Panizo-LLedot, A.; Bello-Orgaz, G.; Gonzalez-Pardo, A.; Cambria, E. The four dimensions of social network analysis: An overview of research methods, applications, and software tools. *Inf. Fusion* **2020**, *63*, 88–120. [CrossRef]
24. Cinque, M.; Cotroneo, D.; Pecchia, A. Challenges and Directions in Security Information and Event Management (SIEM). In Proceedings of the 2018 IEEE International Symposium on Software Reliability Engineering Workshops, Memphis, TN, USA, 18 October 2018.
25. O'Callaghan, D.; Greene, D.; Carthy, J.; Cunningham, P. An analysis of the coherence of descriptors in topic modeling. *Expert Syst. Appl.* **2015**, *42*, 5645–5657. [CrossRef]
26. Hidayatullah, A.F.; Ma'arif, M.R. Road Traffic topic modeling on twitter using Latent Dirichlet Allocation. In Proceedings of the 2017 International Conference on Sustainable Information Engineering and Technology (SIET), Malang, Indonesia, 1 March 2018.
27. Park, G.; An, H. The Topic Modeling Analysis of The DMZ Tours Issues Using Text Mining. *Korean J. Leis. Recreat. Stud.* **2019**, *31*, 143–159.

buildings

MDPI

Article

Managing Safety Risks from Overlapping Construction Activities: A BIM Approach

Armin Rashidi Nasab [1,*], Hassan Malekitabar [2], Hazem Elzarka [1,*], Ala Nekouvaght Tak [3] and Khashayar Ghorab [4]

1 Department of Civil, Architectural Engineering, and Construction Management, University of Cincinnati, Cincinnati, OH 45220, USA
2 School of Civil Engineering, Iran University of Science and Technology, University St., Hengam St., Resalat Square, Tehran 13114-16846, Iran; malekitabar@iust.ac.ir
3 Institute for Creative Technologies, University of Southern California, Playa Vista, CA 90094, USA; antak@ict.usc.edu
4 Department of Civil Engineering, University of Calgary, 2500 University Drive NW, Calgary, AB T2N 1N4, Canada; khashayar.ghorab@ucalgary.ca
* Correspondence: rashidan@mail.uc.edu (A.R.N.); elzarkhm@ucmail.uc.edu (H.E.)

Abstract: Addressing safety risks in construction is an ongoing priority, and integrating safety considerations into construction scheduling is a crucial aspect of this effort. A notable challenge is the safety risk posed by concurrent tasks, which has received limited attention in prior research. This study aims to address this research gap by introducing a novel Building Information Modeling (BIM)-based model that assesses the increased hazardousness resulting from overlapping construction activities. Historically, research has predominantly focused on individual task safety, with less emphasis on the risks associated with overlapping activities. Our innovative approach introduces the concept of a 'source–target' match, which evaluates the degree of hazardousness escalation when activities overlap. Drawing on data from the Occupational Safety and Health Administration (OSHA) and the National Institute for Occupational Safety and Health (NIOSH) fatal accident reports, we extracted 11 hazardous and 9 susceptibility attributes to build a source–target match table. This table reveals the characteristics of activities that generate hazardous conflicts when overlapping. The key contribution of this research is the assessment, prioritization, and visualization of risk levels in a BIM environment. This framework empowers safety managers to proactively address safety risks resulting from overlapping construction activities, ultimately reducing accidents in the construction industry. By shedding light on this overlooked aspect of construction safety, our research highlights the importance of integrating safety considerations into construction scheduling and provides a practical tool for mitigating risks, enhancing workplace safety, and ultimately improving project outcomes.

Keywords: construction safety; overlapping activities; BIM-based model; hazardous conflicts; risk assessment

Citation: Rashidi Nasab, A.; Malekitabar, H.; Elzarka, H.; Nekouvaght Tak, A.; Ghorab, K. Managing Safety Risks from Overlapping Construction Activities: A BIM Approach. *Buildings* **2023**, *13*, 2647. https://doi.org/10.3390/buildings13102647

Academic Editors: Mariusz Szóstak and Marek Sawicki

Received: 25 September 2023
Revised: 17 October 2023
Accepted: 17 October 2023
Published: 20 October 2023

1. Introduction

The construction industry faces a concerning statistic: the rate of fatalities is 50% higher compared to other industries [1]. Despite the stringent safety standards set by organizations like the Occupational Safety and Health Administration (OSHA) and the substantial investments made by construction companies in pursuit of accident-free projects, construction accidents continue to occur at an alarming rate [1]. Addressing the root causes of safety risks demands advanced techniques and a collaborative, multidisciplinary approach [2].

While safety procedures often target high-risk situations associated with specific project activities [3,4], construction sites frequently host numerous small teams, or subcontractors, engaged in diverse tasks in close proximity to each other [5]. However, a critical aspect remains unaddressed: the impact of overlapping conflicts between project activities on safety. When two or more hazardous activities are scheduled concurrently, they not only accumulate risk [6] but also synergistically heighten safety risks by affecting each other. Surprisingly, this increased risk factor has received limited attention in prior research.

In bustling construction environments, it is common for multiple tasks to be scheduled simultaneously, leading to a higher incidence of accidents. Workers may be aware of safety risks associated with their own tasks but lack knowledge or training to manage safety hazards arising from other concurrent activities. These overlapping conflicts, while essential for efficiency, can result in unforeseen safety hazards [7]. For instance, the simultaneous positioning of surveying instruments and excavation by heavy equipment operators may lead to accidents [8]. Similarly, activities like pouring column concrete and installing reinforcement in elevated slabs can overlap, increasing the risk of accidents. Unfortunately, safety evaluations often focus on individual tasks, leaving the risks from overlapping activities unassessed [9].

Behm's study in 2005 revealed that 42% of construction injuries could be linked to design-related factors [7]. These injuries might have been prevented had safety been a design consideration. Designing for safety is now recognized as pivotal in mitigating safety hazards; however, designers do not consistently prioritize safety criteria in their daily work. Construction managers commonly aim to reduce accidents by addressing hazards similar to those previously encountered. Integrating safety planning into construction scheduling can be a proactive approach to identifying potential safety risks before the commencement of each activity, significantly reducing the accident rate [10].

In the complex realm of construction projects, Building Information Modeling (BIM) has become an invaluable tool for safety managers to identify and understand risks more effectively and enhance the dynamic visualization of safety procedures [11,12]. This research introduces a BIM-based framework to identify a previously overlooked safety risk arising from overlapping activities. The process involves three key steps: (1) defining conditions under which overlapping activities intensify the risk, (2) prioritizing risky activities with overlapping conflicts through risk assessment, and (3) automating and visualizing the identification process using BIM to enhance accessibility for safety managers.

This paper is organized as follows:

- Section 2 reviews relevant literature on overlapping construction activities, risk assessment, and the role of BIM in identifying safety issues.
- Section 3 introduces the proposed framework, detailing how overlapping activities are captured and how a BIM model can be implemented.
- Section 4 presents the findings, including a case study illustrating the model's practical application.
- Finally, Sections 5 and 6 provide a discussion of our findings and conclude the paper.

2. Literature Review

A significant body of research has addressed unsafe conditions at construction project sites. Kartam (1997) found construction accidents to be the source of many human tragedies, delays in activities, and lack of worker motivation [5]. He believed an essential step for mitigating construction hazards is considering safety at the planning stage. He introduced a Critical Path Method (CPM)-based system to manage safety activities and predict unsafe risks. Kartam's research is one of the earliest research projects that suggests considering safety in construction scheduling. Zolfagharian et al. (2014) collected information on high-risk construction accidents to evaluate the risk parameters of every project task [8]. They combined this information with the project schedule in an automated module, which anticipated the rate of accidents in each activity to assist safety managers before the start of the activities. Mirzaei et al. (2018) concluded that conflicts frequently occur when the workspace shared by different activities was not evaluated in the construction time–space and proposed a time–space conflict model using 4D-BIM to decrease workspace conflicts among workers' movements during different stages of construction activities [6]. They mainly focused on conflict between two support platform spaces, conflict between two labor workspaces, and conflict between a labor workspace and a support platform space. Moon et al. developed a schedule–workspace system to manage workspaces that have overlaps with each other [13]. They asserted that simultaneous workspaces increase conflict between resources; therefore, they implemented a platform using BIM and a genetic algorithm (GA) to decrease the level of overlap between scheduled workspaces. Yi and Langford considered the issue of scheduled-based safety risks, to minimize accidents [14]. They believed that in a hazardous environment, they should either pay attention to jobs related to the hazards or the hazardous environment should be eliminated. They concluded that to prevent such dangerous activities, they need to be either rescheduled or relocated; then, using risk assessment of activities, they could identify risky situations where and when they can happen. Martinez et al. used the unmanned aerial system to minimize job site conflicts and collisions, after implementing flight tests they could decrease safety risks probabilities such as worker distraction by more than 40% [15]. Khodabandelu et al. (2020) developed an Agent-Based Modeling (ABM) simulation to decrease collision possibilities between cranes [16]. Their model mitigated overlap risks while increasing crane productivity and schedule time. Guo et al. (2018) highlighted the peer pressure effect and developed a Behavior-Based Safety (BBS) program to prevent unsafe behavior [17].

Project managers have felt unable to fully identify the safety risks when the tasks get complicated [18,19]. Mohandes and Zhang asserted there has been always a lack of comprehensive studies in considering risk parameters, so they conducted a Holistic Occupational Health and Safety Risk Assessment Model (HOHSRAM), and they could assess all the crucial risk parameters using the integration of logarithmic fuzzy ANP, interval-valued Pythagorean fuzzy TOPSIS, and grey relational analysis [20]. Cagno et al. (2001) asserted that improvements in safety measures were highly dependent on risk assessments [21]. Sanni-Anibire et al. (2019) confirmed this association by developing a risk assessment approach and establishing risk scores and weights for every construction hazard in various projects [1].

Okpala et al. (2020) surveyed the different technologies for mitigating safety risks throughout the project life cycle. They stated that BIM substantially improves the identification of unsafe situations before the beginning of construction [22]. While asserting that safety management will start from planning to implementation, Chantawit et al. (2005) found it difficult to analyze what, when, where, and why safety measures are needed. They introduced 3D and 4D-CAD visualization systems that made hazard prevention much more convenient for safety managers [23]. At present, growing technologies such as BIM facilitate planning with more detail and fewer weaknesses. Kassem et al. developed a 4D tool for workspace management to identify spatial and temporal conflicts in site activities [24]. Lu et al. developed a quantitative method in order to automatically assess the safety risks using BIM, which helps architects and structure engineers to use alternative designs easily [25]. They calculated risk assessments using three indexes: likelihood, consequences, and exposure. Nekouvaght Tak et al. conducted a 4D crane simulation framework in the BIM environment to detect potential spatial conflicts related to the crane [26]. The frameworks were verified using three case studies in Alberta, Canada. Recent advancements have further expanded this potential through the introduction of the 8D BIM concept [27]. The eighth dimension of BIM revolves around the integration of health, safety, and well-being of the building's occupants into the modeling process. This dimension aids in creating a holistic view of safety, mapping the risks associated with both the construction phase and the end-user phase into the BIM model. By doing so, the entire lifecycle of a construction project, from inception to decommissioning, is taken into consideration, ensuring that every potential risk is accounted for. This way, 8D BIM not only provides an in-depth visualization of potential spatial conflicts but also ensures the building's design is ergonomically sound and safe for its end-users. The emphasis here is on predicting and eliminating risks even before they manifest, making proactive safety measures an integral part of the construction process. Therefore, as the construction industry evolves, 8D BIM is anticipated to play an increasingly pivotal role in merging technological advancements with a strong commitment to safety [28].

Previous studies mainly focused on conflict of either time or space between activities but did not adequately consider the hazardous impact of one activity on another [5,6]. This research introduces a new BIM-based approach to the identification of activities that overlap with each other, occur at the same location, and interfere with each other in a way that intensifies risky accidents in the construction industry.

3. Methodology

3.1. Degree of Hazardousness from Overlapping Activities

The degree of hazardousness from overlapping activities will be first discussed in this section. As shown in Figure 1, a Fault Tree Analysis (FTA) was used to logically describe the anticipated degree of hazardousness from overlapping construction activities. As shown in Figure 1, the three main factors that have an impact on the anticipated degree of hazardousness resulting from two overlapping activities are (1) the activities' spatial overlap, (2) the activities' concurrency, and (3) the activities' source–target match, which is a newly introduced parameter that is described later. It should be noted that the absence of any of these factors will eliminate the hazardousness that may result from the two overlapping activities.

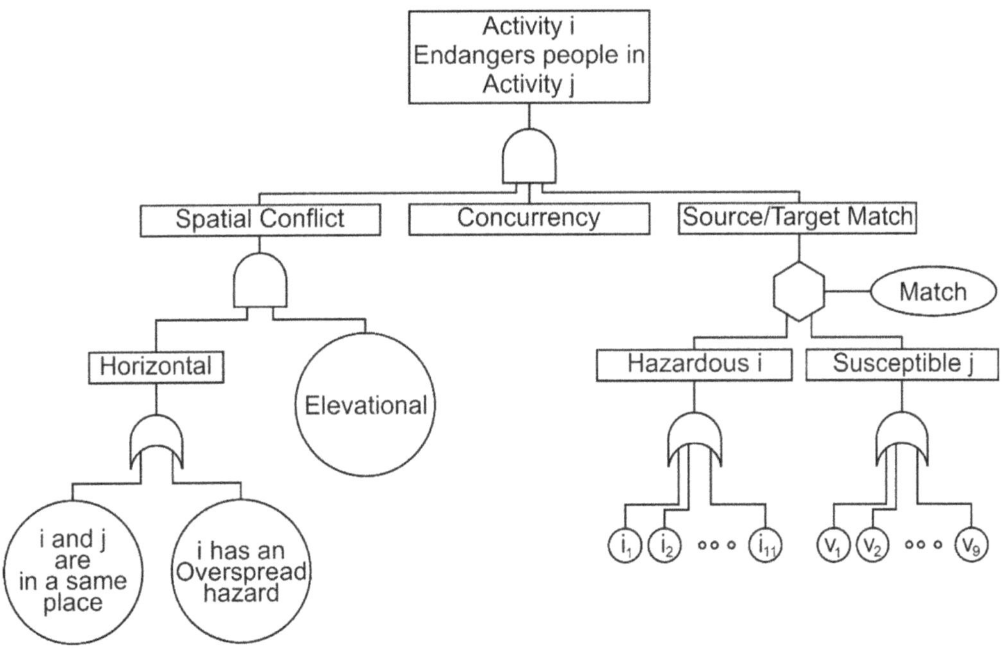

Figure 1. Fault Tree Analysis of overlapping conflicts.

On the other hand, if the three factors co-exist, Equation (1) can be used to determine the degree of hazardousness. An activity that poses risks to workers performing another activity that coincides in time, place, and height generates an overlapping conflict with that activity according to Equation (1), where *F (i, j) = overlapping conflict of activity i and j*. The overlapping conflict is explained in the three steps below.

$$F\ (i.\ j) = Spatial\ overlap \times Concurrency \times Source/Target\ Match \quad (1)$$

3.1.1. Source–Target Match

Source–target match seeks to examine the degree to which two overlapping activities will affect each other. A construction activity has characteristics that may increase the safety risks of all overlapping activities. These characteristics are referred to in this research as "Hazardous" attributes. For example, "exposure to debris and unstable situations" and "possible falling from height" are "Hazardous" attributes associated with the excavation activity. Excavation typically involves loading trucks with debris, which can fall while loading and pose a danger to workers engaged in other close-by activities and passersby. At the same time, construction activity may be susceptible to external factors caused by other activities that increase the safety risks of the activity. These characteristics are referred to in this research as "Susceptibility" attributes. For example, "lack of site space" and "distraction" are "Susceptibility" attributes associated with the construction surveying activity. When there is limited space to work and various distractions are present, the safety of surveyors and other workers can be compromised because of the increased risk of collisions, limited visibility, tripping hazards, workplace stress, and communication challenges. The Source–Target Match concept encompasses a comprehensive table meticulously developed through an exhaustive analysis of OSHA and National Institute for Occupational Safety and Health (NIOSH) reports. This table includes both hazardous and susceptible attributes, providing a thorough overview of potential safety risks in overlapping scenarios. Each attribute is accompanied by an associated severity and probability value, facilitating a more precise assessment of the risk factor associated with overlapping conflicts. Further elaboration on this process will be provided in Section 3.2.

3.1.2. Concurrency of Two Activities

The pivotal chain contributing to increased safety risks among construction activities is the concurrency, which occurs when two or more activities on a construction site take place simultaneously within a shared timeframe. To quantitatively assess concurrency, project management tools, such as Microsoft Project, are often employed to meticulously schedule and organize project activities. This scheduling approach facilitates the determination of the precise start and end dates of each activity, allowing project managers and safety professionals to identify instances of overlap. In the context of our study, we examine two activities, denoted as activities i and j, and classify them as concurrent if they share at least one working day in common. This definition of concurrency provides a foundation for evaluating the potential risks associated with activities that overlap in time.

The duration of overlapping activities plays a crucial role in assessing safety risks. As illustrated in Figure 2, there exists a fundamental correlation between the duration of overlap and the nature of the risks involved. A short period of concurrency between a hazardous activity (i) and a susceptible activity (j) offers limited time for potential mishaps to occur. However, it also implies less time for the workers in the two activities to become familiar with each other's presence and get prepared for potential hazards. Conversely, a longer concurrency duration allows more time for mishaps to potentially take place but also increases the familiarity between the activities, potentially leading to a reduction in safety risks. It is essential to clarify that our current research does not delve into the intricate dynamics of how risk probabilities shift with varying durations of overlap. This question remains open: Does a longer concurrency period significantly elevate the probability of accidents, or does the increased familiarity between tasks effectively mitigate the dangers posed by overlapping activities? These intriguing aspects of safety risk assessment warrant further investigation.

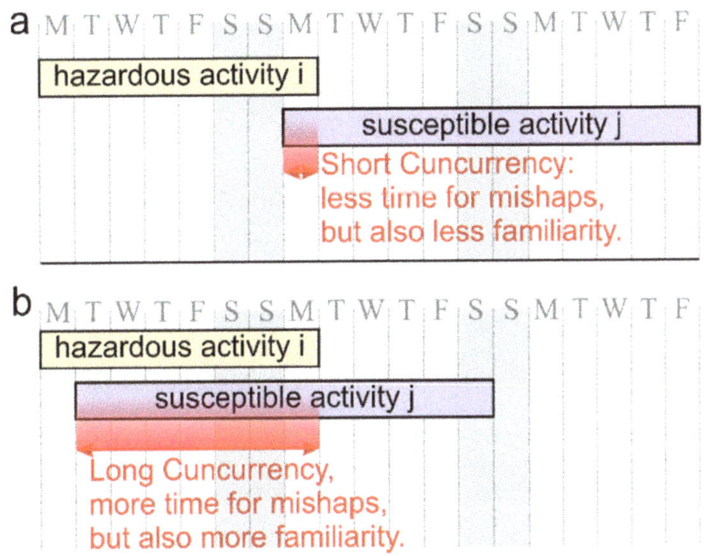

Figure 2. Concurrency of two activities: (**a**) short concurrency (1 day overlap); (**b**) long concurrency (7 day overlap). Letters in grey represent days of the week.

3.1.3. Spatial Conflict

Spatial conflict is another condition that must be present for an overlap conflict to occur, and it includes both elevational and horizontal conflict. The elevational category involves activities such as machinery or equipment operations at varying heights, which can cause distractions or obstructions to other tasks. For instance, elevated operations, such as crane movements, may interfere with ground-level tasks, creating a vertical conflict. This interference can disrupt the workflow and pose safety hazards, particularly when elevated equipment crosses paths with activities on the ground. Elevational conflict does not simply mean being at the same height; two activities with different heights may also influence each other. As it is shown in Figure 3, on a concrete framed building, when the "rebar welding" activity, represented by a red circle, is taking place in level 2, it may cause a distraction for the "rebar installation" activity taking place in level 1 and represented by a blue circle. Thus, it is necessary to consider the elevation of activities and also the height effect of hazardous attributes.

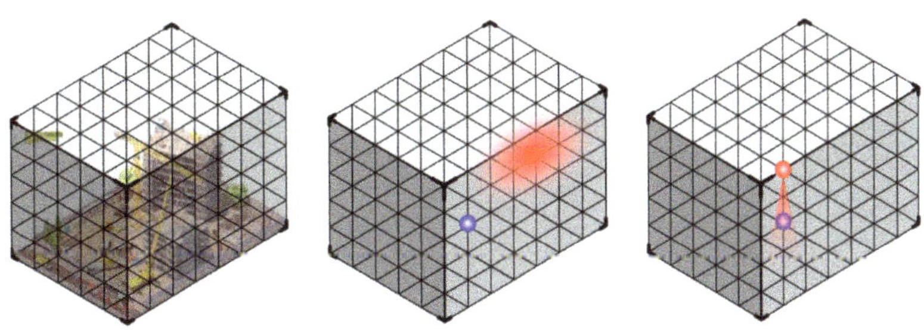

Figure 3. Three-dimensional elevation hazard spread. On a concrete framed building, when the "rebar welding" activity (red circle) is taking place in level 2, it may cause a distraction for the "rebar installation" activity taking place in level 1 (blue circle).

As seen in Equation (2), to identify an elevational conflict, the elevation ranges of the susceptible activity (j) and the hazard produced by the hazardous activity (i) must coincide; that is, the result of Equation (2) must be positive. Otherwise, there will be no interference and thus no conflict. The parameters used in Equation (2) are:

Zmin (j) = minimum height of susceptible activity, Zmax (i) = maximum height of hazardous activity, Zmax (H) = maximum height effect of hazardous attribute, Zmin (H) = minimum height effect of hazardous attribute.

$$Min\ (Zmax\ (j),\ Zmax\ (i) + Zmax(H) - Max\ (Zmin\ (j),\ Zmin\ (i) - Zmin(H) > 0 \quad (2)$$

The second condition of special conflict is horizontal conflict. Horizontal conflicts pertain to activities with extensive spatial demands that conflict with tasks in adjacent zones or overlap workspace, leading to equipment or personnel obstructing neighboring zones. For example, large machinery operations may conflict with tasks in nearby areas, impeding the smooth progress of work and potentially causing safety risks. These horizontal conflicts result from the spatial overlap of activities and the limited physical space available on construction sites. A construction site can be divided into zones to assess the hazard spread from an activity, say activity (i), to which activity (j) is exposed if placed in the same zone as activity (i) or the hazard is able to spread over other zones. The activity shown in zone 7 of Figure 4 spreads its hazard over other zones except zone 13. In fact, not only two activities in the same zone could have horizontal conflict, but hazardous activity that can spread over other zones can be considered as horizontal conflict. Therefore, to assess the horizontal conflict, we need to divide the construction site into different zones.

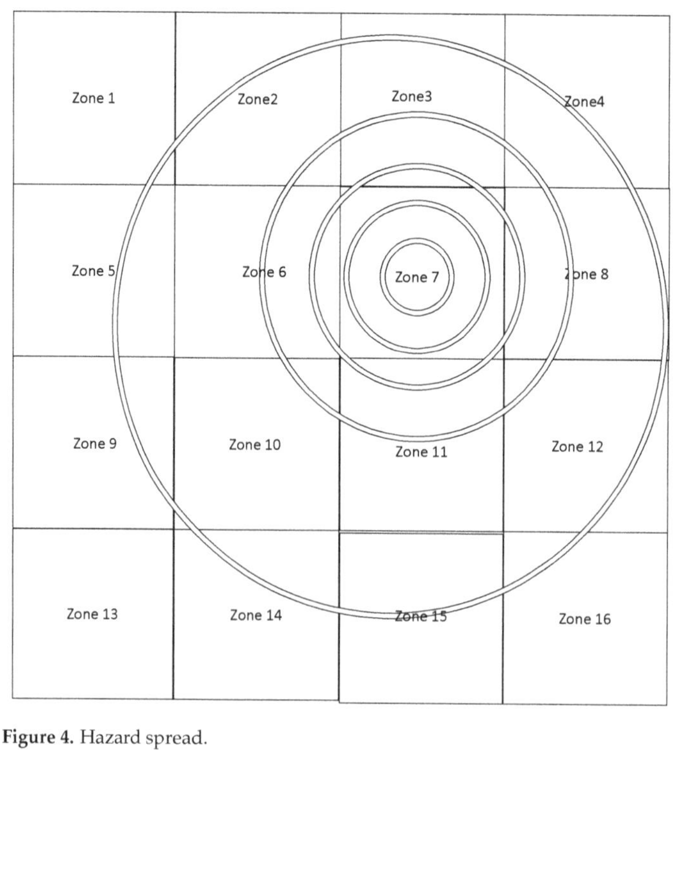

Figure 4. Hazard spread.

3.2. Data Gathering

A database of hazardous and susceptible attributes was extracted from three primary sources: (1) the vivid narrations provided by the National Institute of Occupational Safety and Health (NIOSH) [29], named Fatal Accident Circumstances and Health Epidemiology (FACE), which contains over 200 reports; (2) 70 reports with a similar approach to that of NIOSH FACE which was provided by the Occupational Safety and Health Administration (OSHA) [30], called Fatal Facts; and (3) many scientific papers with accident causation analysis were also reviewed [31–42]. The adequacy of the number of accidents was verified when evidence of data saturation appeared as recommended by Malekitabar et al. [18] and Eisenhardt [43]. An example can clarify how the hazardous and susceptible attributes were extracted from the reports. The NIOSH FACE report number 2005-11 states that on 13 August 2005, in the US state of North Carolina, a worker was hit by a bulldozer traveling in reverse [29], despite all the alarms and warnings sent to the worker and vehicle, the worker is died from multiple blunt force injuries. By reviewing this report, we can extract "machinery with moving parts" as a hazardous attribute and "exposure to heavy materials and equipment" as a susceptible attribute. Another accident took place on 19 April 2012, from NIOSH FACE (2012-02), where a Spanish worker fell from 13.5 feet off the roof onto asphalt; it was reported that the worker had been distracted. From this report, we can extract "distraction" as a hazardous attribute and "exposure to falling from the height" as a susceptible attribute for roofers.

Not all overlapping conflict attributes impose the same, certain level of risk on each other [20]. When excavators are operating, surveyors are exposed to a higher risk level than truck drivers. The truckers, in contrast, are more susceptible to electric shocks as their vehicle might cut a temporary overhead line the electricians have installed. Therefore, the probability and severity of each conflict need to be assessed separately. Considering two activities nominated for a conflict assessment, a source–target match table evaluates the concordance probability of every hazardous attribute in one activity with every susceptible property in the other. In the absence of quantitative measurements, the probability that a hazardous attribute matched a susceptible one was determined using an expert group panel. In this study, ten panelists participated, as Adler and Ziglio suggested that a reasonable result can be achieved with 10–15 experts [44]. This panel included an academician well-versed in construction safety, an expert in Building Information Modeling (BIM), and industry professionals with substantial experience in overseeing safety protocols on construction sites. In addition to these, we also integrated perspectives from site managers, safety inspectors, and construction planners. Each participant brought unique expertise and perspectives, ensuring a well-rounded understanding and robustness of our research findings. All participants possessed four characteristics based on the research of Tersine and Riggs [45]: 1—having a high-performance record, 2—having adequate time, 3—putting in enough effort, and 4—having high rationality. Severity was scored from 0 to 10, where 0 represents the lowest severity and 10 represents the highest severity, and scoring probability from 0 to 5, where 0 represents the lowest probability and 5 represents the highest probability.

3.3. Implementation

As stated by Perlman et al. (2014) [19] and Chantawit et al. (2005) [23], safety risk identification is difficult without new technologies when data are complicated. BIM is thus used to automate the decision process introduced in this research and to visualize the conflicts in the four stages summarized in Figure 5.

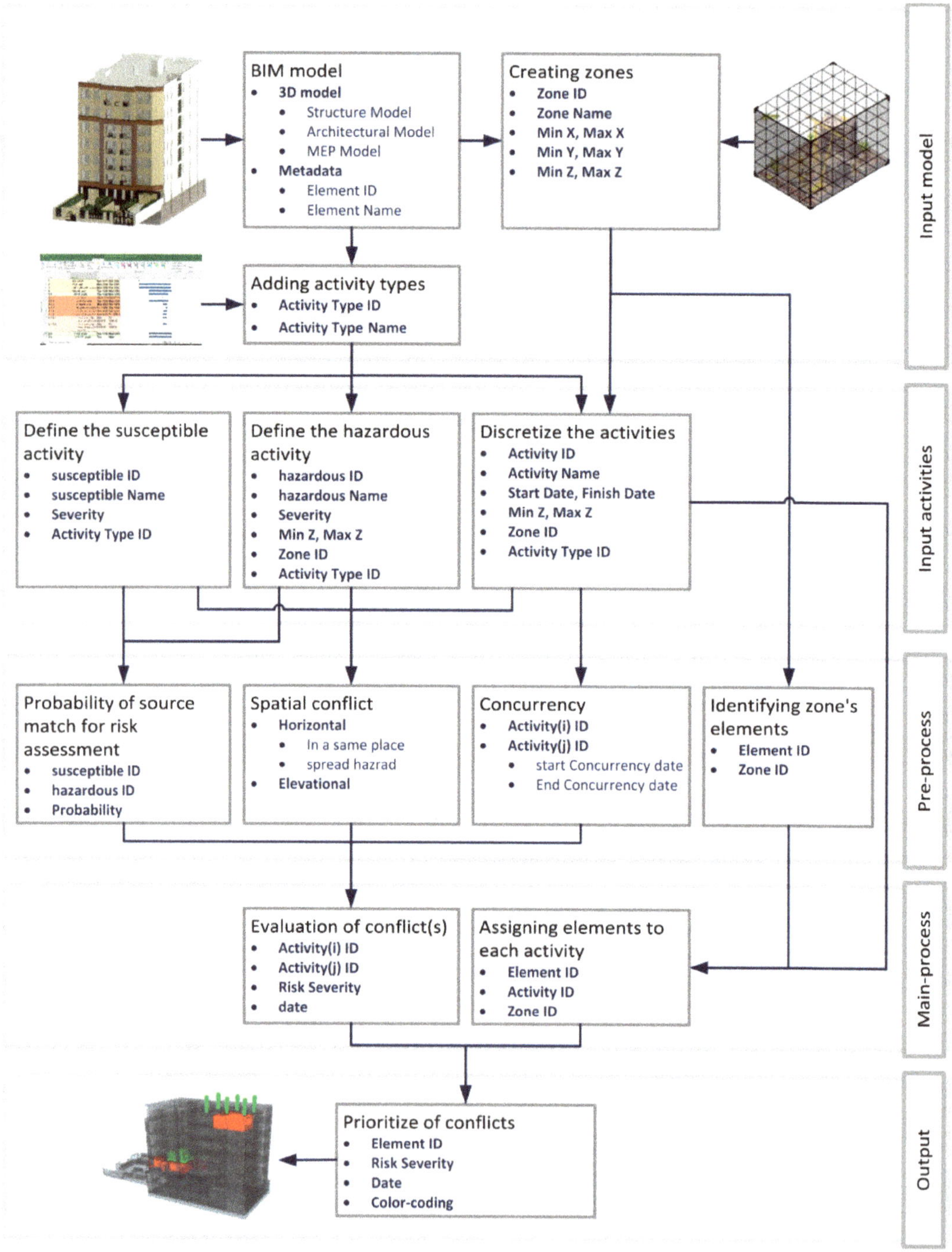

Figure 5. From Data to Output Out: The BIM Workflow.

In the input model stage, a building will be modeled using Autodesk Revit® software (2021 Version) with all construction components including architectural, structural, and Mechanical, Electrical, and Plumbing (MEP) information as well as their element IDs to visualize the conflicts. The model is then imported into Navisworks Manage software (2021 Version) to capture all the activities related to the project; duration, start date, and finish date will be captured as well to evaluate concurrency of activities in the next stages. The other things that need to be captured in this stage to measure the spatial conflict are the height of each activity, which is called Zmin/Zmax (i) and Zmin/Zmax (j) based on Equation (2), and the zone of each activity, which will be defined by min and max of x, y, z using the Application Programming Interface (API) according to their coordinates. Then an operator with general construction knowledge links the 3D components to the activity types. In this way, a 4D model is developed, providing a graphical overview of the involved processes and activity dependencies.

The Pre-Process stage reviews the overlapping conflict essentials that appeared in the first FTA level shown in Figure 1. Firstly, susceptibility and hazardousness attributes, together with their probability and their severity are needed. After extracting them using three sources which are explained in Section 3.1.1, severity and the probability of susceptibility and hazardousness attributes for each activity will be assigned. A hazardous activity imposes one or many hazards, each with severity as an integer, an indicator of whether it spreads over other zones as a Boolean, and relative elevational range above or beneath the host activity as two numbers referred to as Zmax (H) and Zmin (H) based on Equation (2). In fact, all characteristics of overlapping conflicts including spatial conflict, concurrency, and source–target match are assigned at this stage for analysis in the next step.

The main process stage assigns a Risk Factor (RF) to each conflict and couples them with the element ID, activity ID, and zone ID to facilitate the visualization. This process goes through all the activities and calls the routines stated before to check for any overlapping conflicts between the concurrent activities. All calculations are implemented by the API; Figure 5 shows the main module workflow. The evaluation of risk factors depends on the probability and severity of the conflicts, the likelihood of a hazard's potential that results in damage defined as probability, and the extent of the damage which could be related to accidents defined as severity [8,46,47]. Equation (3) measures spatial overlap, concurrency, and source–target match where i = hazardous activity, j = susceptible activity, F (i, j) = overlapping conflict of activity i and j, Xi = severity of hazardous activity, Yj = severity of susceptible activity, Pij = probability of hazardous and susceptible match-based source–target match table, Tij = concurrency of the activities, Eij = elevational exposure, Hij= horizontal exposure. It is worth mentioning that spatial overlap and concurrency are Boolean, and the risk factor rate is measured by using the severity and probability of the source–target match table. Equation (4) calculates and sums the RFs of every ID set where x* = the conflict's score between each two activities. Using the summation risk level of each activity, the model is able to prioritize overlapping conflict using Table 1, where an activity with a scale of 1 and risk level of 0–0.2 is considered "Insignificant" conflict, which means it has minimal impact on the project; however, an activity with the scale of 5 and risk level of 0.8–1.00 is considered a "Catastrophic" conflict which brings a large number of fatalities.

Table 1. Priority of activities' risk levels.

Scale	Risk Level	Description	Priority
1	0–0.2	First aid injuries only and/or minimal impact	Insignificant
2	0.2–0.4	Minor injuries and/or short-term impact	Minor
3	0.4–0.6	Serious injuries and/or significant impact	Moderate
4	0.6–0.8	Fatalities and/or major short-term impact	Major
5	0.8–1.00	Large number of fatalities and/or major long-term impact	Catastrophic

Finally, in the output stage using the total RF, date, element IDs, and a light color code, based on Table 1, of each conflict, to ensure clarity and specificity in visualizing safety risks within the BIM model, a comprehensive visualization protocol is employed. Once the risk factors are meticulously calculated for every activity, the model's elements are color-coded, providing an immediate visual cue to the associated risks. This not only makes it straightforward for stakeholders to immediately discern the severity of potential conflicts but also facilitates swift decision-making. Using a gradient color scheme, risks are depicted on a spectrum, ranging from minimal (green) to significant or catastrophic (red). The color intensity increases with the potential severity of the hazard. Hovering over or selecting a particular element can display a detailed risk analysis tooltip, providing more in-depth information about the associated risks. This dynamic approach ensures that all project stakeholders, from site managers to workers, can intuitively understand and act upon the risk data encapsulated within the model, ensuring enhanced safety protocols and efficient risk management throughout the project lifecycle.

$$\text{Risk Factor} = \text{Probability} \times \text{Severity}$$

$$F\ (i.\ j) = \frac{\text{Source}}{\text{Target}}\text{Match} \times \text{Concurrency} \times \text{Spatial overlap} = \sum i \sum jXi\ Yj \times Pij \times Tij \times Eij \times Hij \quad (3)$$

i = hazardous activity
j = susceptible activity
F (i, j) = overlapping conflict of activity i and j
Xi = severity of hazardous activity
Yj = severity of the susceptible activity
Pij = probability of hazardous and susceptible match-based source–target match table
Tij = concurrency of the activities
Eij = elevational exposure
Hij = horizontal exposure

$$\text{Risk Level of each activity} = \sum_0^n x^* \text{ Risk Level of Each Activity} \quad (4)$$

3.4. Validation of the Model

Given the uncertain and potentially hazardous nature of accidents in the construction industry, conducting direct tests to determine whether the identified high-priority conflicts would lead to actual accidents is both impractical and ethically questionable. As such, our validation approach primarily relied on expert judgment. We consulted with two industry experts, each possessing substantial experience and expertise in construction safety and project management. These experts were presented with our model and the identified high-priority conflicts. Their task was to evaluate the model and confirm whether the conflicts we identified align with the most dangerous events commonly observed in the construction industry. The experts' unanimous agreement that our identified conflicts correspond to the industry's most perilous occurrences serves as a form of face validity [48] and initial validation for our model.

4. Results

4.1. Hazardous and Susceptible Attributes

After reviewing accident reports from OSHA, NIOSH, and reviewed papers, it was found that evidence of data saturation in repeating accidents occurred after the first hundred reports. However, the remaining reports and papers were reviewed to ensure a comprehensive analysis. The main themes found in the reviewed accident literature and reports can be summarized in terms of an activity's "Hazardous attributes," and an activity's "Susceptible attributes" refer to the characteristics of the activity that pose a risk or danger to safety, health, the environment, or other aspects of well-being to workers performing an overlapping activity. We identified 11 "Hazardous attributes" in this research as shown in Table 2, and nine "Susceptible attributes" shown in Table 3, where the first column shows the attribute which is extracted, and the second column shows the description of the attribute. Then, the probability of their match was assessed in Table 4 by the expert group panel as explained in Section 3.2. They ranked the probability from 0 to 5 where 0 represents the lowest probability and 5 represents the highest probability. For example, "site slipperiness" as a hazardous attribute has the highest probability with "exposure to falling from height" as a susceptible attribute, while it has the lowest probability with "exposure to electric shock" as shown in Table 4. Using hazardous and susceptible attributes and their probability match table, we are able to identify the type of conflict an activity has with another one and also calculate the risk factor of each activity containing overlapping conflict.

Table 2. Hazardous attributes.

Attribute	Description	Studies
Construction equipment hazards	Hazards caused by heavy equipment and sharp angles used in construction activity	Li. et al. (2019) [49]
Sparks and lights	Activity generates sparks and lights that reduce visibility and cause distraction	Zhao et al. (2015) [50]
Heavy materials and equipment at site	Heavy materials used in the activity can intentionally or inadvertently fall or fly, posing risks to workers at the same elevations	Li. et al. (2019) [49]
Lack of site space	Activity increases space congestion, causing additional safety risks to workers performing another activity	Mirzaie et al. (2018) [6]
Electric shock	Activity involves electric installation which if not completed correctly can increase risk of electric shock for workers performing another activity	Zhao et al. (2015) [50]
Sharp materials	Activity generates materials cut at sharp angles with no caps or covers	Lee et al. (2020) [2]
Severe impact	Activity involves moving objects that can pose safety risks to workers performing another activity	Lee et al. (2020) [2]
Distraction	Activity generates extra noise, vibration, smell, or any other feature that distracts or unnecessarily attracts workers performing another activity	Hamid et al. (2008) [51]
Heavy materials and equipment at height	Materials or objects at higher level falling on people at lower levels.	Li. et al. (2019) [49]
Displacement of wide and long and large materials	Their installation, transportation, and inertia, and the space they occupy	Li. et al. (2019) [49]
Site slipperiness	Creating slippery surfaces as oil or paint leak	Hamid et al. (2008) [51]

Table 3. Susceptible attributes.

Attribute	Description	Studies
Exposure to distraction	Workers need to be focused and any distractions may result in accidents.	Hamid et al. (2008) [51]
Exposure to slipperiness	Workers can easily experience imbalance or lack of control	Hamid et al. (2008) [51]

Table 3. *Cont.*

Attribute	Description	Studies
Exposure to falling from height	Proximity to the areas that increase the probability of falling, such as working on scaffolding	Kartam and Bouz (1998) [52]
Exposure to debris and unstable situations	Unmanaged debris and poor housekeeping	Tam et al. (2004) [53]
Exposure to heavy materials and equipment	Workers have to work unprotected, or walk or stay close to unstable heavy objects	Li. et al. (2019) [49]
Exposure to light and sparks	Workers are not given light and spark protection equipment as per their standard work procedure	Zhao et al. 2015 [50]
Exposure to displacement of heavy materials	Workers find their way onto the path of moving materials	Li. et al. (2019) [49]
Exposure to sharp equipment	Workers with body parts unprotected against sharp material	Lee et al. (2020) [2]
Exposure to electric shock	Workers untrained or unprotected against electricity	Zhao et al. 2015 [50]

Table 4. Source–target match table.

	Susceptible	**Exposure to Distraction**	**Exposure to Slipperiness**	**Exposure to Falling from Height**	**Exposure to Debris and Unstable Situations**	**Exposure to Heavy Materials and Equipment**	**Exposure to Light and Sparks**	**Exposure to Displacement of Heavy Materials**	**Exposure to Sharp Equipment**	**Exposure to Electric Shock**
Hazardous	Code	A	B	C	D	E	F	G	H	I
Construction equipment hazards	1	4	3	3	2	4	4	5	3	3
Sparks and lights	2	3	3	2	2	4	4	5	2	0
Heavy materials and equipment at site	3	0	0	0	0	5	4	2	0	0
Lack of site space	4	0	0	0	0	4	5	2	0	0
Electric shock	5	0	0	0	0	4	2	0	0	0
Sharp materials	6	0	2	4	5	0	0	4	0	0
Severe impact	7	2	2	0	0	1	1	0	5	0
Distraction	8	0	0	0	0	4	4	0	0	3
Heavy materials and equipment at height	9	2	0	3	0	3	2	3	0	4
Displacement of wide and long and large materials	10	0	3	0	0	4	2	0	0	0
Site slipperiness	11	0	3	5	0	2	2	3	0	0

4.2. Case Study

The proposed model analyzes a seven-story residential concrete building as shown in Figure 6, the construction schedule of which comes from Navisworks, and the AEC and MEP from Revit. The LOD 300 in this framework is sufficient since it allows us to link the work breakdown structure to it. Each roof is divided into four zones, A, B, C, and D, and for simplicity, general activities such as surveying, material handling, demolishing, excavation, paving, formwork installation, reinforcement installation, column concrete pouring, rebar welding, roof installation, MEP installations, and interior finishes have been considered as follows:

- Surveying involves measuring and mapping the land to determine boundaries and elevations, which is crucial for planning and designing.
- Material handling indicates the transporting of the materials, equipment, and waste to and from the construction site.
- Demolishing refers to the process of removal of existing structures and preparing the site for new construction.
- Excavation refers to removing soil and rocks to create a foundation for the building.
- Paving is the process of laying a smooth surface to create driveway and sidewalks
- Formwork installation is a temporary structure used to shape fresh concrete until it gains its strength.
- Reinforcement installation refers to steel bars placed within formwork as part of the roof installation
- Rebar Welding is the process of joining metal bars within roofs by melting them together using heat.
- Column concrete pouring refers to pouring concrete into the formwork for vertical structure columns.
- Roof installation is the process of installing an upper covering of the building using various materials; in this project, a mixture of concrete and metal bars was used
- MEP installation refers to setting up the heating, air conditioning, electrical systems, and plumbing within the building
- Interior finishes refer to painting, ceiling finishing, drywalling, and installing fixtures

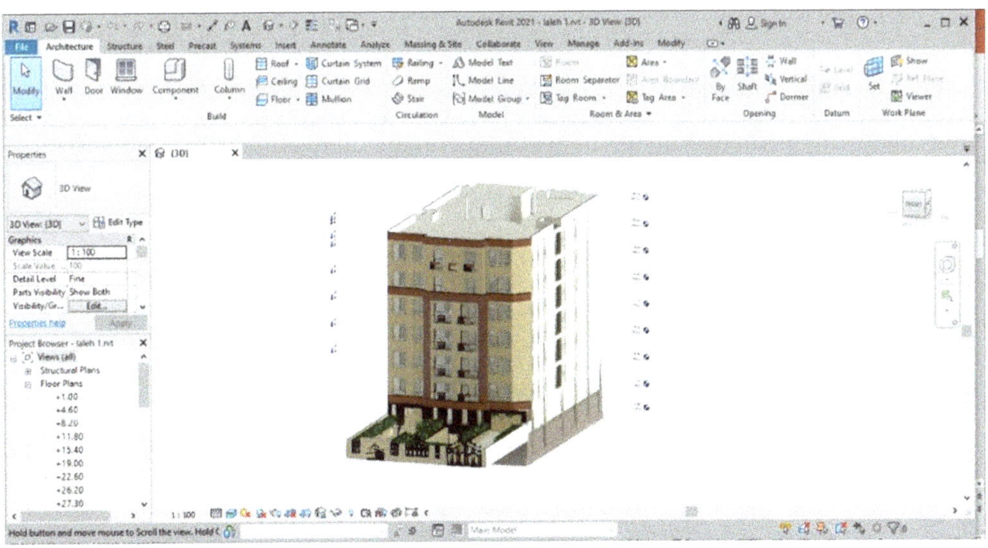

Figure 6. Sample project modeled in Revit.

After running the process using the proposed model, 75 overlapping conflicts among activities were found, as shown in Table 5. Six catastrophic conflicts were marked as priority 5 and 22 insignificant overlapping conflicts were marked as priority 1. Examples of different conflicts are shown in Table 6, indicating that activities such as roof installation, column concrete pouring, and rebar welding, which typically happen at height, are riskier than activities like material handling and paving. This may be due to the high number of accidents that occur due to the falling from height. It is worth mentioning that while the proposed model can be implemented in all kinds of projects, the type of conflict varies based on the type, scheduling, and size of the project. Therefore, overlapping conflicts identified in this case study cannot be considered a rule of thumb for other projects.

Table 5. Risk profile in the case study project.

Risk Level	Priority	Conflicts
0.8–1.0	5	6
0.6–0.8	4	23
0.4–0.6	3	10
0.2–0.4	2	14
0–0.2	1	22
		Sum = 75

Table 6. Some examples of overlapping conflicts.

Activity i	Zone	Activity j	Zone	Risk Level	Priority
Column concrete pouring	A, B, C, D	Roof Installation	A, B, C, D	0.8	5
Formwork installation	C, D	Roof installation	C, D	0.87	5
Excavation	A, B, C, D	Material handling	A, B, C, D	0.6	4
Reinforcement installation	B	Rebar welding	B	0.67	4
Reinforcement installation	D	Rebar welding	C	0.47	3
Formwork installation	A, B	Material handling	A, B	0.28	2
Material handling	A, B, C, D	Reinforcement installation	A, B, C, D	0.17	1

To understand how the RF for each conflict has been calculated, Table 7 represents overlapping conflicts between "column concrete pouring (i)" as a hazardous activity that matches the "roof installation (j)" susceptibilities. A construction activity involving pouring concrete in columns can increase safety hazards for another construction activity involving roof installation in elevated slabs, especially if both activities are happening concurrently or in close proximity. The increased safety hazards can result from falling objects, concrete overflow or spillage, dust, and debris, noise distraction, vibrations, limited workspace, material handling, etc. Using Equation (3) by knowing the parameters of each activity such as scheduling for evaluating the concurrency (Tij), zone, and height to evaluate the elevational and horizontal exposure (Eij, Hij), as well as the probability of each attribute (Pij) from Table 4 and their severity (Xi, Yj), the risk factor for each overlapping conflict can be calculated. The severity of each hazardous and susceptible attribute is in the range of 0 to 10 and their probability of matching is in the range of 0 to 5; therefore, the minimum RF score for each conflict can be zero and the maximum can be 500 ($10 \times 10 \times 5 = 500$). Since all conflicts correspond to 11 hazardous and 9 susceptible attributes, there are 99 types of conflicts that can occur between each pair of activities with the maximum summation RF of 49,500 ($500 \times 99 = 49{,}500$). It is shown that there are 36 different types of conflicts that can occur between column concrete pouring and roof installation activity. The most probable conflict with an RF of 450 out of 500, labeled "1-G" and represented by a yellow box in Table 7, refers to "machinery with moving parts" being hazardous to workers susceptible to "exposure to the displacement of heavy things". After checking concurrency and spatial conflict among these two attributes, a severity of 10 for machinery with moving parts, and a severity of 9 for exposure to the displacement of heavy things, with a probability match of

5, results in an RF of 450. The reason for the high number of hazardous conflicts involving machinery with moving parts could be due to crane buckets used for transferring concrete to columns which can cause many accidents. Other overlapping conflicts include "4-F", that is, "lack of site space", which is hazardous to workers susceptible to "exposure to light and sparks". The sum of all overlapping conflicts in Table 7 using Equation (4), is the RF of column concrete pouring and roof installation activities which is 8030 out of 49,500.

Table 7. Conflict of Column Concrete Pouring and Roof Installation at Zone C.

	Susceptible	**Exposure to Distraction**	**Exposure to Slipperiness**	**Exposure to Falling from Height**	**Exposure to Debris and Unstable Situations**	**Exposure to Heavy Materials and Equipment**	**Exposure to Light and Sparks**	**Exposure to the Displacement of Heavy Things**	**Exposure to Sharp Equipment**	**Exposure to Electric Shock**
Hazardous	Code	A	B	C	D	E	F	G	H	I
Machinery with moving parts	1	0	270	216	144	360	360	450	162	162
Sparks and lights	2	0	120	64	64	160	160	200	48	0
Heavy materials and equipment at site	3	0	0	0	0	400	320	160	0	0
Lack of site space	4	0	0	0	0	320	400	160	0	0
Electric shock	5	0	0	0	0	0	0	0	0	0
Sharp materials	6	0	60	0	0	0	0	0	0	0
Sever impact	7	0	0	0	0	0	0	0	0	0
Distraction	8	0	0	0	0	200	200	0	0	90
Heavy materials and equipment at height	9	0	0	240	0	300	200	300	0	240
Displacement of wide and long and large materials	10	0	240	0	0	320	160	0	0	0
Site slipperiness	11	0	180	240	0	120	120	180	0	0

The nature of the conflicts and how the model captures and evaluates them are better understood through the example in Table 8. A matrix of hazardous (i) and susceptible (j) activities is analyzed to assess the RF of two selected activities for each overlapping conflict. The first column represents hazardous activity (i) with 20 activities which have been divided into four different zones (A, B, C, and D). The first row represents susceptible activities (j), similar to hazardous activity, with 20 activities divided into four zones. Note that the activities listed in Table 8 pertain exclusively to the sixth floor and do not encompass all activities within the entire project. All activities are checked against each other to find the overlapping conflict among them. For this purpose, spatial conflict, concurrency, and source–target match of each pair of activities that have been selected are analyzed to calculate their RF. The sixth floor, for example, reflects 16 overlapping conflicts. Column concrete pouring and roof installation activities are represented by a yellow box shown in Table 8, marked as a priority 5, which indicates "catastrophic" severity. It means safety managers should pay attention to these two activities more than lower priorities; however, activities such as formwork and reinforcement installation in zones A and C have an RF of 4620 marked as "moderate" overlapping conflict with column concrete pouring in zone A and C, so they are in the third place of importance compared to "catastrophic" conflicts; interior finishing activities have least overlapping conflict, and only interior finishing activity in zone C have "moderate" risk with framing and drywall activity at zone C. Because the maximum RF score of conflicts in this case study is 8737 out of 49,500, using Table 1, the first 20% of conflicts are marked as priority 1 and the last 20% RF scores are marked as priority 5. For example, an RF of 8030 out of 8737 falls to the upper 20% of RF levels, so it is given a priority 5, or an RF of 688 out of 49,500 falls to the range of 0–20% risk levels and is given priority 1. It should be noted that missing one of the essential conditions of overlapping conflicts will result in an RF of zero.

The output model shown in Figure 7 visualizes the conflict between column concrete pouring and interior finishing (drywalling) in zone C. The slider below the model area assists the safety manager in performing a day-by-day review of every project step to decide about possible safety conflicts between the activities. It mainly gives the safety reviewer information like hazardous and susceptible activities, date of conflict, zone, and risk level. All possible conflicts have a background assessment accessible as in Table 7, the summary of which shows more details to the safety reviewer than the report.

While this study was limited to identifying conflicts, some safety measures could be implemented to reduce hazards. These measures include:

- Scheduling: If possible, schedule these activities at different times or in separate areas of the construction site. Ensure that the installation of reinforcement in elevated slabs is completed or temporarily halted before concrete pouring begins to avoid potential clashes.
- Safety training: Ensure that all workers involved in these activities receive appropriate safety training and are aware of potential hazards resulting from overlapping activities.
- Zoning: Establish safety zones to prevent workers from entering areas where concrete is being poured.
- Environmental controls: Implement measures to control dust and debris generated during concrete pouring, such as water spraying or dust collection systems.
- Personal protective equipment (PPE): Require all workers to wear appropriate PPE, including helmets, safety glasses, and high-visibility vests, to enhance their safety and visibility on the construction site.
- Temporary Access: Provide safe and clear temporary access routes for workers on the elevated slab to reach their work areas, even during concrete pouring activities.
- Communication: Implement effective communication protocols and systems, such as radios or hand signals, to maintain clear communication between workers in both activities. Emphasize the importance of communication and situational awareness.
- Supervision: Assign competent supervisors to oversee both activities and ensure that safety protocols are followed. Conduct regular safety inspections and audits to identify and address potential hazards.
- Barriers and safety nets: Install physical barriers or safety nets beneath the columns being poured to catch falling objects or concrete spillage. This helps protect workers on the elevated slab

Table 8. Sixth Floor Conflicts.

	Susceptible	Reinforcement Installation Zone A	Reinforcement Installation Zone B	Reinforcement Installation Zone C	Reinforcement Installation Zone D	Column Concrete Pouring Zone A	Column Concrete Pouring Zone B	Column Concrete Pouring Zone C	Column Concrete Pouring Zone D	Roof Installation Zone A	Roof Installation Zone B	Roof Installation Zone C	Roof Installation Zone D	MEP Installation Zone A	MEP Installation Zone B	MEP Installation Zone C	MEP Installation Zone D	Interior Finishes Zone A	Interior Finishes Zone B	Interior Finishes Zone C	Interior Finishes Zone D
Hazardous	Code	1	2	3	4	5	6	7	8	9	10	11	12	13	14	15	16	17	18	19	20
Reinforcement installation zone A	1	0	0	0	0	0	4620	0	0	0	0	0	0	0	0	0	0	0	0	0	0
Reinforcement installation zone B	2	0	0	0	0	0	6686	0	0	0	0	0	0	0	0	0	0	0	0	0	0
Reinforcement installation zone C	3	0	0	0	0	0	4620	6686	0	0	0	0	0	0	0	0	0	0	0	0	0
Reinforcement installation zone D	4	0	0	0	0	0	0	4620	0	0	0	0	0	0	0	0	0	0	0	0	0
Column concrete pouring zone A	5	0	0	0	0	0	0	0	0	0	0	0	0	0	0	0	0	0	0	0	0
Column concrete pouring zone B	6	0	7240	0	0	0	0	0	0	0	0	0	0	0	0	0	0	0	0	0	0
Column concrete pouring zone C	7	0	0	7240	0	0	0	0	0	0	0	8030	0	0	0	0	0	0	0	0	0
Column concrete pouring zone D	8	0	0	0	0	0	0	0	0	0	0	0	8738	0	0	0	0	0	0	0	0
Roof Installation zone A	9	0	0	0	0	0	0	688	0	0	0	0	0	0	0	0	0	0	0	0	0
Roof Installation zone B	10	0	0	0	0	0	0	688	0	0	0	0	0	0	0	0	0	0	0	0	0

Table 8. *Cont.*

Susceptible		Reinforcement Installation Zone A	Reinforcement Installation Zone B	Reinforcement Installation Zone C	Reinforcement Installation Zone D	Column Concrete Pouring Zone A	Column Concrete Pouring Zone B	Column Concrete Pouring Zone C	Column Concrete Pouring Zone D	Roof Installation Zone A	Roof Installation Zone B	Roof Installation Zone C	Roof Installation Zone D	MEP Installation Zone A	MEP Installation Zone B	MEP Installation Zone C	MEP Installation Zone D	Interior Finishes Zone A	Interior Finishes Zone B	Interior Finishes Zone C	Interior Finishes Zone D
Roof Installation zone C	11	0	0	0	0	0	0	3898	0	0	0	0	0	0	0	0	0	0	0	0	0
Roof Installation zone D	12	0	0	0	0	0	0	688	0	0	0	0	0	0	0	0	0	0	0	0	0
MEP installation zone A	13	0	0	0	0	0	0	0	0	0	0	0	0	0	0	0	0	0	0	0	0
MEP installation zone B	14	0	0	0	0	0	0	0	0	0	0	0	0	0	0	0	0	0	5854	0	0
MEP installation zone C	15	0	0	0	0	0	0	0	0	0	0	0	0	0	0	0	0	0	0	0	0
MEP installation zone D	16	0	0	0	0	0	0	0	0	0	0	0	0	0	0	0	0	0	0	0	0
Interior finishes zone A	17	0	0	0	0	0	0	0	0	0	0	0	0	0	0	0	0	0	0	0	0
Interior finishes zone B	18	0	0	0	0	0	0	0	0	0	0	0	0	0	5854	0	0	0	0	0	0
Interior finishes zone C	19	0	0	0	0	0	0	0	0	0	0	6048	0	0	0	0	0	0	0	0	0
Interior finishes zone D	20	0	0	0	0	0	0	0	0	0	0	0	0	0	0	0	0	0	0	0	0

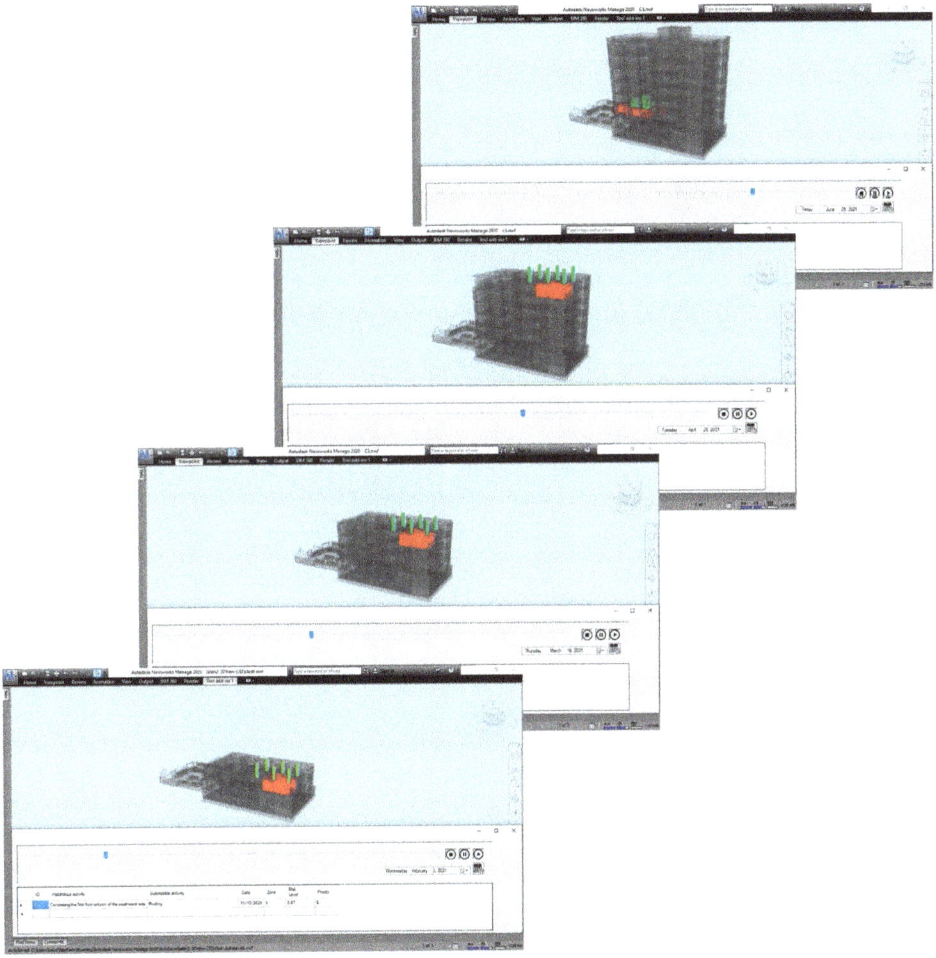

Figure 7. High-risk conflicts highlighted using the Navisworks API.

5. Discussion

While the construction industry has long been aware of the inherent risks associated with individual activities, overlapping construction activities introduce another layer of complexity to safety assessment. Many authors have diagnosed the accident-prone construction industry with a wide variety of working groups from different trades who may be unaware of hazards from neighboring activities [3,5,8]. Authors believe that unfamiliarity of workers with the hazards around them predispose them to higher numbers of accidents. Accident causation models and safety standards substantially evaluate the safe methods for every construction activity, and research is needed to identify how workers are exposed to hazards from activities other than what they have been trained in.

This research developed a model to identify "overlapping conflicts" between activities. We believe overlapping conflicts are expected when simultaneous activities with a physical influence path suffer from a third parameter called the source–target match. A review of 300+ accident synopses led us to a summary of 11 hazardousness and 9 susceptibility attributes a construction activity might have. For example, "distraction" as a hazardous attribute can create extra noise, light, vibration, smell, or any other attribute that distracts or unnecessarily attracts other people, or "exposure from falling from a height" as a susceptible attribute can alert proximate workers to the areas that increase the probability of falling such as working on scaffolding. We used the probability of each hazard matching a susceptibility to assess the probability of every cell of the source–target match table. Using the activities' characteristics such as start and end dates, height, location (zone), and their hazardous and susceptible attributes, conflicts are then given priority based on their total RF. Finally, an automated BIM tool visualizes the model so that a safety expert or manager can decide the most worrying conflicts. It is worth mentioning that the integration of cutting-edge technologies such as BIM, machine learning, and AI has significantly streamed the decision-making process, allowing for more accurate accident prediction in the construction industry, infrastructure management, roads, and bridges [54,55].

The model was implemented on a sample building model, and the results revealed 75 conflicts that were neglected before. These overlooked conflicts can be considered blind spots in traditional evaluations, underscoring the need for a more comprehensive approach like the proposed research. Pouring concrete in columns and roof installation in elevated slabs are two critical construction activities that often occur sequentially or concurrently in building construction. The safety hazards associated with one activity can indeed increase the risks for the other. Activities such as roof installation and column concrete pouring, rebar welding and reinforcement installation, excavating, and material handling are always considered the most dangerous activities that can take place at the same time, and many injuries happen when these activities occur together. Focusing on the hazards pertaining to each activity without considering hazards stemming from neighboring activities can expose workers to unforeseen hazards. For example, column concrete pouring and roof installation activities are two separate activities that can take place at the same time and in the same location. Each activity has its own hazards that need to be prevented, but considering safety risks stemming from both activities showed that machinery with moving parts as a hazardous attribute from column concrete pouring activity can make workers susceptible to exposure to the displacement of heavy things from roof installation activity. This risk could be overlooked without considering overlapping conflicts. Efficiency of visual tools, like BIM, can make the decision much easier for safety managers. Prioritizing and visualizing the overlapping conflicts aid in intuitive understanding of safety risks. As the construction industry progresses, the projects become increasingly intricate and complex [26]. Addressing the challenges of today and anticipating those of tomorrow requires innovative approaches to safety. Our research not only proposes a pioneering method for assessing overlapping conflicts, but also paves the way for a more integrated and technologically advanced future in construction safety, ensuring both improved efficiency and enhanced safety of workers.

After implementing the proposed model, it becomes evident that while numerous studies have delved into identifying and categorizing individual construction risks, few have holistically approached the issue of overlapping conflicts. Kartam (1997) and others [3,5,8] emphasize the importance of understanding individual construction risks, particularly during the planning phase. However, our study uniquely underscores the significance of simultaneous activities and their compound risks. For instance, while Mirzaei et al. (2018) proposed models to decrease workspace conflicts [6], our research extends beyond the singular scope of spatial and concurrent considerations to factor in the hazardous impacts that one activity might have on another. Similarly, while Zolfagharian et al. (2014) and Moon et al. integrated safety measures into project scheduling [8,13], we further this paradigm by integrating overlapping risks into BIM visualizations. The introduction of the "overlapping conflict" concept, as we have proposed, fills a notable gap in the current literature, emphasizing overlapping conflicts where activities intersect in time, space, and hazard potential. Our model not only aligns with the sentiments of Okpala et al. (2020) and Chantawit et al. (2005) regarding the significance of BIM in safety management [22,23], but also pioneers an innovative methodology that enhances the depth and breadth of hazard analysis within the BIM environment.

6. Conclusions

In this study, a specific type of conflict, called "overlapping conflict" was proposed, which had been neglected by previous research and could increase the number of accidents in the construction industry. This research developed a framework for defining overlapping conflict among activities using not only time–space conflict but also source–target matching of those conflicts. Our finding can be summarized as follows:

- This research introduces the concept of overlapping conflicts, which broadens our understanding of safety risks in the construction industry
- A meticulous analysis resulted in identification of eleven hazardous and nine susceptible attributes; moreover, the probability and severity of their matches were assessed.
- To provide a practical dimension to this study, a case study was used to verify the process
- The risk factor associated with each overlapping conflict was assessed and then prioritized to further analyze the conflicts
- This research provides an actionable framework using BIM, proposing a real-world tool for supervision and evaluation of overlapping construction activities

Although this research investigates a new type of construction hazard, its core concept can unveil the main reason for accidents in many other industries. The model presented here equips safety supervisors with a BIM-based monitoring tool whose decision on decoupling or distancing unsafely conflicting activities goes beyond the scope of this research.

This study is limited to identifying and prioritizing the conflicts, while the treatment, safety mitigation strategies, and monitoring updates and communications of which can be a good topic for future research. Integrated management of all overlapping and multiple-node influences that make the nature of construction complex is possibly the next service BIM will offer to the industry. Moreover, as previously noted, this study did not evaluate the duration of concurrent conflicts; therefore, it can be a valuable area for future research. A more refined analysis of hazardous and susceptibility attributes presents a potential topic for future research. We also recognize the need to further explore and integrate a comprehensive range of factors, including types of activities, workforce experience, regulatory compliance, and communication dynamics, to develop holistic safety management strategies that encompass hazard identification and mitigation across the construction industry.

Author Contributions: Conceptualization, A.R.N., H.M. and H.E.; Methodology, A.R.N., H.M. and H.E.; Software, A.N.T. and K.G.; Validation, A.R.N.; Formal analysis, A.R.N.; Investigation, A.R.N.; Resources, A.R.N.; Data curation, A.R.N.; Writing—original draft, A.R.N.; Writing—review & editing, A.R.N., H.M. and H.E.; Visualization, A.N.T. and K.G.; Supervision, A.R.N., H.M. and H.E.; Project administration, A.R.N., H.M. and H.E. All authors have read and agreed to the published version of the manuscript.

Funding: This research received no external funding.

Data Availability Statement: No new original data is generated in this study. Tables and figures represent data processed in an artificial, sample case study. However all the data presented in this study are available on request from the corresponding author.

Conflicts of Interest: The authors declare no conflict of interest.

Abbreviations

pt API	Application Programming Interface
BIM	Building Information Modeling
CPM	Critical Path Method
FACE	Fatal Accident Circumstances and Health Epidemiology
FTA	Fault Tree Analysis
LOD	Level of Development
MEP	Mechanical, Electrical, and Plumbing
NIOSH	National Institute for Occupational Safety and Health
OSHA	Occupational Safety and Health Administration
RF	Risk Factor

References

1. Sanni-Anibire, M.O.; Mahmoud, A.S.; Hassanain, M.A.; Salami, B.A. A risk assessment approach for enhancing construction safety performance. *Saf. Sci.* **2020**, *121*, 15–29. [CrossRef]
2. Lee, Y.-C.; Shariatfar, M.; Rashidi, A.; Lee, H.W. Evidence-driven sound detection for prenotification and identification of construction safety hazards and accidents. *Autom. Constr.* **2020**, *113*, 103127. [CrossRef]
3. Morteza, A.; Ilbeigi, M.; Schwed, J. A Blockchain Information Management Framework for Construction Safety. In Proceedings of the ASCE International Conference on Computing in Civil Engineering 2021, Orlando, FL, USA, 12–14 September 2021; pp. 342–349.
4. Ojha, A.; Seagers, J.; Shayesteh, S.; Habibnezhad, M.; Jebelli, H. Construction safety training methods and their evaluation approaches: A systematic literature review. In Proceedings of the 8th International Conference on Construction Engineering and Project Management, Hong Kong, China, 7–8 December 2020; pp. 188–197.
5. Kartam, N.A. Integrating Safety and Health Performance into Construction CPM. *J. Constr. Eng. Manag.* **1997**, *123*, 121–126. [CrossRef]
6. Mirzaei, A.; Nasirzadeh, F.; Jalal, M.P.; Zamani, Y. 4D-BIM Dynamic Time–Space Conflict Detection and Quantification System for Building Construction Projects. *J. Constr. Eng. Manag.* **2018**, *144*, 04018056. [CrossRef]
7. Behm, M. Linking construction fatalities to the design for construction safety concept. *Saf. Sci.* **2005**, *43*, 589–611. [CrossRef]
8. Zolfagharian, S.; Irizarry, J.; Ressang, A.; Nourbakhsh, M.; Gheisari, M. Automated safety planning approach for residential construction sites in Malaysia. *Int. J. Constr. Manag.* **2014**, *14*, 134–147. [CrossRef]
9. Alirezaei, S.; Taghaddos, H.; Ghorab, K.; Tak, A.N.; Alirezaei, S. BIM-augmented reality integrated approach to risk management. *Autom. Constr.* **2022**, *141*, 104458. [CrossRef]
10. Cao, Y.; Miraba, S.; Rafiei, S.; Ghabussi, A.; Bokaei, F.; Baharom, S.; Haramipour, P.; Assilzadeh, H. Economic application of structural health monitoring and internet of things in efficiency of building information modeling. *Smart Struct. Syst.* **2020**, *26*, 559–573.
11. Omer, M.M.; Mohd-Ezazee, N.M.A.; Lee, Y.S.; Rajabi, M.S.; Rahman, R.A. Constructive and Destructive Leadership Behaviors, Skills, Styles and Traits in BIM-Based Construction Projects. *Buildings* **2022**, *12*, 2068. [CrossRef]
12. Rajabi, M.S.; Radzi, A.R.; Rezaeiashtiani, M.; Famili, A.; Rashidi, M.E.; Rahman, R.A. Key Assessment Criteria for Organizational BIM Capabilities: A Cross-Regional Study. *Buildings* **2022**, *12*, 1013. [CrossRef]
13. Moon, H.; Kim, H.; Kim, C.; Kang, L. Development of a schedule-workspace interference management system simultaneously considering the overlap level of parallel schedules and workspaces. *Autom. Constr.* **2014**, *39*, 93–105. [CrossRef]
14. Yi, K.-J.; Langford, D. Scheduling-Based Risk Estimation and Safety Planning for Construction Projects. *J. Constr. Eng. Manag.* **2006**, *132*, 626–635. [CrossRef]
15. Martinez, J.G.; Albeaino, G.; Gheisari, M.; Issa, R.R.; Alarcón, L.F. iSafeUAS: An unmanned aerial system for construction safety inspection. *Autom. Constr.* **2021**, *125*, 103595. [CrossRef]
16. Khodabandelu, A.; Park, J.; Arteaga, C. Crane operation planning in overlapping areas through dynamic supply selection. *Autom. Constr.* **2020**, *117*, 103253. [CrossRef]
17. Guo, B.H.; Goh, Y.M.; Wong, K.L.X. A system dynamics view of a behavior-based safety program in the construction industry. *Saf. Sci.* **2018**, *104*, 202–215. [CrossRef]
18. Malekitabar, H.; Ardeshir, A.; Sebt, M.H.; Stouffs, R. Construction safety risk drivers: A BIM approach. *Saf. Sci.* **2016**, *82*, 445–455. [CrossRef]
19. Perlman, A.; Sacks, R.; Barak, R. Hazard recognition and risk perception in construction. *Saf. Sci.* **2014**, *64*, 22–31. [CrossRef]
20. Mohandes, S.R.; Zhang, X. Developing a Holistic Occupational Health and Safety risk assessment model: An application to a case of sustainable construction project. *J. Clean. Prod.* **2021**, *291*, 125934. [CrossRef]
21. Cagno, E.; Di Giulio, A.; Trucco, P. An algorithm for the implementation of safety improvement programs. *Saf. Sci.* **2001**, *37*, 59–75. [CrossRef]

22. Okpala, I.; Nnaji, C.; Karakhan, A.A. Utilizing Emerging Technologies for Construction Safety Risk Mitigation. *Pr. Period. Struct. Des. Constr.* **2020**, *25*, 04020002. [CrossRef]
23. Chantawit, D.; Hadikusumo, B.H.; Charoenngam, C.; Rowlinson, S. 4DCAD-Safety: Visualizing project scheduling and safety planning. *Constr. Innov.* **2005**, *5*, 99–114.
24. Kassem, M.; Dawood, N.; Chavada, R. Construction workspace management within an Industry Foundation Class-Compliant 4D tool. *Autom. Constr.* **2015**, *52*, 42–58. [CrossRef]
25. Lu, Y.; Gong, P.; Tang, Y.; Sun, S.; Li, Q. BIM-integrated construction safety risk assessment at the design stage of building projects. *Autom. Constr.* **2021**, *124*, 103553. [CrossRef]
26. Tak, A.N.; Taghaddos, H.; Mousaei, A.; Bolourani, A.; Hermann, U. BIM-based 4D mobile crane simulation and onsite operation management. *Autom. Constr.* **2021**, *128*, 103766. [CrossRef]
27. Sampaio, A.Z.; Constantino, G.B.; Almeida, N.M. 8D BIM Model in Urban Rehabilitation Projects: Enhanced Occupational Safety for Temporary Construction Works. *Appl. Sci.* **2022**, *12*, 10577. [CrossRef]
28. Charef, R. The use of Building Information Modelling in the circular economy context: Several models and a new dimension of BIM (8D). *Clean. Eng. Technol.* **2022**, *7*, 100414. [CrossRef]
29. NIOSH. Fetality Assessment and Control Evaluation (FACE). 2012. Available online: http://www.cdc.com.gov/niosh/face/faceweb.html (accessed on 23 March 2006).
30. OSHA. Fetal Facts. 2014. Available online: http://www.osha.gov/publication/fetalfacts.html (accessed on 23 March 2006).
31. Pinto, A.; Nunes, I.L.; Ribeiro, R.A. Occupational risk assessment in construction industry—Overview and reflection. *Saf. Sci.* **2011**, *49*, 616–624. [CrossRef]
32. Zhou, Z.; Goh, Y.M.; Li, Q. Overview and analysis of safety management studies in the construction industry. *Saf. Sci.* **2015**, *72*, 337–350. [CrossRef]
33. Pillay, M. Accident Causation, Prevention and Safety Management: A Review of the State-of-the-art. *Procedia Manuf.* **2015**, *3*, 1838–1845. [CrossRef]
34. Choi, S.D.; Guo, L.; Kim, J.; Xiong, S. Comparison of fatal occupational injuries in construction industry in the United States, South Korea, and China. *Int. J. Ind. Ergon.* **2019**, *71*, 64–74. [CrossRef]
35. Camino López, M.A.; Ritzel, D.O.; Fontaneda, I.; González Alcantara, O.J. Construction industry accidents in Spain. *J. Saf. Res.* **2008**, *39*, 497–507. [CrossRef] [PubMed]
36. Hamid, A.R.; Noor Azmi, M.R.; Aminudin, E.; Jaya, R.P.; Zakaria, R.; Zawawi, A.M.; Yahya, K.; Haron, Z.; Yunus, R.; Saar, C.C. Causes of fatal construction accidents in Malaysia. *IOP Conf. Ser. Earth Environ. Sci.* **2019**, *220*, 012044. [CrossRef]
37. Shafique, M.; Rafiq, M. An Overview of Construction Occupational Accidents in Hong Kong: A Recent Trend and Future Perspectives. *Appl. Sci.* **2019**, *9*, 2069. [CrossRef]
38. Zhu, R.; Hu, X.; Hou, J.; Li, X. Application of machine learning techniques for predicting the consequences of construction accidents in China. *Process. Saf. Environ. Prot.* **2021**, *145*, 293–302. [CrossRef]
39. Zhou, Z.; Irizarry, J.; Guo, W. A network-based approach to modeling safety accidents and causations within the context of subway construction project management. *Saf. Sci.* **2021**, *139*, 105261. [CrossRef]
40. Saloniemi, A.; Oksanen, H. Accidents and fatal accidents—Some paradoxes. *Saf. Sci.* **1998**, *29*, 59–66. [CrossRef]
41. Winge, S.; Albrechtsen, E. Accident types and barrier failures in the construction industry. *Saf. Sci.* **2018**, *105*, 158–166. [CrossRef]
42. Xu, Q.; Xu, K. Analysis of the Characteristics of Fatal Accidents in the Construction Industry in China Based on Statistical Data. *Int. J. Environ. Res. Public Health* **2021**, *18*, 2162. [CrossRef]
43. Eisenhardt, K.M. Building theories from case study research. *Acad. Manag. Rev.* **1989**, *14*, 532–550. [CrossRef]
44. Adler, M.; Ziglio, E. *Gazing into the Oracle: The Delphi Method and Its Application to Social Policy and Public Health*; Jessica Kingsley Publishers: London, UK, 1996.
45. Tersine, R.J.; Riggs, W.E. The delphi technique: A long-range planning tool. *Bus. Horizons* **1976**, *19*, 51–56. [CrossRef]
46. Carter, G.; Smith, S.D. Safety Hazard Identification on Construction Projects. *J. Constr. Eng. Manag.* **2006**, *132*, 197–205. [CrossRef]
47. Manuele, F.A. Achieving risk reduction, effectively. *Process Saf. Environ. Prot.* **2006**, *84*, 184–190. [CrossRef]
48. Lucko, G.; Rojas, E.M. Research Validation: Challenges and Opportunities in the Construction Domain. *J. Constr. Eng. Manag.* **2010**, *136*, 127–135. [CrossRef]
49. Li, J.; Li, H.; Wang, H.; Umer, W.; Fu, H.; Xing, X. Evaluating the impact of mental fatigue on construction equipment operators' ability to detect hazards using wearable eye-tracking technology. *Autom. Constr.* **2019**, *105*. [CrossRef]
50. Zhao, D.; McCoy, A.P.; Kleiner, B.M.; Smith-Jackson, T.L. Control measures of electrical hazards: An analysis of construction industry. *Saf. Sci.* **2015**, *77*, 143–151. [CrossRef]
51. Hamid, A.R.; Majid, M.Z.; Singh, B. Causes of accidents at construction sites. *Malays. J. Civ. Eng.* **2008**, *20*, 242–259.
52. Kartam, A.N.; Bouz, R.G. Fatalities and injuries in the Kuwaiti construction industry. *Accid. Anal. Prev.* **1998**, *30*, 805–814. [CrossRef]
53. Tam, C.; Zeng, S.; Deng, Z. Identifying elements of poor construction safety management in China. *Saf. Sci.* **2004**, *42*, 569–586. [CrossRef]

54. Rashidi Nasab, A.; Elzarka, H. Predicting Ohio Bridges Deteriorations Using Multi-Target Machine Learning Algorithms, International Conference on Transportation and Development (2023). In Proceedings of the International Conference on Transportation and Development (ICTD 2024), Atlanta, GA, USA, 15–18 June 2024. Scheduled to Be Held.
55. Rashidi Nasab, A.; Elzarka, H. Optimizing Machine Learning Algorithms for Improving Prediction of Bridge Deck Deterioration: A Case Study of Ohio Bridges. *Buildings* **2023**, *13*, 1517. [CrossRef]

buildings

MDPI

Article

Inertial Measurement Unit- and Time Series Approach-Based Motion Trajectory Reconstruction of the Safety Rope Fastening Behaviour

Zixin Han * and Yaowu Wang

School of Civil Engineering, Harbin Institute of Technology, Harbin 150090, China; ywwang@hit.edu.cn
* Correspondence: 17B333002@stu.hit.edu.cn; Tel.: +86-451-8628-2095

Abstract: Monitoring workers' safety compliance is critical to construction accident prevention. However, most of the previous research focuses on checking the entry of workers wearing safety belts and the monitoring of the fastening action of safety rope screw buckles has not yet been considered. In this paper, a reconstruction method for the safety rope fastening behaviour of workers is proposed based on inertial measurement units (IMU) and a time series approach to monitor safety belt use. The proposed method was applied and evaluated through on-site construction experiments. The experimental results show that the acceleration, angular velocity, and magnetic induction intensity data obtained by the inertial measurement unit exhibit clear behavioural characteristics during safety rope fastening. The trajectory of the safety rope can be reconstructed and monitored through inertial measurement units and a time series approach. The results of this study will contribute to the reconstruction and monitoring of safety rope attachment trajectories for scaffolding workers working at heights in order to prevent falls at construction sites.

Keywords: unsafe behaviour; falls from height; inertial measurement unit; quaternions; motion trajectory reconstruction

Citation: Han, Z.; Wang, Y. Inertial Measurement Unit- and Time Series Approach-Based Motion Trajectory Reconstruction of the Safety Rope Fastening Behaviour. *Buildings* **2024**, *14*, 219. https://doi.org/10.3390/buildings14010219

Academic Editors: Marek Sawicki and Mariusz Szóstak

Received: 1 October 2023
Revised: 13 December 2023
Accepted: 11 January 2024
Published: 14 January 2024

1. Introduction

The construction industry is particularly hazardous due to its dynamic work environment [1,2]. According to the Bureau of Labour Statistics (BLS) in the United States, falls, slips, and trips accounted for a substantial portion of the fatalities in construction. There were 370 fatalities in 2021 in construction and extraction occupations due to falls, slips, and trips, accounting for a 7.2% increase from 2020 [3]. From 2017 to 2021, there were 1919 incidents of falling from heights in China, accounting for 53% of the total housing and municipal engineering safety accidents [4]. These casualties not only cause irreparable economic losses but also restrict the sustainable development of the construction industry to a certain extent. The safety belt is a personal protective device for workers to be used at various heights; it can prevent casualties and is praised as lifesaving by construction workers. The safety belt is composed of a lanyard (composed of a strap and metal accessories, which are worn by workers), a safety rope, a buffer bag, and metal accessories, and is generally known as a fall suspension belt. The safety belt must be hung high and used low, and the opposite should be avoided [5,6]. The safety rope should be fastened to a strong body part. However, sometimes, the workers' safety awareness is poor or workers are negligent and fail to attach the belt's safety cord to a fixed structure, which can increase the risk of falling.

In order to reduce risk when scaffolding workers are working at heights and ensure their safety, it is crucial to accurately predict the trajectory of scaffolding workers wearing safety belts. Motion trajectory monitoring plays a key role in achieving this, as it allows collaborative robots to proactively respond and anticipate human movements, reducing safety risks for workers [7,8]. Motion trajectory monitoring refers to the process of forecasting the positions of a human body determined by a previous series of its movements with the highest possible accuracy [9]. Early work on human motion prediction employed methods such as clustering, Kalman filters, linear regression, autoregression, and nonlinear Gaussian processes. However, these methods were typically only effective in laboratory settings and simple scenarios [10]. In contrast, the time series approach is traditionally performed using data-based models that simulate potential human motion using a manually created trajectory. These models construct motion trajectories from data [11], in contrast to the first approach. The primary reason for using data-driven learning methodologies in motion prediction is that it is more practical to forecast an object's movement based on observable data than to use a standard algorithm [11,12]. This is because the movement of an object is varied and unpredictable.

With the advancement of deep learning algorithms and graphic processing units, vision-based behaviour recognition has been applied to recognise unsafe behaviour in construction [13]. Han et al. [14] detected unsafe actions of construction workers using vision-based depth sensors. Detecting personal protection equipment (PPE), such as safety harnesses, using a vision-based approach was also studied by Fang et al. [15]. In addition, motion sensors such as inertial measurement units (IMU) have a high suitability due to their high durability, availability in low-visibility conditions, and low power usage [16–18]. With these advantages, IMU is one of the most widely used sensors in the construction industry [19–21]. Joshua and Varghese [22] showed that IMU data can classify the characteristics of activities. Yoon et al. [23,24] attempted to examine the gait stability of ironworkers based on IMU data. These studies validated the feasibility of using IMU sensors to extract the subtle characteristics of workers' activities. Although these studies focused on classifying each activity using IMU data and extracting unsafe behaviours from their task sequence, it is also essential to identify whether a worker's behaviour follows the safety regulations.

The above literature shows that the unsafe behaviours of scaffold workers working at altitude can be guided and managed. With the help of IMU, we can monitor the safety harness fastening behaviour of scaffold workers, which can effectively reduce the number of safety accidents caused by unsafe human behaviour and greatly promote the automation and intelligence of construction management. Therefore, to ensure the safety of scaffolding workers working at altitude, a method based on IMUs and quaternions [25] is proposed to reconstruct the motion trajectories and buckling behaviours of scaffolding workers working at altitude. In this paper, based on IMUs used to collect safety rope fastening action data for scaffold workers working at altitude and quaternions used to establish a trajectory model of safety rope fastening movement, the safety belt fastening actions of scaffold workers working at altitude are reconstructed to ultimately realize the intelligent monitoring and management of construction safety information.

The remainder of this paper is organized as follows. The specific reconstruction method of the safety harness screw fastening movement of scaffold workers working at altitude based on IMU and quaternions is explained in Section 2. In Section 3, the feasibility of the proposed method is verified. In Section 4, the innovations and limitations of this paper and suggestions for future work are discussed. Section 5 presents the conclusions of this research.

2. Methods

To ensure the safety of scaffolding workers working at height and prevent falling, scaffolding workers must wear safety belts (and straps; Figure 1), attaching one end of the safety rope to the horizontal scaffold pole. According to the operating procedures and on-site observations, the steps of safety rope use can be divided into several stages. First, the safety rope is placed around the waist. Second, the safety harness screw is tightened around the horizontal bar of the scaffold, and the connecting ring is then attached to create a hanging support, which is very important. The safety belt is not functional unless this step is completed. To facilitate workers performing tasks at height and not affect these operations, in this paper, an IMU is fixed at the position where the turnbuckle and the safety rope are connected. The IMU is used to collect action data, and the data are transmitted to a receiver. Quaternions are used to express the attitude changes of the turnbuckle, and a spatial motion trajectory model of tie behaviour is established to monitor tie movement behaviour in combination with the safety belt (Figure 1c).

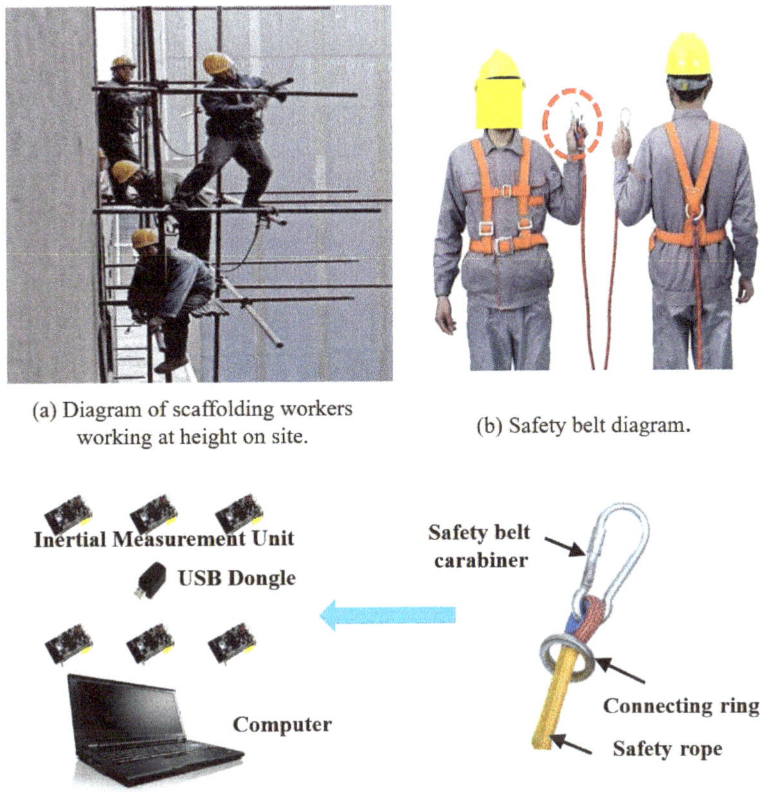

(a) Diagram of scaffolding workers working at height on site.

(b) Safety belt diagram.

(c) Inertial measurement unit layout.

Figure 1. Schematic diagram of IMU-based scaffolding worker safety tethering action data acquisition.

2.1. *Quaternions Represent the Action Position of the Safety Harness Screw*

The safety harness screw is wrapped around the horizontal bar and attached to the connecting ring. The completion of this action confirms the belt fastening behaviour of the scaffolding workers working at altitude. To monitor this behaviour, the first step is to determine how to describe the gesture of the behaviour [26]. Mathematically, attitude refers to the parameters of the angular position of the coordinate system of a rigid body and a reference coordinate system. The action attitude angle represents the change in the spatial angle of rotation in the carrier coordinate system relative to the reference coordinate system. The yaw angle, pitch angle, and roll angle are commonly used to describe the rotation angle of a carrier. A quaternion is an expression of the attitude angle, which can describe the change in the motion angle of a rigid body and can effectively avoid the generation of Euler angle singularity. A quaternion consists of a solid unit and three imaginary numbers $\vec{i}$, $\vec{j}$, and $\vec{k}$, as shown in Equation (1).

$$q = q_0 + q_1\vec{i} + q_2\vec{j} + q_3\vec{k} \tag{1}$$

In this paper, a quaternion expression is used to describe the position of the safety harness screw. The rotation angle is adjusted by changing the quaternion correlation matrix, and attitude variations are based on changes in the rotation angle. In this work, the safety harness screw is regarded as a rigid body, and an IMU is fixed at the turnbuckle position for data acquisition. The IMU and the safety harness screw share a spatial rotation system: the yaw angle ψ is around the z axis, pitch θ is in the y direction, and roll ϕ is in the x direction. The coordinate system is defined as follows. The IMU is fixed in the carrier coordinate system, and the carrier coordinate system is a front–left–up (xyz) right-handed coordinate system. The geographic coordinate system is a west–north–sky coordinate system. If the rotation axis is defined by the directional cosine of the reference coordinate system and the quaternion norm is unitized, the rotation motion of the safety rope hook can be conveniently represented. The cosine angle in each direction is based on the yaw angle ψ, pitch angle θ, and roll angle ϕ, and the rotation angle is δ. $\boldsymbol{n}$ is the unit vector along the rotation axis. Mathematically, any vector $\boldsymbol{u}$ changes the angle around the rotation axis $\boldsymbol{n}$ of the unit vector, and the vector $\boldsymbol{v}$ after rotating an angle δ is determined from Equation (2).

$$\boldsymbol{v} = \boldsymbol{q} \cdot \boldsymbol{u} \cdot \boldsymbol{q}^{-1} \tag{2}$$

where $\boldsymbol{q}^{-1}$ is the inverse matrix of the quaternion $\boldsymbol{q}$.

Based on the IMU of the wearable device, we can acquire the three-axis acceleration sensing data, three-axis gyroscope sensing data and three-axis magnetic sensing data for the same time interval. The above data are used as the inputs of the algorithm, and the attitude angle of the safety harness can be calculated with the attitude angle algorithm and expressed in quadratic form. The attitude angle algorithm is shown below.

In the geographical coordinate system, gravitational acceleration and the geomagnetic field are $\boldsymbol{g}$ and $\boldsymbol{m}$, respectively, as expressed in Equations (3) and (4).

$$\boldsymbol{g} = \begin{bmatrix} 0 & 0 & g_0 \end{bmatrix}^T \tag{3}$$

$$\boldsymbol{m} = \begin{bmatrix} m_x & m_y & m_z \end{bmatrix}^T \tag{4}$$

where g_0 is gravitational acceleration and m_x and m_y are the components of the geomagnetic field corresponding to different coordinate axes. In the carrier coordinate system composed of the safety harness screw and an IMU, gravitational acceleration and the geomagnetic field can be expressed as shown in Equations (5) and (6), respectively.

$$\boldsymbol{g}_{\mathrm{f}} = \boldsymbol{M}_{fg} \cdot \boldsymbol{g} \tag{5}$$

$$\boldsymbol{m}_f = \boldsymbol{M}_{fg} \cdot \boldsymbol{m} \tag{6}$$

where $\boldsymbol{g}_{\mathrm{f}}$ and $\boldsymbol{m}_f$ are the vectors of gravitational acceleration and the geomagnetic field in the carrier coordinate system, respectively. $\boldsymbol{M}_{fg}$ is the coordinate transformation matrix from the geographic coordinate system to the carrier coordinate system, as shown in Equation (7).

$$\boldsymbol{M}_{fg} = \begin{bmatrix} \cos\theta\cos\psi & \cos\theta\sin\psi & -\sin\theta \\ \sin\phi\sin\theta\cos\psi - \cos\phi\sin\psi & \sin\phi\sin\theta\sin\psi + \cos\phi\cos\psi & \sin\phi\cos\theta \\ \cos\phi\sin\theta\cos\psi + \sin\phi\sin\psi & \cos\phi\sin\theta\sin\psi - \sin\phi\cos\psi & \cos\phi\cos\theta \end{bmatrix} \tag{7}$$

After obtaining the initial geomagnetic field $\boldsymbol{m}$ and gravitational acceleration $\boldsymbol{g}$ data at a construction site, Equations (3), (5) and (7) are applied. By combining the results with $\boldsymbol{g}_f$ and $\boldsymbol{m}_f$ obtained by the IMU in the carrier coordinate system, Equation (8) can be established, as shown below.

$$\begin{bmatrix} g_{xf} \\ g_{yf} \\ g_{zf} \end{bmatrix} = \begin{bmatrix} -\sin\theta \\ \sin\phi\cos\theta \\ \cos\phi\cos\theta \end{bmatrix} g_0 \tag{8}$$

where g_{xf}, g_{yf}, and g_{zf} are the acceleration readings in different axis directions in the carrier coordinate system of the IMU fixed on the safety harness screw. Equation (8) can be used to determine the pitching angle θ and roll angle ϕ, as shown in Equations (9) and (10), respectively.

$$\theta = \arcsin(-\frac{g_{xf}}{g_0}) \tag{9}$$

$$\phi = \arctan 2(g_{yf}, g_{zf}) \tag{10}$$

By applying Equations (4), (6) and (7) with m_{xf}, which is the reading from the magnetic induction sensor (on the IMU fixed on the safety harness screw) along the x-axis of the carrier coordinate system, as the input, Equation (11) is obtained. Additionally, the yaw angle is determined as shown in Equation (12).

$$m_{xf} = m_x \cos\theta \cos\psi + m_y \cos\theta \sin\psi - m_z \sin\theta \tag{11}$$

$$\psi = \arcsin(\frac{m_{xf} + mz\sin\theta}{\cos\theta \cdot \sqrt{{m_x}^2 + {m_y}^2}}) - \arctan 2(m_x, m_y) \tag{12}$$

Through the above derivation, Equations (9), (10) and (12) are used to determine the attitude angles (pitch, roll, and heading) of the safety harness screw, and the attitude angles are expressed in the form of a quaternion, as shown in Equation (13). The quaternion norm above is unitized and expressed with a trigonometric function, as shown in Equation (14). The quaternion represents the rotation change in the direction of each axis of the carrier coordinate system based on the IMU; that is, it represents the active attitude of the safety rope buckle when it is attached.

$$q = \begin{bmatrix} q_0 \\ q_1 \\ q_2 \\ q_3 \end{bmatrix} = \begin{bmatrix} \cos\frac{\psi}{2}\cos\frac{\theta}{2}\cos\frac{\phi}{2} - \sin\frac{\psi}{2}\sin\frac{\theta}{2}\sin\frac{\phi}{2} \\ \sin\frac{\psi}{2}\sin\frac{\theta}{2}\cos\frac{\phi}{2} + \cos\frac{\psi}{2}\cos\frac{\theta}{2}\sin\frac{\phi}{2} \\ \cos\frac{\psi}{2}\sin\frac{\theta}{2}\cos\frac{\phi}{2} - \sin\frac{\psi}{2}\cos\frac{\theta}{2}\sin\frac{\phi}{2} \\ \sin\frac{\psi}{2}\cos\frac{\theta}{2}\cos\frac{\phi}{2} + \cos\frac{\psi}{2}\sin\frac{\theta}{2}\sin\frac{\phi}{2} \end{bmatrix} \tag{13}$$

$$q = \cos\frac{\delta}{2} + x\sin\frac{\delta}{2}\vec{i} + y\sin\frac{\delta}{2}\vec{j} + z\sin\frac{\delta}{2}\vec{k} \tag{14}$$

where $n = \left\{x\vec{i}, y\vec{j}, z\vec{k}\right\}$ represents the rotation axis of the unit vector and δ represents the rotation angle of an action.

2.2. Spatial Motion Trajectory Model of Safety Rope Buckle Fastening Behaviour

In the above subsection, the action attitude of the safety rope buckle when it is attached is described based on a quaternion approach. With the aim of increasing the representativeness of quaternion action data, the next step is to establish a spatial motion trajectory model of the safety rope fastening behaviour of workers working at height.

The specific steps used for safety rope tethering behaviour monitoring are as follows.

(1) A virtual space coordinate system is created, and one part of the human body is set as the coordinate origin. Considering the action characteristics of the safety rope hanging behaviour and that the relative position of the waist of a worker performing actions barely changes, we set the waist as the origin of the spatial coordinate system.

(2) The initial coordinates of the IMU fixed to the safety harness screw are (a, b, c), and the corresponding coordinate system is established (in a right hand Cartesian coordinate system, the front of the person is the x direction, and from the foot to the head is the z direction).

(3) Then, through the quaternion rotation algorithm in three-dimensional spatial coordinates, the relative coordinates of the IMU during the safety rope fastening movement are obtained, and the motion trajectory with the worker's waist position as the origin is recorded.

Starting from the initial state, we monitor and track the fastening of the safety harness screw and represent the process in a time series [27,28]. By collecting data at each discrete time point, a series of safety rope fastening actions for scaffold workers working at heights is established.

The algorithmic flow of the time series approach is as follows.

(1) The natural state point A_0 when the IMU is first fixed on the safety harness screw is regarded as the initial state of action, and the waist position of the scaffold worker working at altitude is selected as the origin of the reference coordinate system, i.e., $(0,0,0)$. Then, the initial position of the wearable IMU is A_0 $(0,1,1)$ (assuming that the distance from the waist to the shoulder is 1 and that the length of a worker's arm is 1).

(2) A_0 is converted to a quaternion, expressed as $\boldsymbol{A_0} = 0 + 0\vec{i} + 1\vec{j} + 1\vec{k}$.

(3) The quaternion at each discrete time point of the time series $\{1,2,3,\ldots,t\}$ is expressed as $\{\boldsymbol{B_1}, \boldsymbol{B_2}, \boldsymbol{B_3}, \ldots, \boldsymbol{B_t}\}$. B_t represents the rotation angle of the IMU in this state at time t, that is, the change in the attitude angle (the change is represented by a quaternion).

(4) At time t, the coordinate of the IMU relative to the initial point A_0 is the imaginary part of the quaternion $\boldsymbol{A_t} = \boldsymbol{B_t} \cdot \boldsymbol{A_0} \cdot \boldsymbol{B_t}^{-1}$.

Based on the complex method of quaternions for matrix operations, the spatial coordinates of the IMU of the safety rope screw fastener fixed at a given moment in the time series are obtained. Based on the constructed spatial coordinate system and the quaternion data, the spatial motion trajectory of the fastening action of the safety rope buckle is determined.

2.3. Evaluation of Model Performance

For the evaluation of motion trajectory reconstruction, the final displacement error (FDE) is often used as an evaluation metric [9,29]. The FDE takes into account the difference between the predicted point and the true final point. The FDE metric is computed as Equation (15).

$$FDE = \frac{\sum_{i=1}^{k} \sqrt{\left(\widehat{Y}_i^t - Y_i^t\right)^2}}{k} \tag{15}$$

where k is the number of target points, Y_i^t and $\widehat{Y}_i^t$ are the true and predicted locations of target point i.

$$Y_i = \sqrt{x_i^2 + y_i^2 + z_i^2} \tag{16}$$

Since in this study the prediction is in 3D space, the location is calculated by distance formula as Equation (16) using corresponding x, y, and z .

3. Experiment

3.1. Experimental Design

To verify the feasibility of the reconstruction method of safety rope attachment for scaffold workers based on an IMU and quaternions, the research team conducted field tests involving field workers and fastening operations. The IMU was fixed at the lower end of the safety harness screws, and turnbuckle movement data were transmitted to a computer application that used a practical timestamp to store the data. The test was conducted at the Harbin Daowai Hospital construction site operated by the China Construction Eighth Engineering Bureau. There are many scaffolds used in this construction project. Scaffold workers usually need to climb high to work, so they are exposed to more danger than others working at ground level. In order to ensure that the individual behaviours of the different workers were taken into account, 20 workers participated in this field trial, each of them following the safety rope procedure to attach the safety rope. In this experiment, the scaffold workers fastened their safety belts according to the actual standard operation process. When the workers wearing the safety belt were working at height on the scaffold, they uncoiled the safety rope at their waist, lifted the end of the safety rope with the buckle slightly higher than the horizontal bar of the scaffold, wound the rope around the horizontal bar, and then returned the rope down to hook the connecting ring on the safety harness. Then, the buckle was pushed up near the lower part of the horizontal scaffold bar of the scaffold. Finally, the safety rope fastening action was completed (Figure 2). It was observed in the field test that the workers tied the safety rope to the scaffolding, and one of the most important actions is to wrap the safety rope with a screw buckle on the horizontal pole, and this action indicates that the safety rope is in a safe state. An IMU was used to collect and transmit acceleration, angular velocity, and magnetic induction strength data, and the action data for the safety rope fastening action were stored in an application on a computer with a USB receiver.

Figure 2. Field test of safety rope fastening.

3.2. Data Generation

The collection of original data is the most important step in this research, and all data in this study were collected with the IMU. The frequency of data collection was 50 Hz. Workers maintained a normal attitude and acted naturally during the process of data collection. Additionally, they did not deliberately change their behaviour due to the test, which would affect the authenticity of the measured data. Based on the IMU data, scaffold workers' safety belt use when working at heights was monitored, and some collected test data are shown in Table 1.

Table 1. Some of the raw data collected using the IMU. (Time (ms), Acc(0) [G], Gyr(0) [deg/s], Mag(0) [uT]).

Time	AccX(0)	AccY(0)	AccZ(0)	GyrX(0)	GyrY(0)	GyrZ(0)	MagX(0)	MagY(0)	MagZ(0)
0	−0.14	−0.412	0.37	−79.6	20.5	−74.1	24.5	72.3	−10.2
0.02	−0.198	−0.213	0.423	76.9	−5.1	−56.2	23.2	73	−8.6
0.038	−0.262	−0.172	0.363	−98.8	−21.2	−39.4	22.6	73.6	−7.2
0.058	−0.263	−0.304	0.316	−107.1	−19.8	−27.7	21.7	74	−6.2
0.079	−0.199	−0.471	0.361	−98.2	−9.7	−14.8	20.9	74	−4.4
0.099	−0.259	−0.562	0.414	−78.2	−1.6	−26	19.6	73.9	−2.9
0.12	−0.319	−0.573	0.437	−98.2	16.6	−35.4	18.5	73.7	−1.5
0.138	−0.314	−0.523	0.478	−138.1	28.6	−40.2	16.9	73.5	0.3

Table 1. *Cont.*

Time	AccX(0)	AccY(0)	AccZ(0)	GyrX(0)	GyrY(0)	GyrZ(0)	MagX(0)	MagY(0)	MagZ(0)
0.159	−0.238	−0.655	0.516	−85.7	51.9	−6.5	15.8	73.2	2.3
0.179	−0.318	−0.981	0.598	−3.4	66.2	32.7	15.5	72.8	3.1
0.2	−0.393	−1.127	0.53	−22.3	54.2	73.2	15.1	72.2	3.4
0.22	−0.363	−1.085	0.408	10.8	55.9	67.6	15.7	71.7	3.7
0.24	−0.583	−0.958	0.516	4.4	0.2	31	16.8	71.4	3.6
0.26	−0.479	−0.639	0.595	−59.2	−37.5	−29.1	17	71.6	3.8
0.28	−0.065	−0.277	0.61	−131.7	−2	−6.9	16.4	71.3	5.4
0.29	−0.09	−0.114	0.614	−111.6	17.3	71.8	17.2	69.7	7.6

3.3. Data Analysis

The safety rope fastening behaviours of scaffold workers working at heights varied. Data collected by the IMU were visualized. Figure 3 shows the three-axis acceleration change curve for the safety rope buckle fastening action, and Figure 4 shows the change in the angular velocity monitored by the three-axis gyroscope during the hooking action of the safety rope. Figure 5 shows the three-axis magnetic induction intensity curve of the safety rope hook during the fastening action.

The abscissas in Figures 3–5 are time, and the ordinates are physical components corresponding to the three axes of the carrier coordinate system, where blue represents the x-axis component, red represents the y-axis component, and black represents the z-axis component. Figure 3 shows that the change in the three-axis acceleration data collected by the IMU is relatively stable in the initial stage, and a variation occurs at approximately 1.5 s. The curve fluctuates significantly between 1.5 s and 3 s, producing two prominent peaks, and then enters a relatively stable state at approximately 3.5 s. The curve of the three-axis angular velocity change measured by the gyroscope (Figure 4) and the curve of the three-axis magnetic induction change (Figure 5) display similar characteristics as the three-axis acceleration curve. First, the curves are relatively stable; then, they fluctuate sharply between 1.5 s and 3.5 s before returning to a stable state.

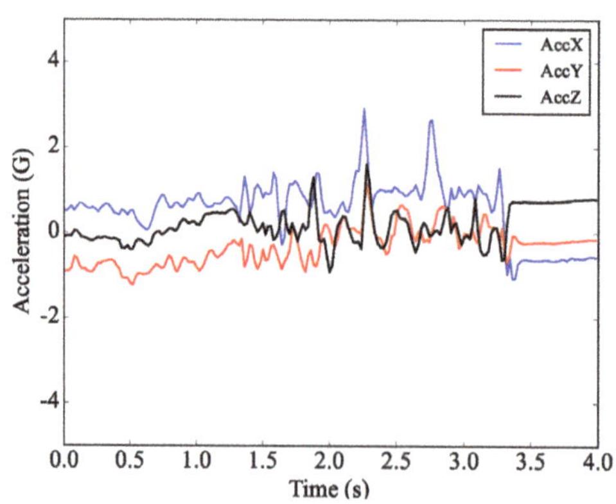

Figure 3. Three−axis acceleration curve.

In order to reduce the influence of external factors, e.g., the deployment location of the IMU and the data transmission process on the raw collected data, the Euclidean distance of 3D data, which enables the provision of the most intuitive characterisation of the safety rope tightening action data, is utlized. Figure 6 shows the three-axis curve of the acceleration change during the safety rope buckle fastening action, Figure 7 shows the three-axis curve of the angular acceleration change during the safety rope fastening action, and Figure 8 shows the curve of the three-axis composite magnetic induction intensity during the safety rope hook fastening action. Figures 6–8 illustrate curves with similar trends; the curves are initially stable, fluctuate sharply from between 1.5 s to and 3.5 s, and finally return to a stable state.

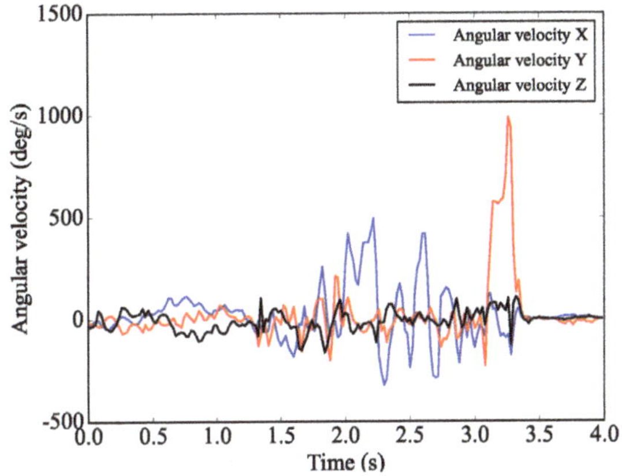

Figure 4. Angular velocity curve based on three−axis gyroscope data.

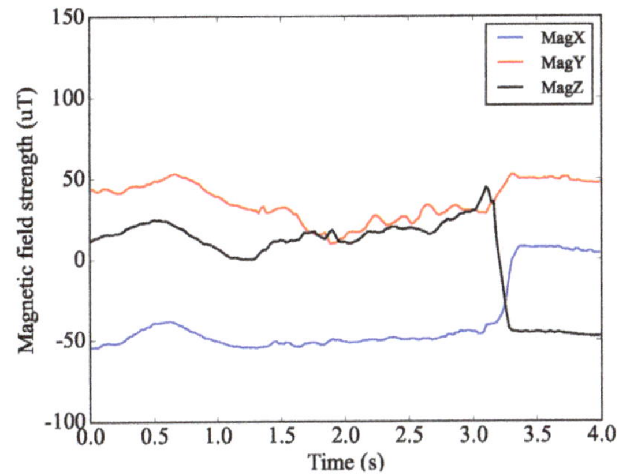

Figure 5. Three−axis magnetic induction curve.

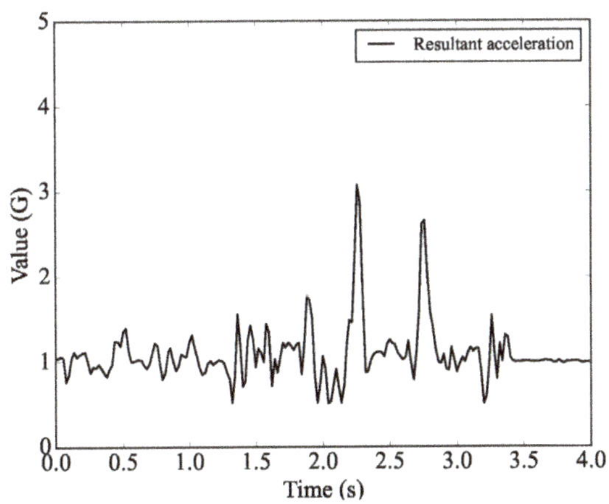

Figure 6. Three−axis resultant acceleration.

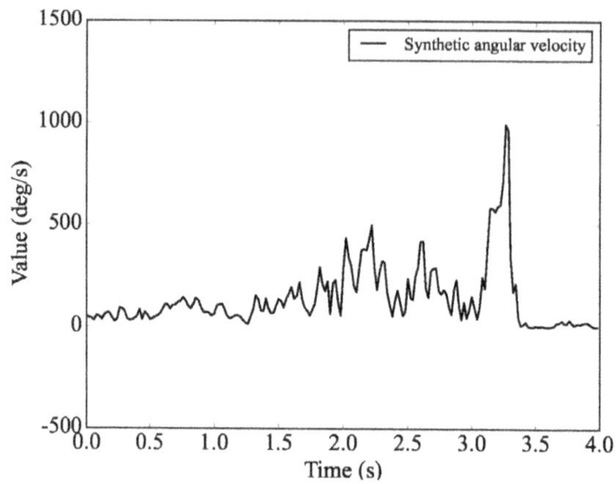

Figure 7. Three−axis resultant angular velocity.

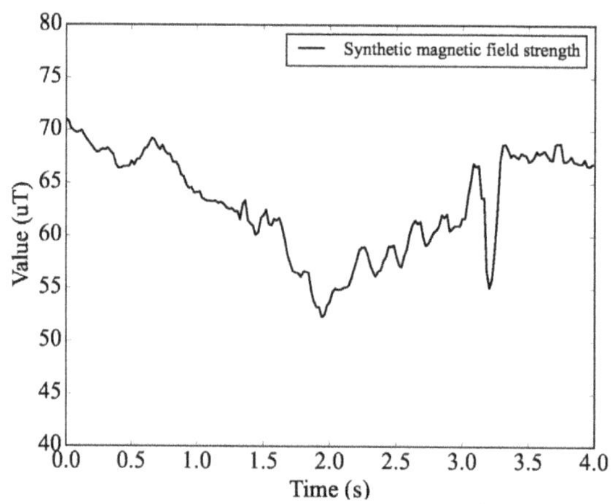

Figure 8. Three−axis composite magnetic induction.

4. Results and Discussions

This section provides an overview of the trajectory results of scaffolding worker safety rope screw buckle fastening based on a time series approach and experimental data, and explores the advantages and limitations of this study.

In Section 2.1, the use of quadratic expressions was proposed to describe the tethering action and position of the safety rope screw buckle. This approach can express the change in the rotation angle based on the change in the quaternion correlation matrix and express attitude based on rotation angle change. The acceleration data and magnetic induction intensity data can be used to obtain the attitude angle for the safety harness screw by using the attitude angle calculation formulas (Equations (9), (10) and (12)); the result can be expressed in the form of quadratic numbers (Equation (13)). Figure 9 shows the curve of the attitude angle corresponding to the safety rope fastening action data from construction workers; specifically, the attitude angle changes over time, as demonstrated based on the data collected by the IMU on the buckle of the safety belt. Figure 10 shows the curve of the safety harness screw rotation vector, that is, the curve of the rotation vector based on quaternions. Figure 10 shows the change in the rotation angle in the established coordinate system during the movement of the safety harness screw. Figure 11 shows the projection changes along the three axes during unit vector rotation in the quaternion approach. As shown in Figures 9 and 11, the data fluctuations in the range of approximately 1.5 to 3.5 s are prominent, with obvious characteristics of fastening action. Figure 10 shows the curve changes associated with the screw buckle fastening process.

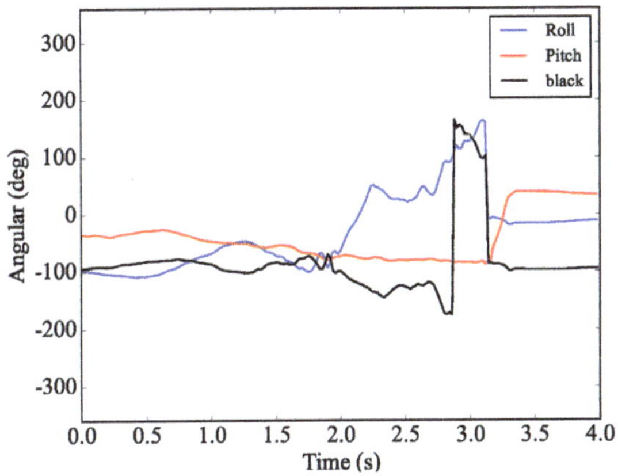

Figure 9. Attitude angle of the safety rope buckle fastening action.

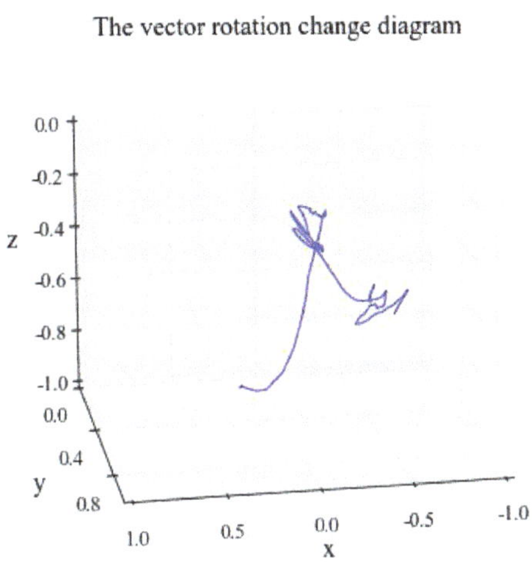

Figure 10. Changes in the safety harness screw rotation vector.

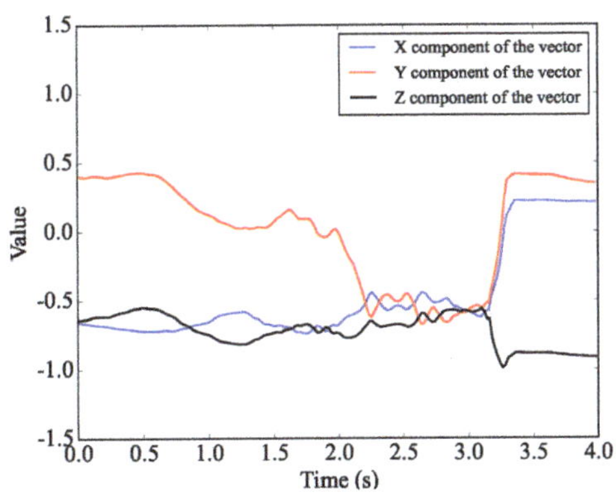

Figure 11. Projection changes of the unit vector along three axes during the rotation process.

Through the collection of the above data, although it is found that the safety rope screw buckle fastening action and position display obvious fluctuations and characteristics, the quaternion-based characterization of the initial screw buckle angle changes is clear, but other detailed actions are difficult to differentiate. To monitor specific actions and not just assess fluctuations in the data, it is necessary to correlate the data features with detailed actions. The acquired IMU data were processed and transformed based on a quadratic correlation approach from raw data containing a single fluctuation feature to three-dimensional spatial coordinate data with multiple feature attributes. Then, they were plotted with the spatial motion trajectories in the safety rope screw buckle attachment process, as shown in Figure 12. Figure 12 shows the trajectory of the entire safety rope screw buckle fastening action. The safety rope screw buckle moves upwards, and a loop-like trajectory appears at the uppermost end of the upwards trajectory, which characterizes the action of the safety rope screw buckle being wound around the horizontal bar of the scaffold. Such an action trajectory indicates that the scaffold worker has attached the safety rope and that the worker is in a safe state. The safety rope attachment trajectory provides real-time monitoring information for construction site safety management. The final displacement error (FDE) was used to evaluate the model for performing kinematic trajectory reconstruction, and the FDE value was 3.8%, and the accuracy of kinematic trajectory reconstruction reached 96.2%, which proved the reliability of the model. Figure 13 shows the schematic diagram of the safety rope attachment motion trajectory in each dimension. Similar data fluctuation characteristics as described previously were again observed in the range of 1.5 to 3.5 s.

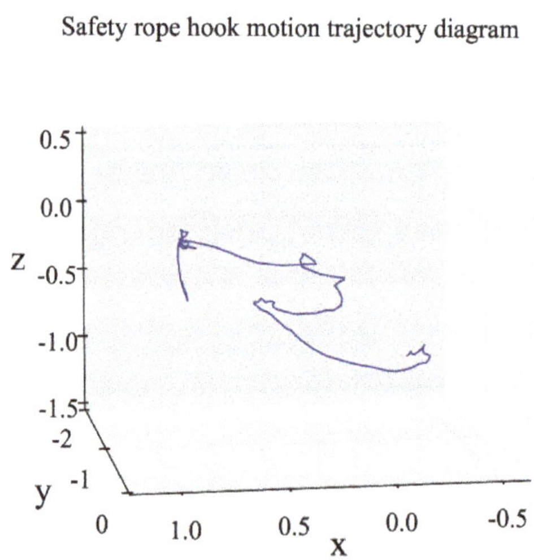

Figure 12. Movement diagram for safety rope attachment.

Existing research mostly focuses on identifying the behaviour of workers wearing safety belts on construction sites, with little attention paid to monitoring this behaviour during high-altitude operations. Through the analysis of experimental data, it can be determined that the acceleration, angular velocity, and magnetic induction intensity data obtained by the inertial measurement unit exhibit identifiable characteristics during the safety rope hanging action. In order to reduce the influence of external conditions, such as the fixed position of the inertial measurement unit and data transmission, on the collected raw data, three-dimensional acceleration, angular velocity, and magnetic induction intensity data can be summed by square root processing and then synthesized into one-dimensional data. These data can intuitively describe the characteristics of the safety rope hanging action. On the other hand, the attitude angle changes corresponding to the data of the safety rope tying action can also be used to indicate the completion of the tying action.

A spatial motion trajectory model of the safety rope screw buckle fastening behaviour was established based on a time series approach, and the reconstruction of the safety rope attachment actions of scaffold workers was achieved.

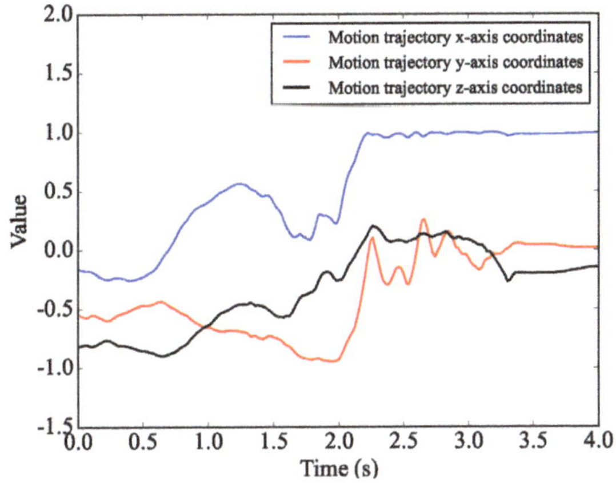

Figure 13. Diagram of safety rope attachment trajectories in each dimension.

The above results show that it is possible to monitor the safety harness fastening action of scaffold workers with the help of IMU and a time series approach. The current research, which involves a state-of-the-art approach for monitoring the movements of workers at height, has some limitations, mainly regarding the insufficient amount of data in the studied sample. Additionally, the physical conditions of construction workers are not the same, and workers' operation behaviours, although based on set standards, can vary in practice [30]. In future studies, additional physical and operational data associated with the actions of high-altitude scaffold workers will be obtained to reveal the characteristics of safety rope fastening and support construction management and worker safety and security. In the future, the combination of artificial intelligence and machine learning will be explored, and unsafe action recognition and real-time behaviour assessments will be performed, thus facilitating improved construction safety management through data safety management measures involving wearable devices.

5. Conclusions

To improve the efficiency of the safety inspection process and effectively avoid working-at-height accidents, a method based on a combination of IMU and a time series approach was proposed to establish a spatial movement trajectory model of safety rope screw buckle fastening. The research results show that (1) the IMU approach is simple to apply, and an IMU can be fixed directly to the safety rope hook. Additionally, the safety rope screw buckle fastening action displayed obvious data fluctuation characteristics in the construction site test. (2) A quaternion approach can be used to describe the attitude of the safety harness screw fastening action and to develop an algorithm for modelling and reconstructing the spatial movements of safety harness screw fastening. (3) The reconstruction method for scaffold worker fastening actions based on IMU and a time series approach is feasible. By analysing the relationship between the data collected by the IMU and the worker fastening actions, a spatial movement trajectory of the fastening behaviour is established. Ultimately, scaffold workers' fastening behaviour is reconstructed and a coupled movement technique is provided for monitoring safety belt fastening, thus contributing to construction safety management in practice and theory.

Author Contributions: Conception, Z.H. and Y.W.; Methodology, Z.H.; Investigation: Z.H.; Validation, Z.H. and Y.W.; Writing—original draft preparation, Z.H.; Writing—review and editing, Z.H.

and Y.W. Project administration, Z.H. All authors have read and agreed to the published version of the manuscript.

Funding: This research was supported by grants from the National Natural Science Foundation of China (No. 52108279) and Heilongjiang Provincial Key Laboratory of Polar Environment and Ecosystem (No. 2021007). This support is gratefully acknowledged.

Data Availability Statement: The data presented in this study are available on request from the corresponding author.

Conflicts of Interest: The authors declare no conflicts of interest.

References

1. Uddin, S.M.J.; Albert, A.; Alsharef, A.; Pandit, B.; Patil, Y.; Nnaji, C. Hazard Recognition Patterns Demonstrated by Construction Workers. *Int. J. Environ. Res. Public Health* **2020**, *17*, 7788. [CrossRef] [PubMed]
2. Karamoozian, A.; Luo, C.; Wu, D. Risk assessment of occupational safety in construction projects using uncertain information. *Hum. Ecol. Risk Assess. Int. J.* **2023**, *29*, 1134–1151. [CrossRef]
3. Paul, M.L.; Blasko, M. Bureau of Labor Statistics Releases 2021 Construction Industry Fatality Injury Data. Available online: https://www.jacksonlewis.com/insights/bureau-labor-statistics-releases-2021-construction-industry-fatality-injury-data (accessed on 30 March 2023).
4. Li, Z. Analysis of the Situation of Construction Safety in China. Available online: http://www.hywit.com/portal/Article/index/id/9986/cid/207.html (accessed on 8 September 2022).
5. Goh, Y.M. An Empirical Investigation of the Average Deployment Force of Personal Fall Arrest Energy Absorbers. *J. Constr. Eng. Manag.* **2014**, *141*, 04014059. [CrossRef]
6. Evan, N.; Carol, H.; Bo, X.; Ian, S.; Dongping, F. Falls from Height in the Construction Industry: A Critical Review of the Scientific Literature. *Int. J. Environ. Res. Public Health* **2016**, *13*, 638. [CrossRef]
7. Zhou, T.; Zhu, Q.; Shi, Y.; Du, J. Construction Robot Teleoperation Safeguard Based on Real-Time Human Hand Motion Prediction. *J. Constr. Eng. Manag.* **2022**, *7*, 148. [CrossRef]
8. Terzioglu, T.; Polat, G. Formwork System Selection in Building Construction Projects Using an Integrated Rough AHP-EDAS Approach: A Case Study. *Buildings* **2022**, *12*, 1084. [CrossRef]
9. Heravi, M.Y.; Jang, Y.; Jeong, I.; Sarkar, S. Deep learning-based activity-aware 3D human motion trajectory prediction in construction. *Expert Syst. Appl.* **2024**, *239*, 122423. [CrossRef]
10. Hirakawa, T.; Yamashita, T.; Tamaki, T.; Fujiyoshi, H. *Survey on Vision-Based Path Prediction*; Springer International Publishing: Cham, Switzerland, 2018; pp. 48–64. [CrossRef]
11. Du, H.; Manns, M.; Herrmann, E.; Fischer, K. Joint Angle Data Representation for Data Driven Human Motion Synthesis. *Procedia CIRP* **2016**, *41*, 746–751. [CrossRef]
12. Kim, K.; Cho, Y.K. Effective inertial sensor quantity and locations on a body for deep learning-based worker's motion recognition. *Autom. Constr.* **2020**, *113*, 103126. [CrossRef]
13. Fang, W.; Zhong, B.; Zhao, N.; Love, P.E.; Luo, H.; Xue, J.; Xu, S. A deep learning-based approach for mitigating falls from height with computer vision: Convolutional neural network. *Adv. Eng. Inform.* **2019**, *39*, 170–177. [CrossRef]
14. Han, S.; Lee, S.; Peña-Mora, F. Vision-Based Detection of Unsafe Actions of a Construction Worker: Case Study of Ladder Climbing. *J. Comput. Civ. Eng.* **2013**, *27*, 635–644. [CrossRef]
15. Fang, W.; Ding, L.; Luo, H.; Love, P.E. Falls from heights: A computer vision-based approach for safety harness detection. *Autom. Constr.* **2018**, *91*, 53–61. [CrossRef]
16. Sanhudo, L.; Calvetti, D.; Martins, J.P.; Ramos, N.M.; Mêda, P.; Gonçalves, M.C.; Sousa, H. Activity classification using accelerometers and machine learning for complex construction worker activities. *J. Build. Eng.* **2021**, *35*, 102001. [CrossRef]
17. Akhavian, R.; Behzadan, A.H. Smartphone-based construction workers' activity recognition and classification. *Autom. Constr.* **2016**, *71*, 198–209. [CrossRef]
18. Tian, Y.; Zhang, J.; Chen, L.; Geng, Y.; Wang, X. Single Wearable Accelerometer-Based Human Activity Recognition via Kernel Discriminant Analysis and QPSO-KELM Classifier. *IEEE Access* **2019**, *7*, 109216–109227. [CrossRef]
19. Ahn, C.R.; Lee, S.; Sun, C.; Jebelli, H.; Yang, K.; Choi, B. Wearable Sensing Technology Applications in Construction Safety and Health. *J. Constr. Eng. Manag.* **2019**, *145*, 03119007. [CrossRef]
20. Bangaru, S.S.; Wang, C.; Busam, S.A.; Aghazadeh, F. ANN-based automated scaffold builder activity recognition through wearable EMG and IMU sensors. *Autom. Constr.* **2021**, *126*, 103653. [CrossRef]
21. Barraza Madrigal, J.A.; Contreras Rodríguez, L.A.; Cardiel Pérez, E.; Hernández Rodríguez, P.R.; Sossa, H. Hip and lower limbs 3D motion tracking using a double-stage data fusion algorithm for IMU/MARG-based wearables sensors. *Biomed. Signal Process. Control* **2023**, *86*, 104938. [CrossRef]
22. Joshua, L.; Varghese, K. Accelerometer-Based Activity Recognition in Construction. *J. Comput. Civ. Eng.* **2011**, *25*, 370–379. [CrossRef]

23. Yoon, J.; Lee, B.; Chun, J.; Son, B.; Kim, H. Investigation of the relationship between Ironworker's gait stability and different types of load carrying using wearable sensors. *Adv. Eng. Inform.* **2022**, *51*, 101521. [CrossRef]
24. Hong, S.; Yoon, J.; Ham, Y.; Lee, B.; Kim, H. Monitoring safety behaviors of scaffolding workers using Gramian angular field convolution neural network based on IMU sensing data. *Autom. Constr.* **2023**, *148*, 104748. [CrossRef]
25. Terze, Z.; Zlatar, D.; Pandža, V. Aircraft attitude reconstruction via novel quaternion-integration procedure. *Aerosp. Sci. Technol.* **2020**, *97*, 105617. [CrossRef]
26. Pellois, R.; Brüls, O. An inertial human upper limb motion tracking method for robot programming by demonstration. *Robot. Auton. Syst.* **2022**, *156*, 104201. [CrossRef]
27. Roy, A.; Nelson, T.; Turaga, P. Functional data analysis approach for mapping change in time series: A case study using bicycle ridership patterns. *Transp. Res. Interdiscip. Perspect.* **2023**, *17*, 100752. [CrossRef]
28. Ligęza, P. Reconstructing the trajectory of the object's motion on the basis of measuring the components of its velocity. *Measurement* **2023**, *221*, 113546. [CrossRef]
29. Alahi, A.; Goel, K.; Ramanathan, V.; Robicquet, A.; Li, F.; Savarese, S. Social LSTM: Human Trajectory Prediction in Crowded Spaces. In Proceedings of the 2016 IEEE Conference on Computer Vision and Pattern Recognition (CVPR), Las Vegas, NV, USA, 27–30 June 2016; pp. 961–971. [CrossRef]
30. Karamoozian, A.; Wu, D.; Luo, C. Green Supplier Selection in the Construction Industry Using a Novel Fuzzy Decision-Making Approach. *J. Constr. Eng. Manag.* **2023**, *149*, 04023033. [CrossRef]

buildings

MDPI

Article

Causal Model Analysis of the Effect of Policy Formalism, Equipment Insufficiency and COVID-19 Fear on Construction Workers' Job Burnout, and Insomnia during the Epidemic

Tsung-Lin Wu [1], Tsai-Feng Chu [2] and Hsiang-Te Liu [3,*]

1 Department of Sports Technology and Leisure Management, I-Shou University, Kaohsiung 84001, Taiwan
2 Office of Occupational Safety and Health, Taipei City Hospital, Taipei 108, Taiwan
3 Department of Public Affairs and Administration, Ming Chuan University, Taoyuan 333, Taiwan
* Correspondence: lback@mail.mcu.edu.tw

Abstract: During the epidemic, construction site workers suffered from COVID-19 fear, job burnout, and job insecurity due to insufficient personal protective equipment. This study mainly explores the effect of policy formalism, equipment insufficiency, COVID-19 fear, and job insecurity on construction workers' job burnout and insomnia during the epidemic. The main contribution of this article is to introduce policy formalism into the exploration of the causes of insomnia among construction site workers during the epidemic. This study collected 733 valid samples of construction site workers. We tested the established hypotheses using confirmatory factor analysis and structural equation modeling. The research results found that equipment insufficiency positively affects COVID-19 fear. Policy formalism and COVID-19 fear positively affect job burnout. Social support negatively affects job burnout. Job burnout and job insecurity positively affect insomnia. The government should establish clear policy evaluation standards and implement policies to reduce construction site workers' perceptions of policy formalism. The government should prepare sufficient personal protective equipment so that workers on construction sites can cope with the spread of the epidemic.

Keywords: policy formalism; equipment insufficiency; COVID-19 fear; job burnout; insomnia

Citation: Wu, T.-L.; Chu, T.-F.; Liu, H.-T. Causal Model Analysis of the Effect of Policy Formalism, Equipment Insufficiency and COVID-19 Fear on Construction Workers' Job Burnout, and Insomnia during the Epidemic. *Buildings* **2024**, *14*, 265. https://doi.org/10.3390/buildings14010265

Academic Editors: Mariusz Szóstak and Marek Sawicki

Received: 30 November 2023
Revised: 9 January 2024
Accepted: 16 January 2024
Published: 18 January 2024

1. Introduction

During the epidemic, construction sites were considered places where the virus spread rapidly [1]. There were many workers at construction sites, including directors, engineers, supervisors, administrators, technical staff, and workers. This study refers to all the categories listed above as construction site workers. Many workers at construction sites are at risk of COVID-19 infection [2]. Social distancing policies, personal protective equipment, and video conferencing have all changed worksite operations [3,4].

Construction site workers are prone to industrial safety accidents due to negligence and fatigue when operating machinery and equipment. During the pandemic, construction workers at primary sites faced mounting fears, insecurities, job burnout, and insomnia. The COVID-19 fear, job burnout, and job insecurity of construction site workers during the epidemic are worth exploring. The negative emotions of these construction site workers profoundly affect the effective operation and safety of construction sites. Insomnia is characterized by interrupted sleep, difficulty falling asleep, and poor sleep quality, which negatively affect normal body functions and health and in turn affect daily life routines [5,6]. Insomnia can be caused by stress, anxiety, and traumatic events [7]. Studies have found that stressors related to work, family, and health—including work overload and lack of support—are the main causes of insomnia; more recently, COVID-19 anxiety has also been confirmed as a cause [8]. The current study explored whether job burnout and insecurity have an impact on insomnia among workers.

Many studies have confirmed the occurrence of severe sleep disorders during the COVID-19 pandemic [9,10]. Severe insomnia led to subsequent physical and psychological problems among frontline workers [11]. Many researchers have identified hyperarousal as a major cause of insomnia [12]. Insomnia is mostly caused by stressful situations [13]; when individuals are unable to cope with the challenges of stressful events, it can be difficult to fall asleep. The COVID-19 outbreak inevitably also contributed to job burnout and insomnia among construction site workers.

Job burnout is a symptom of prolonged work stress [14]. It can cause emotional exhaustion, depersonalization, and diminished personal accomplishment. Emotional exhaustion causes employees to feel unmotivated to work [15]. Depersonalization occurs when an individual is blamed for problems at work and therefore feels negative emotions and attitudes [16]. Diminished personal accomplishment is a negative evaluation of one's ability to perform tasks and interact with people, along with feelings of dissatisfaction and unhappiness [17]. Employees who suffer from job burnout are prone to anxiety and depression [18]. The sudden onset of COVID-19 affected many aspects of the economy, and it impacted many part-time and full-time workers [8]. Although most natural disasters are short-lived, the COVID-19 pandemic lasted 2–3 years, and during that time, one out of every five frontline workers was confirmed to have experienced job burnout [19].

Zhang found that group support reduced job burnout for psychological counselors [20]. In a study of healthcare workers, Zhao et al. found that social support enhanced workers' self-efficacy, which in turn reduced job burnout [21]. A nursing study in South Korea found that social support was beneficial for frontline personnel to face the hardships of the pandemic [22]. Multiple studies have noted that frontline workers faced anxiety, depression, and burnout during the pandemic [23]. Job demands require employees to be committed to their work, which creates psychological and physical stress [24].

From the perspective of the conservation of resources theory, individuals who have social support resources are more likely to effectively cope with stress and reduce burnout [25]. Conversely, when individuals lack personal, social, and material resources, it is more difficult for them to cope with stress [26]. According to the conservation of resources theory, individuals will strive to acquire and conserve the physical and psychological resources they value [27]. The loss of these valuable resources can cause individuals to feel stressed. Economic crises, natural disasters, and the spread of disease can all cause individuals to feel a sense of loss in terms of resources.

According to the job demand–resources model, job demands are job requirements that tend to create job stress for workers, whereas job resources are the provision of job skills and resources that help to reduce job burnout [28]. The job demand–resources model stipulates that when job demands exceed job resources, employees are prone to burnout and health problems [29].

The outbreak of the pandemic in 2019 led to changes in working hours and workloads for many frontline workers, in addition to contributing to job burnout [30]. Furthermore, the lockdown policy during the pandemic containment period disrupted social support for workers. Some studies have demonstrated that social support during outbreaks can reduce job burnout [31]; conversely, a lack of social support can cause anxiety, depression, and insomnia in workers [32].

Many studies have emphasized the importance of personal protective equipment during a pandemic. A study by Gallop et al. revealed that the greatest fear among caregivers for HIV patients was that of being infected [33]. During the severe acute respiratory syndrome (SARS) and Middle East respiratory syndrome (MERS) outbreaks, the WHO found that failure to ensure the proper use of personal protective equipment contributed to the spread of the diseases [34]. Inadequate personal protective equipment also led to a high risk of infection during the 2014 Ebola virus disease (EVD) outbreak [35]. Therefore, the lack of personal protective equipment increased the fear of COVID-19 among frontline workers [36]. Frontline workers were also concerned about transmission to their families as a result of their work [37]. Construction workers also faced fears of COVID-19 infection while working at construction sites.

Job insecurity is the feeling of powerlessness to maintain a stable job in dangerous situations [38]. Job insecurity is recognized as a source of stress in the work environment, and it has a negative impact on the physical and mental health of workers [39]. The COVID-19 pandemic created a sense of uncertainty and insecurity about future construction site work. In addition to feeding insecurities about future job stability, the COVID-19 outbreak necessitated increased safety behavior in the workplace, i.e., the need for workers to comply with protocols in the face of risky situations [40]. Being asked to conduct temperature checks, wear masks, and practice social distancing at the worksite during the pandemic were all causes of stress for workers. At the same time, high environmental stress, work overload, uncertainty, and excessive demands are known to affect the safety behavior of workers [41].

Job security is the degree to which an existing job is stable or consistently threatened [42]. Individuals' feelings of uncertainty about job security can increase their stress levels [43]. The perception of job insecurity can lead to insomnia, and the self-regulatory resource theory suggests that insomnia affects self-recovery [44]. When individuals lack sufficient resources to cope with hazardous environments, they can develop sleep disorders and sleep impairments. Economic depression, travel restrictions, and material shortages caused by the COVID-19 pandemic all affected individuals and negatively impacted the work and life aspects of construction site workers.

The main contribution of this study is to introduce policy formalism into the study of job burnout of construction site workers during the epidemic. Riggs' formalism asserts that culture has a profound impact on institutions and administration [45]. The introduction of Western laws and systems will be affected by local culture [45]. Policy formalism is an important feature of Riggs' ecological administration. The formalism in the formulation and implementation of epidemic prevention policies has caused more job burnout for construction site workers.

This study mainly explores the fear of COVID-19 caused by insufficient equipment among construction site workers during the epidemic. COVID-19 fears among construction site workers have further caused job burnout, job insecurity, and insomnia. How job burnout among construction site workers is affected by social support and policy formalism is investigated. This study mainly explores how construction site workers' social support, policy formalism, and COVID-19 fear affect their job burnout, job insecurity, and insomnia.

2. Literature Review and Hypothesis Development

2.1. Theoretical Basis

2.1.1. Policy Formalism

The main contribution of the current study is the incorporation of Riggs' formalism into the COVID-19 study of construction site workers. When there is a discrepancy between pandemic prevention regulations and their implementation, it can lead to anxiety and job burnout among workers. Riggs described the formalism of administrative agencies mainly in the context of developing countries [45]. His criterion of using American society as a diffracted society has been challenged by scholars who argue that, although the United States is the benchmark for developed and industrialized societies, it is a mistake to claim that American society is totally devoid of formalism [46]. Some scholars have argued that prismatic societies not only exist in underdeveloped countries but are also common in countries with different levels of development [46]. Taiwan is also gradually emerging from its status as a developing country; however, due to the influence of Asian and Chinese cultures in the past, studies have also confirmed the existence of formalism in Taiwan [47–49].

Formalism as proposed by Riggs involves ritualistic methods, a lack of authorization, and centralization. Ritualistic methods create a gap between the norms of law and their effective implementation, along with a gap between administrative norms and realities.

The absence of objective evaluation standards and the lack of pressure on civil servants to implement programs allow formalism to develop easily [45]. Riggs therefore concluded that there is a difference between administrative ritualistic procedures and the rationalistic procedures of developed countries. Ritualistic procedures are not implemented in practice, and thus leave a gap between administrative requirements and actual implementation performance. When epidemic prevention measures are not implemented in practice, workers' psychological exhaustion and burnout increase. In countries with higher levels of formalism, the lack of shared values at the administrative level and inconsistency between governmental and social values creates a gap between administrative planning and people's expectations [50]. Workers' job burnout therefore increases due to dissatisfaction with epidemic prevention measures at work.

2.1.2. Conservation of Resources Theory

Conservation of resources (COR) is a psychological theoretical framework proposed by Stevan E. Hobfoll [27]. Stress occurs when individuals perceive a net loss of their own resources. Individuals use coping mechanisms to deal with these stressors. They also seek social support to reduce stress and burnout [25,26]. Construction site workers feel the loss of resources under the pressure of COVID-19 and will look for social support to reduce work burnout.

2.1.3. Job Demand–Resources Model

The job demand–resources (JD-R) model is a theoretical framework in organizational psychology. JD-R is used to explain the impact of job characteristics on employee health [28,29]. Job demands include high workload, time pressure, complex tasks, and uncertainty. These job demands can lead to employee stress and burnout [28,29]. Sufficient job resources will help improve employee happiness and job satisfaction. Construction site workers do face many job requirements during the COVID-19 pandemic, including wearing personal protective equipment, virus testing, and social distancing policies. This study also applies the job demand–resources model to explain the relationship between job resources and job burnout during the epidemic period for construction site workers.

2.2. Hypotheses Development

2.2.1. The Relationship between Equipment Insufficiency and COVID-19 Fear

During the first 2 years of the COVID-19 pandemic, rapid transmission and ease of infection led to fear among frontline workers, who were at high risk of infection due to their exposure to patients and inadequate personal protective equipment [51,52]. Early in the outbreak, frontline workers had yet to determine the route of transmission; therefore, many became infected [53]. In Spain, the lack of personal protective equipment at the beginning of the outbreak increased the risk of infection, leading to a fear of contact between frontline workers and patients [54,55]. Studies at the time confirmed the link between infection and lack of personal protective equipment for frontline workers [56], so the lack of effective and adequate personal protective equipment for workers during the pandemic naturally led to increased fear.

One study noted a relationship between personal protective equipment and psychological stress and fear [57]. Wearing personal protective equipment has been recognized as a key measure to prevent the spread of COVID-19. Personal protective equipment can protect frontline workers from droplet infection [58]. However, when personal protective equipment becomes contaminated, frontline workers discontinue its use, which increases the risk of infection for patients and workers alike [59]. Workers at construction sites experienced increased fear due to inadequate protective equipment and infections.

Hypothesis 1. *Equipment insufficiency positively affects COVID-19 fear.*

2.2.2. The Relationship between Social Support and Job Burnout

Many studies have confirmed that burnout is a negative outcome of prolonged exposure to stressful events [60]. Job burnout can be summarized as emotional exhaustion caused by physical and psychological exhaustion. Maslach and Jackson defined job burnout as a syndrome of ineffectiveness, exhaustion, and ridicule that occurs in the workplace [61]. Heavy workloads and high levels of work stress can lead to job burnout [62].

Social support is recognized as a resource for dealing with stressful events [63]. Individuals who perceive higher levels of social support demonstrate more optimism and motivation when faced with difficult situations [64]. Therefore, social support is negatively correlated with job burnout [65]. Social support is characterized as interpersonal interactions and relationships that provide care and emotional attachment [66]. Individuals feel social support when they are helped by others. Social supporters are mostly coworkers, family members, and friends. Some studies have demonstrated that social support can alleviate the stressful feelings of frontline workers [67]. A study by Dir et al. suggested that social support can mediate the relationship between job burnout and stigmatization [68]. Anxiety and burnout caused by frontline work at construction sites necessitates immediate social support.

Social support is the moral or material support provided by organizations, supervisors, and coworkers. Social support reflects the quality of an individual's social relationships and affects their physical and mental health [69]. The more social support an individual receives, the more likely it is that they will be able to cope with difficult problems. Lee et al. discovered a negative correlation between social support and job burnout [70]. In a study on transportation industry employees, Feldman et al. found that family support reduced job burnout [71]. The job demand–resources (JD-R) model is a theory that was developed to explain job burnout. Job stress and social support are important factors in the JD-R model [72]. Job stress can cause internalized tension and frustration in individuals, which in turn can lead to job burnout. Social support is a work resource that can reduce job stress and job burnout.

Frontline workers faced many job demands related to outbreak screening and control. Many frontline workers felt weary due to outbreak screening [73]. The workflows at many worksites were altered due to the outbreak, resulting in job burnout due to reduced autonomy [74]. Job resources are believed to assist in achieving goals and reducing the physical and psychological stress caused by job demands. Job resources are also thought to increase work engagement while reducing health threats caused by job demands [75]. Job demands at construction sites during the pandemic were higher than usual, and the social distancing, protective equipment, disinfection, and temperature-taking protocols implemented at construction sites due to the pandemic increased the job demands of workers.

Hypothesis 2. *Social support negatively affects job burnout.*

2.2.3. The Relationship between COVID-19 Fear and Job Burnout

Fear is a functional emotion that occurs when an individual perceives a threat. However, fear of COVID-19 infection is a dysfunctional fear that can be physically and psychologically harmful at high levels [76,77]. Furthermore, fear of COVID-19 is thought to increase job burnout, as indicated by the first responders who experienced higher levels of job burnout during the pandemic [78].

Fear is a psychological response to environmental stressors. The social distancing policy and demands for virus screening during the pandemic increased individuals' perceived fear [79]. Quarantine, job insecurity, lifestyle changes, and the death of loved ones all contributed to this heightened sense of fear. Fear of COVID-19 is thought to be associated with job burnout as well [80]. At construction sites, workers' first job was to face the fear of COVID-19 infection. When individuals have more resources and experience in facing situations of uncertainty, their feelings of fear can be reduced. Conversely, when individuals have low tolerance for uncertainty, they are prone to fear and anxiety [81].

COVID-19 is a novel coronavirus, having only been discovered in 2019, and this led to a high sense of uncertainty.

The lack of personal protective equipment, uncertainty surrounding the outbreak, and the lack of an effective vaccine all contributed to the fear felt by frontline workers [82]. The most worrying aspect of the COVID-19 virus was its spread by asymptomatic carriers, which was a source of stress and fear for frontline workers. At construction sites, as well, fear among workers further contributed to job burnout [82]. During the COVID-19 pandemic, the rising number of infections and the lack of resources resulted in job burnout. The risk of death from COVID-19 infection led to many psychological problems including stress, anxiety, and fear [83]. Many frontline workers experienced these negative physical and psychological problems [80]. Increased workloads, deaths of colleagues due to infection, and lack of social support all contributed to anxiety and burnout.

Hypothesis 3. *COVID-19 fear positively affects job burnout.*

2.2.4. The Relationship between Policy Formalism and Job Burnout

The gap between formal power and effective power has resulted in a highly formalistic administration that centralizes power but lacks administrative efficiency. The COVID-19 pandemic came on quickly, and government agencies were often unable to respond to it accordingly, leaving construction site workers with a higher sense of job burnout. Thompson found that developing countries had a relatively large number of generalists [84], and that they emphasized hierarchy and procedures while forgetting their purely instrumental origins; generalist civil servants would incorporate programmatic procedures into the law, forgetting that they were merely tools. Even scholars have pointed out that the generalists in developing countries are superior to the professionals [85].

Bureaucracies with a high degree of formalism are full of pathological behavior, such as lack of authorization, over-emphasis of control, red tape, indifference, and fear of innovation. Most civil servants wait for orders from the top and lack a sense of personal security [86]. When faced with the inaccurate implementation of epidemic policies, workers are prone to feel a higher level of job burnout. Burns and Stalker argued that in organizations with an organic model, the responsibilities and duties of professionals are not clearly defined; work content is generated through continuous interaction with colleagues, and not everything is decided by the supervisor [87]. In a mechanistic system, the supervisor decides whether the work of the professionals aligns with the goals of the organization. Most formalist government agencies belong to the mechanistic system. During the pandemic, the influence of administrative generalists exceeded that of professionals, which made it easier for a gap to form between the pandemic prevention laws and their actual implementation. It is difficult for a developing country to implement organizations with an organic model [88] because the flexible design causes anxiety for executives [87]. Organic systems require a sense of trust among members, which developing countries often lack [89].

Countries with a low degree of formalism encourage creativity and motivate civil servants to achieve their goals. However, countries with a high degree of formalism are characterized by authoritarianism and inherent work practices that inhibit creativity [90]. Administrations in developing countries are characterized by "irrational management". Due to the lack of authorization, many minor decisions must be made by administrative executives. Many civil servants in developing countries set themselves apart from the general public—they are mainly concerned with their own interests, and care little about the public interest [90]. During the COVID-19 outbreak, many delays in handling the crisis were attributable to civil servants waiting for orders from their superiors. When construction site workers feel higher policy formalism, this will increase their job burnout.

Hypothesis 4. *Policy formalism positively affects job burnout.*

2.2.5. The Relationship between COVID-19 Fear and Job Insecurity

For the general public, concerns about COVID-19 infection could be attributed to the high infection and mortality rates [76]. Later studies also associated fear of COVID-19 with job insecurity, psychological stress, and anxiety [76]. Job security is an important tool for securing personal resources [91]; therefore, job insecurity can easily lead to fear of poverty, stigmatization, and social exclusion [92]. Literature has confirmed the association of job insecurity with anxiety, depression, and mental health [39].

Frone found that in times of economic downturn, individuals experience job insecurity through pay cuts, reduced hours, and physical and mental anguish [93]. During the early stages of the pandemic, the rapid spread of COVID-19 led to urban lockdowns and increased remote work. Company closures caused by the pandemic caused employees to experience job insecurity [94]. Furthermore, uncertainty about the future of the economy at the time caused many people to worry about their family finances.

Sudden destabilizing events can cause social, economic, and psychological disruptions to an individual's life. The COVID-19 pandemic affected paychecks and employment and sparked family–work conflicts [95]. It led to a decrease in job opportunities, and the resulting financial stress caused individuals to become anxious. Indeterminate stressors increased job insecurity, which in turn created anxiety, restlessness, and depression in individuals. The economic downturn caused by the pandemic also contributed to psychological depression among workers [96].

Depression and anxiety caused by the pandemic affected the mood in the workplace. Work tasks and priorities were altered following the escalation of the outbreak. Fear of the pandemic, insecurity, and loss of income all contributed to the loss of psychological resources [97]. From the perspective of the conservation of resources theory, job insecurity makes it necessary to expend energy to cope with stressful situations. Herzberg asserted that job security is the combination of a stable work environment and employment protection, i.e., retirement security, stable salary, and opportunities for self-development and advancement [98]. Therefore, job security encompasses the provision of financial, economic, and social security within an organization [98]. Fear of COVID-19 is thought to influence the relationship between job insecurity and depression [99].

Hypothesis 5. *COVID-19 fear positively affects job insecurity.*

2.2.6. The Relationship between Job Burnout and Insomnia

High levels of emotional and physical exhaustion tend to cause sleep problems. Cognitive and emotional demands in the workplace make it difficult for workers to fall asleep [100]. COVID-19 posed new challenges and threats to the workplace. Frontline workers faced a high risk of contracting the virus during the pandemic [101]. Their stress factors included lack of personal protective equipment, high workloads, quarantine, and loss of loved ones [102].

Prolonged anxiety about the pandemic led to insomnia, stress disorders, and job burnout [103]. Furthermore, long shifts and working hours were required during this period, something known to cause work stress, insomnia, and fatigue problems. Working late shifts can result in poor sleep quality and difficulty falling asleep. Prolonged exposure to emotionally challenging work can lead to physical and mental health problems [104]. Job burnout occurs when individuals experience chronic stress that is difficult to cope with in the workplace. The relationship between job burnout and insomnia has been explored, and it has been determined that increased job burnout causes insomnia and psychological exhaustion [105]. Workers at construction sites during the pandemic often suffered from insomnia due to job burnout.

Hypothesis 6. *Job burnout positively affects insomnia.*

2.2.7. The Relationship between Job Insecurity and Insomnia

Davy, Kinicki, and Scheck defined job insecurity as an individual's expectation of job continuity [106]. Perceived job insecurity is heightened when the job is threatened by uncertainty. Insecurity is a precursor to long-term personal stress and leaves employees feeling that they are powerless to determine their own job continuity. Workers' fear and dread about their jobs is job insecurity [42]. Such worries and fears can lead to insomnia, depriving workers of the sleep needed to maintain energy and motivation. Poor sleep quality makes it difficult to relieve fatigue, and inability to concentrate due to insomnia can lead to workplace safety problems [107].

Multiple studies have pointed to job insecurity as a source of stress that exacerbates insomnia problems [108]. Worrying about lack of job continuity leads to stress and insomnia. Sleep is a way for individuals to regain energy and cognitive clarity, as well as a means to recover self-regulating resources. Additionally, sleep grants individuals with sufficient resources to respond to safety regulations [109]. Studies have indicated that people with sleep problems are more likely to have work accidents [109]. Work stress, work overload, and role conflicts due to job demands are all factors that contribute to insomnia, and workers were subjected to greater job demands during the pandemic.

Hypothesis 7. *Job insecurity positively affects insomnia.*

We drew out the research framework based on the above hypothesis argument (Figure 1).

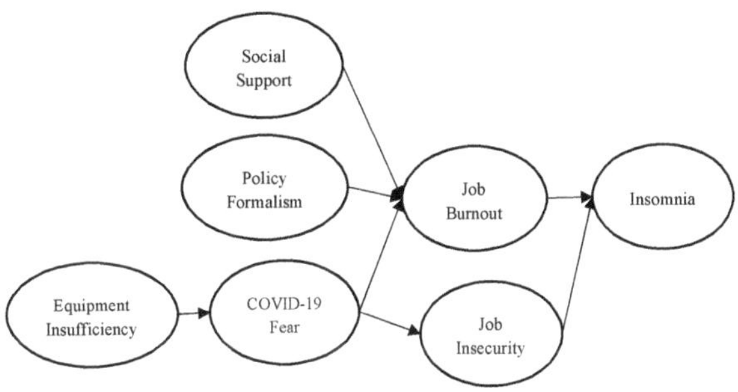

Figure 1. Research Framework.

3. Materials and Methods

3.1. Samples, Tools, and Procedure

This study takes construction site workers in Taiwan as the research subjects. The Taiwanese government announced on 1 May 2023 that the epidemic prevention threat had been lowered. According to statistics from Taiwan's Ministry of Labor in November 2023, the number of construction site workers in Taiwan was 11,977 [110]. This study adopted regional stratified sampling; we obtained a total of 733 valid construction site worker samples from 2022 to early 2023, with a recovery rate of 48.3%. This study used G*Power3.1.9.7 to calculate the sample size required for the study. We set α err prob = 0.05, Power (1-β err prob) = 0.95, and G*Power calculates the total sample size = 146. This confirms that 733 valid construction site worker samples are sufficient for inference. Demographic information of respondents: males account for 72.0%. In terms of occupation category, directors account for 3.0%, engineers account for 3.1%, supervisors account for 3.3%, administrators account for 17.7%, technical staff accounts for 14.9%, and workers account for 58.00% (see Table 1). This study advocates that all construction site workers are affected by epidemic prevention policies during the epidemic. Among the respondents, 49.8% of construction site workers have more than 8 years of experience. A total of 63.6% of the respondents are married. In terms of informed consent, respondents were informed that responses to the questionnaire would be anonymous and unidentifiable. Research data will be stored in the project leader's research room until June 2024 and then deleted. Respondents were free to decide whether to join the study by filling out the questionnaire. Respondents can quit filling in at any time without feeling pressured.

Table 1. Sample basic information.

Gender	Percentage (%)	Seniority	Percentage (%)
Male	72.0%	1–3 years	34.1%
Female	28.0%	4–7 years	16.1%
Age		8–11 years	13.5%
20–29 years old	25.0%	12–15 years	5.7%
30–39 years old	43.0%	16 years or more	30.6%
40–49 years old	9.1%		
50 years old or older	22.9%		
Occupation		**Marriage**	
Director	3.0%	Unmarried	31.2%
Engineer	3.1%	Married	63.6%
Supervisor	3.3%	Other	5.2%
Administrator	17.7%		
Technical staff	14.9%		
Worker	58.00%		

3.2. Measures

The measurement items in this study all refer to scales with good reliability and validity in the past. In this study, Cronbach's α, composite reliability, average variance extraction, and heterotrait–monotrait (HTMT) were analyzed. Social support was modified with reference to the scale of Sarason et al. [111]. Example items are as follows: I can share my most private worries and fears with colleagues. There is someone I can turn to for advice about handling problems with my job. When I need suggestions on how to deal with a personal problem, I know someone I can turn to. If a family crisis arose, it would be difficult to find someone who could give me good advice about how to handle it. Job burnout refers to the scale of Maslach et al. [14]. Example items are as follows: I feel burned out from my work. I have become more callous toward people since I took this job. COVID-19 fear refers to the scale developed by Ahorsu et al. [76]. Example items are as follows: I am most afraid of coronavirus-19. It makes me uncomfortable to think about coronavirus-19. I worry a lot about coronavirus-19. Coronavirus-19 is almost always terminal.

Equipment insufficiency refers to the Cohen and Rodgers article [112] and includes the following items: There are not enough personal face masks at the construction site. I cannot assign eye protection to every job. Gloves are not easy to obtain. Personal epidemic prevention equipment is a scarce resource. Job insecurity refers to the scale developed by Vander Elst and others [113] and includes the following items: Chances are, I will soon lose my job. I am sure I can keep my job (R). I feel insecure about the future of my job. I think I might lose my job in the near future. Insomnia refers to the Pittsburgh Sleep Quality Index [114] and includes the following items: I could not fall asleep within 30 min. I would wake up suddenly in the middle of the night or early in the morning. I would have nightmares while sleeping.

Based on the definition of policy formalism and questionnaires used in the past, the following items were included: I believe that the actual implementation of epidemic prevention regulations and construction sites will not be exactly the same. I think it is sometimes difficult to implement anti-epidemic protocols on construction sites. I think many anti-epidemic plans are not so easy to implement. It is believed that there are differences between anti-epidemic regulations and current construction site practices. The Cronbach's α value of each factor in this study ranges from 0.92 to 0.96 (Table 2), which is higher than the minimum reliability standard of 0.60 set by Nunnally [115].

Table 2. Item loading and reliability.

Variables	Items	Lambda	Z Values	Composite Reliability	Cronbach's Alpha
Policy formalism	Policy formalism 1	0.90	-	0.95	0.95
	Policy formalism 2	0.92	109.8		
	Policy formalism 3	0.93	108.2		
	Policy formalism 4	0.90	104.9		
Social support	Social support 1	0.80	-	0.94	0.94
	Social support 2	0.91	45		
	Social support 3	0.88	45.5		
	Social support 4	0.91	45.5		
	Social support 5	0.81	43.4		
Job burnout	Job burnout 1	0.89	-	0.96	0.96
	Job burnout 2	0.85	180.4		
	Job burnout 3	0.94	191.1		
	Job burnout 4	0.87	191.7		
	Job burnout 5	0.90	188.8		
	Job burnout 6	0.93	193.7		
Job insecurity	Job insecurity 1	0.83	-	0.92	0.92
	Job insecurity 2	0.90	173.3		
	Job insecurity 3	0.86	173		
	Job insecurity 4	0.87	180.7		
COVID-19 fear	COVID-19 fear 1	0.77	-	0.92	0.92
	COVID-19 fear 2	0.82	149.1		
	COVID-19 fear 3	0.76	147.8		
	COVID-19 fear 4	0.88	158.2		
	COVID-19 fear 5	0.88	162.1		
	COVID-19 fear 6	0.76	154.1		
Equipment insufficiency	Equipment insufficiency 1	0.72	-	0.82	0.93
	Equipment insufficiency 2	0.83	155.6		
	Equipment insufficiency 3	0.67	143.4		
	Equipment insufficiency 4	0.68	146.3		
Insomnia	Insomnia 1	0.80	-	0.93	0.93
	Insomnia 2	0.86	190.3		
	Insomnia 3	0.88	196.9		
	Insomnia 4	0.81	178.6		
	Insomnia 5	0.89	189.3		

Note: The first item of each variable is set to 1, so there is no Z value.

3.3. Detection of Common Method Variance (CMV)

Common method variance (CMV) is considered the variance caused by measurement error; the internal consistency bias caused by CMV should be controlled [116,117]. The self-administered questionnaire used in this study is prone to CMV problems. This study used an anonymous method to fill in the questionnaire and used a mix of 5- and 7-point Likert scales [116]. The questionnaire was designed according to standard operating procedures, and the items were designed to be as simple and easy to understand as possible. Items that were difficult to understand, easy to misunderstand, and difficult to answer were deleted.

Harman's one-factor test is considered a tool for subsequent testing [118]. In the exploratory factor analysis without rotation, the explained variance of the first factor extracted was 46.79%. The explained variance of the first factor does not exceed 50%, indicating that the CMV problem in this study is not serious.

3.4. Validity and Reliability Analysis

Confirmatory factor analysis (CFA) is used to measure the relationship between latent factors and observed variables. Structural equation modeling (SEM) can simultaneously estimate complex relationships among latent factors [119]. This study used confirmatory factor analysis with SEM software lavaan version 0.6-17 to test the reliability and validity of constructs [119]. Among overall model fit measures, the SRMR of the conceptual model was 0.081. Although the SRMR is higher than the critical value of 0.05, it is still within the acceptable range. GFI = 0.98 is higher than 0.90. The NNFI = 0.97, NFI = 0.97, CFI = 0.97, IFI = 0.97, and RFI = 0.97, all higher than the specified 0.90. These indicators confirm that the hypothetical model of this study is good. In the model parsimonious fit measures, PNFI = 0.89, and PGFI = 0.81; both are higher than the critical value of 0.50. This confirms that the conceptual model and data of this study fit.

The factor loading λ values of all the constructs in this study ranged from 0.67 to 0.94, which is higher than the 0.5 recommended by Hair, Anderson, Tatham, and Black [120]. This means that the individual items in this study have good reliability and validity. The item loading t-values of all factors have reached statistically significant levels, which also confirms the construct validity and convergent validity of this research construct. The composite reliability (CR) of the latent constructs can measure the consistency of the items in the construct. Previous scholars have proposed that the CR value must be greater than 0.7 [120]. The CR values of the latent constructs in this study ranged from 0.92–0.96. This indicates that the latent constructs in this study have good internal consistency.

The average variance extraction (AVE) is the percentage of latent constructs that can be measured by observed items. AVE can test the reliability, discriminant validity, and convergent validity of research variables. The AVE values of latent constructs in this study range from 0.66 to 0.84, all greater than 0.5. It means that the latent constructs in this study have good discriminant and convergent validity.

The square root of the average variance extracted (AVE) must be higher than the correlation coefficient between constructs and is considered to have discriminant validity [121]. The lower left corner of Table 3 is the matrix of correlation coefficients between the constructs, and the diagonal is the square root of the constructs' AVEs. The square roots of the constructs' AVEs in this study ranged from 0.53 to 0.83, which are all higher than the correlation coefficients between constructs. The upper right half of the table is the heterotrait–monotrait (HTMT) ratio of correlations. The HTMT values of this study are lower than 0.90, which also confirms the discriminant validity between constructs [122]. AVEs are larger than MSVs and ASVs, which again confirms the discriminant validity of this study [115].

Table 3. Square root of AVE and inter-correlations.

	1	2	3	4	5	6	7	ASV	MSV	AVE
Policy Formalism (1)	(0.91)	0.12	0.46	0.43	0.39	0.78	0.41	0.20	0.54	0.83
COVID-19 Fear (2)	0.44	(0.87)	0.08	0.06	0.10	0.13	0.10	0.39	0.65	0.75
Social Support (3)	0.16	0.08	(0.82)	0.82	0.88	0.62	0.79	0.01	0.02	0.67
Job Burnout (4)	0.42	0.78	−0.02	(0.90)	0.69	0.64	0.82	0.36	0.60	0.81
Job Insecurity (5)	0.37	0.81	0.11	0.65	(0.86)	0.53	0.67	0.31	0.65	0.75
Equipment Insufficiency (6)	0.73	0.58	0.15	0.61	0.49	(0.73)	0.60	0.30	0.53	0.53
Insomnia (7)	0.38	0.74	−0.08	0.77	0.62	0.55	(0.85)	0.33	0.60	0.73

Note: The figures in parentheses indicate the square roots of AVEs of the study constructs. The lower left table on the diagonal is the Pearson correlation coefficient, and the upper right table is the heterotrait–monotrait (HTMT) ratio of correlations. MSV = maximum share variance, ASV = average share variance.

The correlation matrix allows for observing preliminary relationships between constructs. Policy formalism is negatively related to job burnout and insomnia, and their coefficients are 0.42 and 0.38. COVID-19 fear negatively affects job burnout and job insecurity; the coefficients are 0.78 and 0.81. This indicates that when the fear of COVID-19 among construction site workers is high, their job burnout and job insecurity will also increase. Social support negatively affects job burnout (−0.02). This means that high social support for workers can reduce their job burnout. The correlation coefficients of job burnout, job insecurity, and insomnia are 0.77 and 0.62. This represents the high degree of job burnout and job insecurity of workers, which will cause insomnia.

This study first defines the research problem and then conducts literature review and formulates research hypotheses (see Figure 2). This study collects raw data, performs statistical analysis, and finally draws conclusions and suggestions.

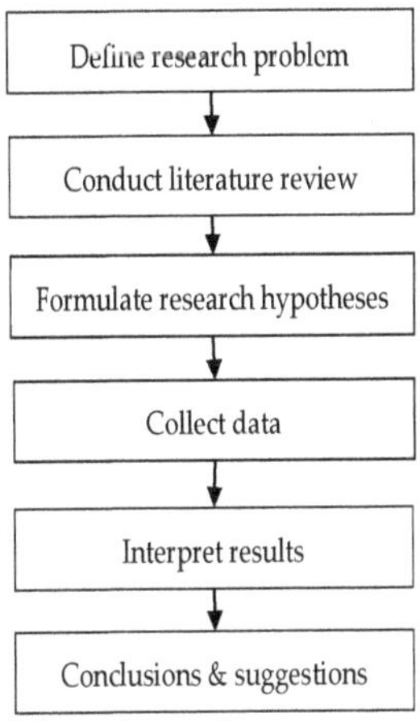

Figure 2. Research Process Flowchart.

4. Results

This study uses lavaan (latent variable analysis) in the R statistical programming language for path coefficient analysis and hypothesis testing [119]. The lavaan package version 0.6-17 is a package for structural equation modeling (SEM) in the statistical software R. It can be easily integrated with other statistical packages in R. It can be seen from Table 4 that equipment insufficiency positively affects COVID-19 fear and path coefficient = 0.63, which supports the argument of Hypothesis 1. Insufficient personal protective equipment can easily put workers on construction sites in fear of infection. Such findings are similar to previous research results conducted by Zhan et al. [8].

Table 4. Path coefficients.

	Causal Path			Path Coefficient	Standard Error	Z Value	*p* Value
H1	Equipment Insufficiency	->	COVID-19 Fear	0.63 ***	0.00	113.40	<0.001
H2	Social Support	->	Job Burnout	−0.17 ***	0.01	−33.80	<0.001
H3	COVID-19 Fear	->	Job Burnout	0.80 ***	0.01	118.90	<0.001
H4	Policy Formalism	->	Job Burnout	0.16 ***	0.01	29.00	<0.001
H5	COVID-19 Fear	->	Job Insecurity	0.84 ***	0.01	148.50	<0.001
H6	Job Burnout	->	Insomnia	0.77 ***	0.01	82.00	<0.001
H7	Job Insecurity	->	Insomnia	0.13 ***	0.01	14.30	<0.001

Note: *** represent statistical significance at $p < 0.001$.

Social support negatively affects job burnout, with a causal coefficient of −0.17, supporting Hypothesis 2. As observed from the JD-R model, social support is regarded as a resource that can reduce job burnout. During the epidemic, many virus tests, process changes, and work requirements can easily cause job burnout among construction site workers. This empirical finding is similar to the research results of Psychiatrist and Bearman [73,74]. Social support can reduce job burnout among construction site workers.

COVID-19 fear positively affects job burnout, with a causal coefficient of 0.80, supporting Hypothesis 3. Fear of COVID-19 infection is considered dysfunctional fear during the epidemic [76]. There were no existing drugs to treat COVID-19 in the early stages of its spread. Increased work demands, infections, and deaths of relatives and friends can easily cause job burnout among construction site workers.

This study found that policy formalism positively affects job burnout, with a causal coefficient of 0.16, which supports Hypothesis 4. When workers experience disparities in epidemic prevention policies and implementation at work sites, they are exposed to long-term work fear and stress. Countries with high formalism are full of pathological behavior, such as lack of authorization, over-emphasis of control, red tape, indifference, and fear of innovation. These irrational management characteristics can easily lead to gaps between epidemic prevention policies and implementation [87]. Construction site workers suffer from job burnout due to inaccurate epidemic prevention.

COVID-19 fear positively affects job insecurity, with a causal coefficient of 0.84, which supports Hypothesis 5. Construction site workers' fear of COVID-19 makes them also worried about the continuation of work due to the epidemic. Construction site workers' fear of COVID-19 also makes them worried about the continuation of work due to the epidemic. The COVID-19 epidemic has caused disruptions in social, economic, and personal lives [95]. COVID-19 fear not only causes physical signs of fear among construction site workers but also makes them feel unsafe at work.

This study confirms that job burnout positively affects insomnia, with a causal coefficient of 0.77, which supports Hypothesis 6. Continuous emotional exhaustion among construction site workers can easily lead to sleep disruption and insufficient sleep. Such empirical findings are similar to the research results obtained by Lee [103]. Changes in shift systems, increased worksite safety inspections, and social distancing policies have all caused worksite workers to experience burnout. Long-term stress and work burnout can cause insomnia problems.

Job insecurity positively affects insomnia, with a causal coefficient of 0.13, which supports Hypothesis 7. Past research has found that job insecurity contributes to insomnia. Social isolation, economic disruption, and job instability caused by the epidemic can easily cause insomnia among construction site workers.

5. Discussion

This study used path coefficient analysis and structural equation modeling (SEM) to verify all research hypotheses. First, this study confirms that insufficient protective equipment will increase COVID-19 fear among construction site workers. The problem of insufficient personal protective equipment has arisen during the COVID-19, severe acute respiratory syndrome (SARS), and Middle East respiratory syndrome (MERS) epidemics. These diseases all have the characteristics of sudden appearance, rapid spread, and lack of vaccines and drugs. These characteristics can easily cause COVID-19 fear among construction site workers when they lack adequate personal protective equipment [55,57]. Construction sites are inherently workplaces where dangerous accidents occur. Workers' fear due to insufficient personal protection is more likely to cause industrial safety incidents.

This study confirms that social support can reduce job burnout among construction site workers. From the perspective of the conservation of resources theory, social support can help construction site workers cope with work stress and burnout [25]. The job demand–resources model regards job resources as a kind of job skills and resources, which can reduce job burnout caused by job demand [29]. Construction site workers during the epidemic face many work demands and new work processes, and they need more social support to reduce work burnout. According to the JD-R model, social support can alleviate job burnout caused by excessive job demand for construction site workers during the epidemic.

COVID-19 fear among construction site workers has caused physical and psychological burdens on construction site workers [77]. COVID-19 fear among construction site workers further contributes to burnout [78]. Insufficient personal protective equipment at construction sites creates greater fear and burnout among construction site workers [82]. The COVID-19 fear scale is a measurement questionnaire developed in recent years due to the COVID-19 epidemic. This study confirms that COVID-19 fear among construction site workers causes job burnout.

Natural disasters and epidemics require government agencies to make quick decisions and effectively implement policy plans. However, institutions in developing countries will make people feel that policies and implementation are inconsistent. Agencies in developing countries often view compliance with work processes as the primary goal, rather than using work processes as a tool to achieve policy goals [84]. Pathological behaviors with high formalism include: over-emphasis on process, red tape, lack of concern for public interests, and fear of innovation [90]. Construction site workers feel that the failure to implement epidemic prevention policies will increase their emotional exhaustion. In the past, research on policy formalism was all qualitative discussion and analysis [45]. This study not only explores the formalism of the epidemic policy through quantitative research but also finds that the formalism of the policy will increase the job burnout of construction site workers.

The high infection and mortality rates of COVID-19 can create a sense of fear among construction site workers [76]. The epidemic has caused factory closures, economic stagnation, and remote work, all of which have increased job insecurity among construction site workers [94]. Reductions in work and financial pressure can create perceptions of job insecurity among workers on the job site. Job insecurity is the pressure that construction site workers have felt due to the epidemic in recent years.

Long-term emotional and physical exhaustion among construction site workers can easily lead to insomnia [103]. Shift changes and increased epidemic prevention work during the epidemic can easily cause burnout and insomnia among construction site workers. This study confirms that job insecurity of construction site workers affects their insomnia. The self-regulatory resource theory suggests that sleep allows individuals to restore energy. Insomnia can lead to more accidents and injuries among construction site workers [109]. It can be seen that the job insecurity caused by the epidemic has a negative impact on the safety of construction site workers.

6. Conclusions

6.1. Theoretical Implications

The greatest contribution of this study is to introduce policy formalism into the physical and psychological exploration of construction workers during the epidemic. Construction site workers feel that failure to implement epidemic prevention policies will increase their job burnout. This article uses empirical data to confirm the negative impact of formalism on construction site workers.

Secondly, this study applies and supplements the conservation of resources theory and job demand–resources model. This study confirms that policy formalism is a negative antecedent of job burnout. Construction site workers' formalistic perceptions of public policies will increase their job burnout. The physical and psychological cognition of construction site workers during the epidemic has rarely been studied. It is even rarer to incorporate the macro-psychological perceptions of public policy into research on construction site workers.

6.2. Practical Implications

If government agencies want to avoid construction site workers' awareness of the gaps in epidemic prevention policies and implementation, they should create a fair and transparent performance evaluation system that links civil servants' performance to rewards. This can motivate civil servants to engage in job performance rather than just respond to external pressures. Secondly, government agencies should actively establish a positive and pragmatic organizational culture. Let civil servants truly care about the interests of the people while shortening the gap between policy and implementation. Some studies in the past have confirmed that formalism has a negative impact on both civil servants and the public [47–49].

After the COVID-19 epidemic, government agencies can foresee and plan the masks, gloves, protective clothing, and inspection equipment needed during future epidemic disasters. Obtaining trustworthy epidemic information, using social media cautiously, and seeking professional psychological support can all reduce the COVID-19 fear of construction site workers. During the epidemic, construction site workers need to stay in touch with their organizations, families, and friends and also seek support from medical and psychological professionals. Worksite workers also need to adapt to shift changes and uncertainty about working hours during the pandemic. Construction site workers should try to maintain a regular schedule, limit the use of electronic products, and continue healthy exercise. These activities can reduce the insomnia problem of construction site workers during the epidemic.

7. Further Study

Most of the past research on policy formalism was qualitative, resulting in insufficient citations of causal relationships. This study did not adopt a longitudinal approach, which requires a lot of time and cost. It is recommended that future researchers perform longitudinal studies to discover causal relationships in depth. This study has collected 733 samples for analysis, and it is recommended that future researchers collect more samples for exploration. The self-reported questionnaire used in this study may have problems with recall bias, social desirability, question understanding, and selective responses. We look forward to researchers adopting more diverse research designs and methods in the future. The administrative agencies of many developing countries have obvious policy formalism attributes. This quantitative research result on policy formalism should have many implications for these countries. However, this study still has problems of cultural differences and bias. Extrapolating the empirical research in Taiwan to the world may still have problems with external validity. We look forward to more researchers from different regions around the world collecting different samples to explore in the future.

Author Contributions: Conceptualization, T.-L.W., T.-F.C. and H.-T.L.; methodology, H.-T.L.; software, H.-T.L.; validation, T.-L.W., T.-F.C. and H.-T.L.; formal analysis, T.-L.W. and H.-T.L.; investigation, T.-L.W.; resources, T.-L.W., T.-F.C. and H.-T.L.; data curation, T.-L.W., T.-F.C. and H.-T.L.; writing—original draft preparation, T.-L.W., T.-F.C. and H.-T.L.; writing—review and editing, T.-L.W. and H.-T.L.; visualization, H.-T.L.; supervision, T.-L.W., T.-F.C. and H.-T.L. All authors have read and agreed to the published version of the manuscript.

Funding: This research received no external funding.

Data Availability Statement: The data presented in this study are available on request from the corresponding author. The data are not publicly available due to privacy.

Conflicts of Interest: The authors declare no conflicts of interest.

References

1. Liu, X.; Zhang, M.; Cheng, Q.; Zhang, Y.; Ye, G.; Huang, X.; Kang, M. Dengue fever transmission between a construction site and its surrounding communities in China. *Parasites Vectors* **2021**, *14*, 22. [CrossRef] [PubMed]
2. Koh, D. Migrant workers and COVID-19. *Occup. Environ. Med.* **2020**, *77*, 634–636. [CrossRef] [PubMed]
3. Penaloza, G.A.; Saurin, T.A.; Formoso, C.T. Monitoring complexity and resilience in construction projects: The contribution of safety performance measurement systems. *Appl. Ergon.* **2020**, *82*, 102978. [CrossRef]
4. Selleck, R.; Hassall, M.; Cattani, M. Determining the Reliability of Critical Controls in Construction Projects. *Safety* **2022**, *8*, 64.
5. Doi, Y. An epidemiologic review of occupational sleep research among Japanese workers. *Ind. Health* **2005**, *43*, 3–10. [CrossRef] [PubMed]
6. Sateia, M.J.; Doghramji, K.; Hauri, P.J.; Morin, C.M. Evaluation of chronic insomnia. An American academy of sleep medicine review. *Sleep* **2000**, *15*, 243–308.
7. Alvaro, P.K.; Roberts, R.M.; Harris, J.K. A Systematic Review Assessing Bidirectionality between Sleep Disturbances, Anxiety, and Depression. *Sleep* **2013**, *36*, 1059–1068. [CrossRef]
8. Zhang, H. China's employment stabilization policies in response to the impact of the COVID-19 pandemic. *Int. J. Sociol. Soc. Policy* **2022**, *42*, 201–209. [CrossRef]
9. Gupta, R.; Grover, S.; Basu, A.; Krishnan, V.; Tripathi, A.; Subramanyam, A.; Nischal, A.; Hussain, A.; Mehra, A.; Ambekar, A.; et al. Changes in sleep pattern and sleep quality during COVID-19 lockdown. *Indian J. Psychiatry* **2020**, *62*, 370.
10. Voitsidis, P.; Gliatas, I.; Bairachtari, V.; Papadopoulou, K.; Papageorgiou, G.; Parlapani, E.; Syngelakis, M.; Holeva, V.; Diakogiannis, I. Insomnia during the COVID-19 pandemic in a Greek population. *Psychiatry Res.* **2020**, *289*, 113076. [CrossRef]

11. Fernandez-Mendoza, J.; Vgontzas, A.N. Insomnia and its impact on physical and mental health. *Curr. Psychiatry Rep.* **2013**, *15*, 418. [PubMed]
12. Bonnet, M.H.; Arand, D.L. Hyperarousal and insomnia: State of the science. *Sleep Med. Rev.* **2010**, *14*, 9–15. [CrossRef] [PubMed]
13. Drake, C.L.; Friedman, N.P.; Wright, K.P., Jr.; Roth, T. Sleep reactivity and insomnia: Genetic and environmental influences. *Sleep* **2011**, *34*, 1179–1188. [PubMed]
14. Maslach, C.; Jackson, S.E. The measurement of experienced burnout. *J. Organ. Behav.* **1981**, *2*, 99–113. [CrossRef]
15. Chiara, P.; Luca, C.; Annalisa, P.; Chiara, R. Emotional exhaustion among healthcare professionals: The effects of role ambiguity, work engagement and professional commitment. *Acta Bio Med. Atenei Parm.* **2019**, *90*, 60.
16. Morgantini, L.A.; Naha, U.; Wang, H.; Francavilla, S.; Acar, Ö.; Flores, J.M.; Crivellaro, S.; Moreira, D.; Abern, M.; Eklund, M. Factors contributing to healthcare professional burnout during the COVID-19 pandemic: A rapid turnaround global survey. *PLoS ONE* **2020**, *15*, e0238217. [CrossRef]
17. Rotenstein, L.S.; Torre, M.; Ramos, M.A.; Rosales, R.C.; Guille, C.; Sen, S.; Mata, D.A. Prevalence of burnout among physicians: A systematic review. *JAMA* **2018**, *320*, 1131–1150. [CrossRef]
18. Bakker, A.B.; Demerouti, E.; Sanz-Vergel, A.I. Burnout and work engagement: The JD–R approach. *Ann. Rev. Organ. Psychol. Organ. Behav.* **2014**, *1*, 389–411. [CrossRef]
19. Roslan, N.; Yusoff, M.; Razak, A.; Morgan, K. Burnout prevalence and its associated factors among Malaysian healthcare workers during COVID-19 pandemic: An embedded mixed-method study. *Healthcare* **2021**, *9*, 90. [CrossRef]
20. Zhang, M. Research on correlation of psychological counselor's job burnout with self-efficacy and social support and professional identity. *Hunan Agric. Univ.* **2016**, *8*, 1–55.
21. Zhao, Y.G.; Li, X.X.; Cui, Y. Analysis of the impact of social support on job burnout among doctors and nurses: Based on mediation effect of self-efficacy. *Chin. Hosp. Manag.* **2019**, *39*, 48–50.
22. Kim, Y. Nurses' experiences of care for patients with Middle East respiratory syndrome-coronavirus in South Korea. *Am. J. Infect. Control* **2018**, *46*, 781–787. [CrossRef] [PubMed]
23. Chirico, F.; Ferrari, G.; Nucera, G.; Szarpak, L.; Crescenzo, P.; Ilesanmi, O. Prevalence of anxiety, depression, burnout syndrome, and mental health disorders among healthcare workers during the COVID-19 pandemic: A rapid umbrella review of systematic reviews. *J. Health Soc. Sci.* **2021**, *6*, 209–220.
24. Hu, Q.; Schaufeli, W.B.; Taris, T.W. The job demands–resources model: An analysis of additive and joint effects of demands and resources. *J. Vocat. Behav.* **2011**, *79*, 181–190. [CrossRef]
25. Leitão, J.; Pereira, D.; Gonçalves, Â. Quality of work life and contribution to productivity: Assessing the moderator effects of burnout syndrome. *Int. J. Environ. Res. Public Health* **2021**, *18*, 2425. [CrossRef] [PubMed]
26. Ojo, A.O.; Fawehinmi, O.; Yusliza, M.Y. Examining the predictors of resilience and work engagement during the COVID-19 pandemic. *Sustainability* **2021**, *13*, 2902. [CrossRef]
27. Hobfoll, S.E. Conservation of resources: A new attempt at conceptualizing stress. *Am. Psychol.* **1989**, *44*, 513–524. [CrossRef]
28. Demerouti, E.; Bakker, A.; Nachreiner, F.; Ebbinghaus, M. From mental strain to burnout. *Eur. J. Work. Organ. Psychol.* **2002**, *11*, 423–441. [CrossRef]
29. Schaufeli, W. The job demands-resources model: A 'how to'guide to increase work engagement and prevent burnout. *Organ. Dyn.* **2017**, *46*, 120–132. [CrossRef]
30. Woon, L.S.-C.; Mansor, N.S.; Mohamad, M.A.; Teoh, S.H.; Leong Bin Abdullah, M.F.I. Quality of life and its predictive factors among healthcare workers after the end of a movement lockdown: The salient roles of COVID-19 stressors, psychological experience, and social support. *Front. Psychol.* **2021**, *12*, 1164. [CrossRef]
31. Gleason, F.; Malone, E.; Wood, L.; Baker, S.J.; Hollis, R.H.; Richman, J.S.; Chu, D.I.; Lindeman, B. The job demands-resources model as a framework to identify factors associated with burnout in surgical residents. *J. Surg. Res.* **2020**, *247*, 121–127. [CrossRef] [PubMed]
32. Arafa, A.; Mohammed, Z.; Mahmoud, O.; Elshazley, M.; Ewis, A. Depressed, anxious, and stressed: What have healthcare workers on the frontlines in Egypt and Saudi Arabia experienced during the COVID-19 pandemic? *J. Affect. Disord.* **2021**, *278*, 365–371. [CrossRef] [PubMed]
33. Gallop, R.M.; Lancee, W.J.; Taerk, G.; Coates, R.A.; Fanning, M. Fear of contagion and AIDS: Nurses' perception of risk. *AIDS Care* **1992**, *4*, 103–109. [PubMed]
34. World Health Organization. Clinical Management of Human Infection with Avian Influenza A (H5N1) Virus. Available online: https://cdn.who.int/media/docs/default-source/wpro---documents/emergency/surveillance/avian-influenza/ai_20231110.pdf?sfvrsn=5f006f99_122 (accessed on 17 December 2023).
35. Reidy, P.; Fletcher, T.; Shieber, C.; Shallcross, J.; Towler, H.; Ping, M.; Kenworthy, L.; Silman, N.; Aarons, E. Personal protective equipment solution for UK military medical personnel working in an Ebola virus disease treatment unit in Sierra Leone. *J. Hosp. Infect.* **2017**, *96*, 42–48. [CrossRef]
36. Blanco-Donoso, L.M.; Moreno-Jiménez, J.; Amutio, A.; Gallego-Alberto, L.; Moreno-Jiménez, B.; Garrosa, E. Stressors, Job Resources, Fear of Contagion, and Secondary Traumatic Stress Among Nursing Home Workers in Face of the COVID-19: The Case of Spain. *J. Appl. Gerontol.* **2021**, *40*, 244–256. [PubMed]
37. Pincha Baduge, M.S.; Moss, C.; Morphet, J. Emergency nurses' perceptions of emergency department preparedness for an Ebola outbreak: A qualitative descriptive study. *Australas. Emerg. Care* **2017**, *20*, 69–74. [CrossRef]

38. Greenhalgh, L.; Rosenblatt, Z. Job insecurity: Toward conceptual clarity. *Acad. Manag. Rev.* **1984**, *9*, 438–448.
39. Llosa, J.A.; Menéndez-Espina, S.; Agulló-Tomás, E.; Rodríguez-Suárez, J. Job insecurity and mental health: A meta-analytical review of the consequences of precarious work in clinical disorders. *An. Psicol.* **2018**, *34*, 211–223.
40. Zhou, Z.; Goh, Y.M.; Li, Q. Overview and analysis of safety management studies in the construction industry. *Saf. Sci.* **2015**, *72*, 337–350.
41. Wang, D.; Wang, X.; Xia, N. How safety-related stress affects workers' safety behavior: The moderating role of psychological capital. *Saf. Sci.* **2018**, *103*, 247–259. [CrossRef]
42. Sverke, M.; Hellgren, J.; Näswall, K. No security: A meta-analysis and review of job insecurity and its consequences. *J. Occup. Health Psychol.* **2002**, *73*, 242–264.
43. Giunchi, M.; Emanuel, F.; Chambel Maria, J.; Ghislieri, C. Job insecurity, workload and job exhaustion in temporary agency workers (TAWs): Gender differences. *Career Dev. Int.* **2016**, *21*, 3–18. [CrossRef]
44. Barnes, C.M.; Hollenbeck, J.R. Sleep Deprivation and Decision-Making Teams: Burning the Midnight Oil or Playing with Fire? *Acad. Manag. Rev.* **2009**, *34*, 56–66.
45. Riggs, F.W. An ecological approach: The 'Sala' model. In *Papers in Comparative Administration*; Heady, F., Stokes, S., Eds.; University of Michigan Press: Ann Arbor, MI, USA, 1962; pp. 19–36.
46. Amaechi, O.; Ayeni, E.; Madu, C.P. The Prismatic Sala Model and its Relevance to Nigeria's Public Administrative System. *CCU J. Humanit.* **2019**, *1*, 113–122.
47. Liu, H.T. Analysis of formalism's moderating effect on the relationships between role stressors and work anxiety—Viewpoints from oriental public administration. *Int. J. Inf. Manag. Sci.* **2015**, *26*, 85–101.
48. Liu, H.-T. The Influence of Public Servant's Perceived Formalism and Organizational Environmental Strategy on Green Behavior in Workplace. *Sustainability* **2021**, *13*, 11020.
49. Huang, F.F.-Y.; Liu, H.-T. Path Model Analysis of the Effects of Perceived Formalism, and Fear of COVID-19 on Police Officers' PTSD and Insomnia. *Behav. Sci.* **2023**, *13*, 867. [CrossRef]
50. Braibanti, R. Transnational inducement of administrative reform: A survey of scope and critique of issues. In *Approaches to Development*; Montgomery, J.D., Sffin, W.J., Eds.; McGraw-Hill: New York, NY, USA, 1966; pp. 133–183.
51. Liu, Q.; Luo, D.; Haase, J.E.; Guo, Q.; Wang, X.Q.; Liu, S.; Xia, L.; Liu, Z.; Yang, J.; Yang, B.X. The experiences of health-care providers during the COVID-19 crisis in China: A qualitative study. *Lancet Glob. Health* **2020**, *8*, e790–e798. [CrossRef]
52. Jin, Y.-H.; Huang, Q.; Wang, Y.-Y.; Zeng, X.-T.; Luo, L.-S.; Pan, Z.-Y.; Yuan, Y.-F.; Chen, Z.-M.; Cheng, Z.-S.; Huang, X.; et al. Perceived infection transmission routes, infection control practices, psychosocial changes, and management of COVID-19 infected healthcare workers in a tertiary acute care hospital in Wuhan: A cross-sectional survey. *Mil. Med. Res.* **2020**, *7*, 24. [CrossRef]
53. Wei, X.-S.; Wang, X.-R.; Zhang, J.-C.; Yang, W.-B.; Ma, W.-L.; Yang, B.-H.; Jiang, N.-C.; Gao, Z.-C.; Shi, H.-Z.; Zhou, Q. A cluster of health care workers with COVID-19 pneumonia caused by SARS-CoV-2. *J. Microbiol. Immunol. Infect.* **2020**, *54*, 54–60. [CrossRef]
54. Lázaro-Pérez, C.; Martínez-López, J.A.; Gómez-Galán, J.; López-Meneses, E. Anxiety about the risk of death of their patients in health professionals in Spain: Analysis at the peak of the COVID-19 pandemic. *Int. J. Environ. Res. Public Health* **2020**, *17*, 5938. [PubMed]
55. Zhan, M.; Anders, R.L.; Lin, B.; Zhang, M.; Chen, X. Lesson learned from China regarding use of personal protective equipment. *Am. J. Infect. Control.* **2020**, *48*, 1462–1465. [CrossRef] [PubMed]
56. Suárez-García, I.; de Aramayona, M.; López, M.J.; Sáez Vicente, A.; Lobo Abascal, P. SARS-CoV-2 infection among healthcare workers in a hospital in Madrid, Spain. *J. Hosp. Infect.* **2020**, *106*, 357–363. [CrossRef]
57. Martínez-Ales, G.; Domingo-Relloso, A.; Arribas, J.R.; Quintana-Díaz, M.; Hernan, M. Critical Care Requirements Under Uncontrolled Transmission of SARS-CoV-2. *Am. J. Public Health* **2021**, *111*, 923–926. [CrossRef] [PubMed]
58. Verbeek, J.H.; Rajamaki, B.; Ijaz, S.; Sauni, R.; Toomey, E.; Blackwood, B.; Tikka, C.; Ruotsalainen, J.H.; Kilinc Balci, F.S. Personal protective equipment for preventing highly infectious diseases due to exposure to contaminated body fluids in healthcare staff. *Cochrane Database Syst. Rev.* **2020**, *4*, Cd011621.
59. Okamoto, K.; Rhee, Y.; Schoeny, M.; Lolans, K.; Cheng, J.; Reddy, S.; Weinstein, R.A.; Hayden, M.K.; Popovich, K.J. Impact of doffing errors on healthcare worker self-contamination when caring for patients on contact precautions. *Infect. Control. Hosp. Epidemiol.* **2019**, *40*, 559–565. [CrossRef]
60. Demerouti, E.; Bakker, A.B.; Nachreiner, F.; Schaufeli, W.B. The job demandsresource model of burnout. *J. Appl. Psychol.* **2001**, *86*, 499–512. [CrossRef]
61. Maslach, C.; Jackson, S.E. *The Maslach Burnout Inventory: Research Edition*; Consulting Psychologists Press: Palo Alto, CA, USA, 1981.
62. Sun, H.P.; Liu, J. Research on the influencing factors of professional social worker retention: Based on qualitative comparative analysis method (QCA). *J. Soc. Work* **2017**, *4*, 77–85.
63. Heaney, C.A.; Israel, B.A. Social networks and social support. *Health Behav. Health Educ. Theory Res. Pract.* **2008**, *4*, 189–210.
64. Schulz, R.; Decker, S. Long-term adjustment to physical disability: The role of social support, perceived control, and self-blame. *J. Pers. Soc. Psychol.* **1985**, *48*, 1162.
65. Russell, S.; Subramanian, B.; Russell, P.S.; Nair, M.K.C. Psychopathology, traumatic life events, and coping skills among patients attending a primary-care adolescent clinic. *Indian J. Pediatr.* **2012**, *79*, 52–59.
66. Hobfoll, S. *The Ecology of Stress*; Hemisphere: Washington DC, USA, 1988.

67. Lee, J.-S.; Koeske, G.F.; Sales, E. Social support buffering of acculturative stress: A study of mental health symptoms among Korean international students. *Int. J. Intercult. Relat.* **2004**, *28*, 399–414. [CrossRef]
68. Dir, A.L.; Saldana, L.; Chapman, J.E.; Aalsma, M.C. Burnout and mental health stigma among juvenile probation officers: The moderating effect of participatory atmosphere. *Adm. Policy Ment. Health* **2019**, *46*, 167–174. [CrossRef]
69. Blumenthal, J.A.; Burg, M.M.; Barefoot, J.; Williams, R.B.; Haney, T.; Zimet, G. Social support, type A behavior, and coronary artery disease. *Psychosom. Med.* **1987**, *49*, 331–340.
70. Lee, C.; Ashford, S.J.; Bobko, P. Interative effects of type A behavior and perceived control on worker performance, job satisfaction, and somatic complaints. *Acad. Manag. J.* **1990**, *33*, 870–881. [CrossRef]
71. Feldman, C.B.; Brondolo, E.; Dayan, D.B. Sources of social support and burnout, job satisfaction and productivity. *J. Occup. Health Psychol.* **2002**, *7*, 84–93.
72. Schaufeli, W.B.; Bakker, A.B.; Van Rhenen, W. How changes in job demands and resources predict burnout, work engagement, and sickness absenteeism. *J. Organ. Behav.* **2009**, *30*, 893–917. [CrossRef]
73. . Eftekhar Ardebili, M.; Naserbakht, M.; Bernstein, C.; Alazmani-Noodeh, F.; Hakimi, H.; Ranjbar, H. Healthcare providers experience of working during the COVID-19 pandemic: A qualitative study. *Am. J. Infect. Control* **2020**, *49*, 547–554. [CrossRef]
74. Bearman, G.; Hota, S.S.; Haessler, S.D. Physician burnout and healthcare epidemiology: Dual implications worthy of greater scrutiny. *Infect. Control Hosp. Epidemiol.* **2020**, *41*, 250–251. [CrossRef]
75. Brauchli, R.; Jenny, G.J.; Füllemann, D.; Bauer, G.F. Towards a job demandsresources health model: Empirical testing with generalizable indicators of job demands, job resources, and comprehensive health outcomes. *BioMed Res. Int.* **2015**, *2015*, 959621.
76. Ahorsu, D.K.; Lin, C.Y.; Imani, V.; Saffari, M.; Griffiths, M.D.; Pakpour, A.H. The fear of COVID-19 scale: Development and initial validation. *Int. J. Ment. Health Addict.* **2020**, *20*, 1537–1545. [CrossRef] [PubMed]
77. Sloan, M.M.; Haner, M.; Graham, A.; Cullen, F.T.; Pickett, J.T.; Jonson, C.L. Pandemic emotions: The extent, correlates, and mental health consequences of fear of COVID-19. *Sociol. Spectr.* **2021**, *41*, 369–386. [CrossRef]
78. Dimitriu, M.C.; Pantea-Stoian, A.; Smaranda, A.C.; Nica, A.A.; Carap, A.C.; Constantin, V.D.; Socea, B.; Cirstoveanu, C.; Bacalbasa, N.; Bratu, O.G.; et al. Burnout syndrome in Romanian medical residents in time of the COVID-19 pandemic. *Med. Hypotheses* **2020**, *144*, 109972.
79. Hu, D.; Kong, Y.; Li, W.; Han, Q.; Zhang, X.; Zhu, L.X.; Wan, S.W.; Liu, Z.; Shen, Q.; Yang, J.; et al. Frontline nurses' burnout, anxiety, depression, and fear statuses and their associated factors during the COVID-19 outbreak in Wuhan, China: A large-scale cross-sectional study. *Eclin. Med.* **2020**, *24*, 100424.
80. Yıldırım, M.; Solmaz, F. COVID-19 burnout, COVID-19 stress and resilience: Initial psychometric properties of COVID-19 burnout scale. *Death Stud.* **2022**, *46*, 524–532. [PubMed]
81. Raja, U.; Jahanzeb, S.; Malik, M.A.R.; Baig, M.U.A. Dispositional causes of burnout, satisfaction, and performance through the fear of COVID-19 during times of pandemic. *Appl. Psychol.* **2022**, *72*, 998–1019. [CrossRef]
82. Zare, S.; Kazemi, R.; Izadi, A.; Smith, A. Beyond the outbreak of COVID-19: Factors affecting burnout in nurses in Iran. *Ann. Glob. Health* **2021**, *87*, 51.
83. Brooks, S.K.; Webster, R.; Smith, L.; Woodland, L.; Wessely, S.; Greenberg, N.; Rubin, G.J. The psychological impact of quarantine and how to reduce it: Rapid review of the evidence. *Lancet* **2020**, *395*, 912–920. [CrossRef]
84. Thompson, V.A. Administrative objectives for development administration. *Adm. Sci. Q.* **1964**, *9*, 91–108. [CrossRef]
85. Guyot, J.F. Bureaucratic transformation in Burma. In *Asian Bureaucratic Systems Emergent from the British Imperial Tradition*; Braibanti, R., Ed.; Duke University Press: Durham, NC, USA, 1966; pp. 354–443.
86. Pye, L.W. *Politics, Personality, and Nation Building: Burma's Search for Identity*; Yale University Press: New Haven, CT, USA, 1962.
87. Burns, T.; Stalker, G.M. *The Management of Innovation*; Tavistock: London, UK, 1961.
88. Milne, R. Mechanistic and organic models of public administration in developing countries. *Adm. Sci. Q.* **1970**, *15*, 57.
89. Bennis, W.G. A new role for the behavioral sciences: Effecting organizational change. *Adm. Sci. Q.* **1963**, *8*, 125–165. [CrossRef]
90. Fox, G.; Joiner, C. Perceptions of the Vietnamese public administration system. *Adm. Sci. Q.* **1964**, *8*, 443–481. [CrossRef]
91. Huang, J.Z.; Han, M.F.; Luo, T.D.; Ren, A.K.; Zhou, X.P. Mental health survey of medical staff in a tertiary infectious disease hospital for COVID-19. *Zhonghua Lao Dong Wei Sheng Zhi Ye Bing Za Zhi* **2020**, *38*, 192–195.
92. Rafi, M.; Mamun, M.A.; Hsan, K.; Hossain, M.; Gozal, D. Psychological implications of unemployment among bangladesh civil service job seekers. *Front. Psychiatry* **2019**, *10*, 578.
93. Frone, M.R. What happened to the employed during the Great Recession? A U.S. Population study of net change in employee insecurity, health, and organizational commitment. *J. Vocat. Behav.* **2018**, *107*, 246–260. [CrossRef] [PubMed]
94. Spurk, D.; Straub, C. Flexible employment relationships and careers in times of the COVID-19 pandemic. *J. Vocat. Behav.* **2020**, *119*, 103–435. [CrossRef]
95. Kawohl, W.; Nordt, C. COVID-19, unemployment, and suicide. *Lancet Psychiatry* **2020**, *7*, 389–390. [CrossRef]
96. Dozois, D.J. Anxiety and depression in Canada during the COVID-19 pandemic: A national survey. *Can. Psychol. Psychol. Can.* **2020**, *62*, 136. [CrossRef]
97. Jofre-Bonet, M.; Serra-Sastre, V.; Vandoros, S. The impact of the Great Recession on health-related risk factors, behaviour and outcomes in England. *Soc. Sci. Med.* **2018**, *197*, 213–225. [CrossRef]
98. Herzberg, F. *Motivation to Work*; Routledge: New York, NY, USA, 2017.

99. Gasparro, R.; Scandurra, C.; Maldonato, N.M.; Dolce, P.; Bochicchio, V.; Valletta, A.; Marenzi, G. Perceived Job Insecurity and Depressive Symptoms among Italian Dentists: The Moderating Role of Fear of COVID-19. *Int. J. Environ. Res. Public Health* **2020**, *17*, 5338. [CrossRef]
100. Loft, M.; Cameron, L. The importance of sleep: Relationships between sleep quality and work demands, the prioritization of sleep and pre-sleep arousal in day-time employees. *Work Stress* **2014**, *28*, 289–304.
101. Pappa, S.; Ntellac, V.; Giannakasc, T.; Giannakoulisc, V.; Papoutsic, E.; Katsaounouc, P. Prevalence of depression, anxiety, and insomnia among healthcare workers during the COVID-19 pandemic: A systematic review and meta-analysis. *Brain Behav. Immun.* **2020**, *88*, 901–907. [CrossRef]
102. Abdulah, D.M.; Musa, D.H. Insomnia and stress of physicians during COVID-19 outbreak. *Sleep Med.* **2020**, *2*, 100017. [CrossRef] [PubMed]
103. Lee, S.A.; Mathis, A.A.; Jobe, M.C. Mental health characteristics associated with dysfunctional Coronavirus Anxiety. *Psychol. Med.* **2020**, *51*, 1403–1404. [PubMed]
104. Leiter, M.P.; Maslach, C. Burnout: 35 years of research and practice. *Career Dev. Int.* **2009**, *14*, 204–220.
105. Ekstedt, M.; Söderström, M.; Akerstedt, T.; Nilsson, J.; Søndergaard, H.P.; Aleksander, P. Disturbed sleep and fatigue in occupational burnout. *Scand. J. Work Environ. Health* **2006**, *32*, 121–131.
106. Davy, J.A.; Kinicki, A.J.; Scheck, C.L. A test of job security's direct and mediated effects on withdrawal cognitions. *J. Organ. Behav.* **1997**, *18*, 323–349.
107. Nahrgang, J.D.; Morgeson, F.P.; Hofmann, D.A. Safety at work: A meta-analytic investigation of the link between job demands, job resources, burnout, engagement, and safety outcomes. *J. Appl. Psychol.* **2011**, *96*, 71–94. [CrossRef]
108. Piccoli, B.; Reisel, W.; De Witte, H. Understanding the Relationship Between Job Insecurity and Performance: Hindrance or Challenge Effect? *J. Career Dev.* **2019**, *48*, 089484531983318. [CrossRef]
109. Kao, K.Y.; Spitzmueller, C.; Cigularov, K.; Wu, H. Linking insomnia to workplace injuries: A moderated mediation model of supervisor safety priority and safety behavior. *J. Occup. Health Psychol.* **2016**, *21*, 91–104. [CrossRef]
110. Labor Statistics Inquiry Network of the Ministry of Labor. Available online: https://statfy.mol.gov.tw/statistic_DB.aspx (accessed on 5 January 2024).
111. Sarason, I.G.; Levine, H.M.; Basham, R.B.; Sarason, B.R. Assessing social support: The Social Support Questionnaire. *J. Personal. Soc. Psychol.* **1983**, *44*, 127–139. [CrossRef]
112. Cohen, J.; Rodgers, Y.V.M. Contributing factors to personal protective equipment shortages during the COVID-19 pandemic. *Prev. Med.* **2020**, *141*, 106263. [CrossRef]
113. Vander Elst, T.; De Witte, H.; De Cuyper, N. The Job Insecurity Scale: A psychometric evaluation across five European countries. *Eur. J. Work. Organ. Psychol.* **2014**, *23*, 364–380. [CrossRef]
114. Buysse, D.J.; Reynolds, C.F., 3rd; Monk, T.H.; Berman, S.R.; Kupfer, D.J. The Pittsburgh Sleep Quality Index: A new instrument for psychiatric practice and research. *Psychiatry Res.* **1989**, *28*, 193–213. [PubMed]
115. Nunnally, J.C. *Psychometric Theory*; McGraw-Hill: New York, NY, USA, 1967.
116. Podsakoff, P.M.; MacKenzie, S.B.; Lee, J.Y.; Podsakoff, N.P. Common method biases in behavioral research: A critical review of the literature and recommended remedies. *J. Appl. Psychol.* **2003**, *88*, 879–903.
117. Murray, J.Y.; Kotabe, M.; Zhou, J.N. Strategic alliance-based sourcing and market performance: Evidence from foreign firms operating in China. *J. Int. Bus. Stud.* **2005**, *36*, 187–208.
118. Fuller, C.M.; Simmering, M.J.; Atinc, G.; Atinc, Y.; Babin, B.J. Common methods variance detection in business research. *J. Bus. Res.* **2016**, *69*, 3192–3198. [CrossRef]
119. Jöreskog, K.G.; Söbom, D. *LISREL 8: Structural Equation Modeling with the SIMPLIS Command Language*; Scientific Software International: Chicago, IL, USA, 1993.
120. Hair, J.; Black, W.; Babin, B.; Anderson, R. *Multivariate Data Analysis*; Prentice-Hall, Inc.: Upper Saddle River, NJ, USA, 2010.
121. Hulland, J.S. Use of partial least squares (PLS) in strategic management research: A review of four recent studies. *Strateg. Manag. J.* **1999**, *20*, 195–204. [CrossRef]
122. Henseler, J.; Ringle, C.M.; Sarstedt, M. A new criterion for assessing discriminant validity in variance-based structural equa-tion modeling. *J. Acad. Mark. Sci.* **2015**, *43*, 115–135. [CrossRef]

buildings

MDPI

Article

Roles and Autonomous Motivation of Safety Officers: The Context of Construction Sites

Kritsada Khun-anod [1,*], Tsunemi Watanabe [2] and Satoshi Tsuchiya [3]

[1] Graduate School of Engineering, Kochi University of Technology, Tosayamada, Kami City 782-8502, Japan
[2] Faculty of Society and Design, Bond University, Gold Coast, QLD 4226, Australia; twatanab@bond.edu.au
[3] School of Economics and Management, Kochi University of Technology, 2-22 Eikokuji, Kochi City 780-8515, Japan; tsuchiya.satoshi@kochi-tech.ac.jp
* Correspondence: 256006k@gs.kochi-tech.ac.jp

Abstract: Safety officers have been underlined as key individuals in the implementation of safety programmes at construction sites. However, previous research mentioned that some project managers predominantly focus on other aspects, such as the productivity of construction and the management of time and cost. Such emphases may potentially demotivate safety officers from fully engaging in safety initiatives for construction projects. Moreover, scholars have devoted piecemeal discussions to the motivation of such practitioners. These problems were addressed in the current work through the development of a conceptual paradigm that captures the actual situation between project managers and safety officers. To this end, four constructs were extensively examined: (1) autonomy-oriented support from project managers; (2) the motivation of safety officers; (3) the engagement of safety officers in safety programmes (covering four major categories of safety-related tasks); and (4) safety performance. From August to September 2022, valid data from 195 safety officers working in construction projects were considered, after which the proposed paradigm was analysed via structural equation modelling. The results showed that the autonomous motivation of safety officers was activated by autonomy-oriented support from project managers (β = 0.520, sig. = 0.000). Such motivation significantly affected their safety performance (β = 0.231, sig. = 0.007) and levels of engagement with safety initiatives (β = 0.529, sig. = 0.000). These findings indicate that in the implementation of safety programmes, the autonomous motivation of safety officers serves as the engine, while autonomy-oriented support from project managers functions as the ignition key. Policymakers in construction companies can use the results as a reference for decision-making on initiating safety policy that highlights methods of training project managers in supporting safety officers.

Keywords: autonomous motivation; self-determination theory; safety management; safety performance; structural equation modelling

Citation: Khun-anod, K.; Watanabe, T.; Tsuchiya, S. Roles and Autonomous Motivation of Safety Officers: The Context of Construction Sites. *Buildings* **2024**, *14*, 460. https://doi.org/10.3390/buildings14020460

Academic Editors: Mariusz Szóstak and Marek Sawicki

Received: 28 December 2023
Revised: 30 January 2024
Accepted: 5 February 2024
Published: 7 February 2024

1. Introduction

The implementation of safety programmes in the construction industry has improved to some extent over the last few decades, but it continues to rank the lowest among all sectors in terms of safety performance. For example, the US construction industry revealed that accidents among construction practitioners have gradually increased since 2016. A report released by the Bureau of Labour Statistics indicated that the industry was plagued with a nearly 50% incidence of fatal occupational injuries from 2016 to 2018, earning it a place alongside agriculture, mining, and manufacturing as one of the riskiest workplaces in commerce [1]. These problems occur equally in Asia, with the Chinese construction industry accounting for 35% of work-related incidents in 2015 [2] and the Thai construction industry accounting for the highest number of accidents among 131 industries from 2001 to 2011 [3]. In Thailand as well, the construction industry accounted for the highest rate of injury and illness (26.40%) among the most dangerous sectors in the country in 2016, according to the Social Security Office's historical data on occupational injury and illness [4].

To resolve the abovementioned issues, researchers have proposed essential strategies for helping stakeholders enhance safety performance in construction projects [5–16], such as cultivating sufficient commitment among management, providing safety training, and setting clear safety goals. Many analyses have also been conducted on success and failure factors, that is, what should and should not be done. However, limited research has focused on the individuals responsible for supporting these success factors to ensure the effective management of construction safety. An important challenge in addressing the 'who' and 'how' of construction safety appears to be the disparity between espoused values and the assumptions that characterise the culture of construction organisations [17]. Another problem is that the stakeholders of construction projects sometimes operate without a uniform agreement on site safety responsibilities [18]. Furthermore, although the slogan 'safety first' is commonly heard on construction sites, its practical implementation is often denied the emphasis that it deserves, with many project managers predominantly concentrating on matters such as construction productivity. This predilection leads to the occasional oversight of safety programmes.

The difficulties described above are poised to be resolved by safety officers, who are one of the primary stakeholders in construction initiatives. Their integral role in the execution of safety programmes, which are crucial to project success, was emphasised in [5]. This research suggested that safety officers have the opportunity to influence the various dimensions constituting the quality of safety programmes, including, but not limited to, empowering workers to safely perform construction activities, clearly communicating safety concerns to involved parties, and continuously monitoring the performance metrics of safety initiatives [5]. Nevertheless, the aforementioned propensity to prioritise facets such as construction productivity gives rise to the potential to inadvertently demotivate safety officers from wholeheartedly pursuing safety.

In a study that underscored the essentiality of autonomous motivation among safety officers involved in construction projects [19], the researchers differentiated these professionals based on their motivational profiles. The authors found that highly motivated practitioners consistently manifest a substantially greater degree of organisational commitment than their counterparts [19]. However, research focusing on both the contributions of safety officers and their autonomous motivation has been fragmented. Thus, there is an exigent need for a comprehensive consideration of the contributions of safety officers to construction safety management. Specifically, the issues that remain inadequately examined are as follows:

(1) The contributions of safety officers to safety performance at construction sites;
(2) The motivation of safety officers to ensure safety in these workplaces and its influencing factors;
(3) The interrelationship between items 1 and 2.

With regard to item 1, because safety officers are responsible for carrying out various tasks related to safety, it is necessary to collectively and exhaustively describe such duties, but accurately representing safety performance at construction sites remains challenging. Moreover, enquiring into the strengths and limitations of previous analyses and proposing new methods by which to improve existing safety approaches necessitate the consideration of various related aspects. As for item 2, noteworthy issues include autonomous motivation and support for autonomy from project managers. Additionally, well-educated safety officers are required to work diligently in challenging situations. On these bases, two hypotheses can be put forward: safety officers are likely to have high autonomous motivation, and maintaining such motivation requires autonomy support from project managers. With respect to item 3, if the aforementioned hypotheses are valid and the features highlighted therein are contributory to excellent safety performance, then various stakeholders can enjoy the following benefits: increased confidence among safety officers, opportunities to change and refine contractors' perceptions of safety officers, and more effective development and implementation of safety laws and regulations.

In consideration of the matters discussed above, this research developed a novel paradigm that captures the actual situation between project managers and safety officers, as well as the essential roles of the latter and their autonomous motivation. To this end, four constructs were extensively examined: the autonomy-supportive behaviours of project managers, the motivation of safety officers, their levels of engagement in safety initiatives, and safety performance. The proposed paradigm was used to quantitatively explore the issues of interest to satisfy three objectives: (1) to assess the effects of autonomy-oriented support from project managers on the motivation of safety officers and safety performance in the construction industry; (2) to evaluate the impact of safety officers' motivation on engagement in safety-related tasks and safety performance; and (3) to assess the relationship between safety officers' levels of engagement in safety tasks and safety performance. The findings are expected to empirically improve the understanding of safety officers' autonomous motivation in the safety programmes currently implemented in the construction industry. They can also help company management and project managers formulate strategies for enhancing safety performance by superseding conventional views in favour of a new approach: developing an environment in which safety officers can work with a tremendous sense of value, responsibility, and enjoyment instead of having their activities constrained.

2. Hypothesis Development

2.1. *Self-Determination Theory (SDT)*

SDT is a broad meta-theory maintaining that humans possess three types of motivation: amotivation (lack of motivation), controlled motivation, and autonomous motivation [20]. Individuals with high amotivation lack the drive to perform or engage in activities. People with controlled motivation take part in an activity to receive an external reward, avoid punishment, and protect their egos. Autonomously motivated individuals engage in a particular activity because of their interest in it and their perception of its importance.

2.2. *Basic Psychological Needs*

The theory of basic psychological needs states that human beings have three fundamental psychological requirements [21] to live in society: autonomy, competence, and relatedness. Autonomy concerns perceived feelings of volition, congruence, and integration, which enhance a person's sense of independence. Competence pertains to the desire of individuals to recognise that they can capably engage in a given activity, and relatedness refers to a person's perception of social connection and belonging. Theoretically, fulfilling these needs enhances the autonomous motivation of individuals to participate in a particular activity.

Studies have examined the positive relationship between autonomy-supportive leadership and the autonomous motivation of followers in other contexts, such as child education (e.g., the autonomous motivation to do homework) [22], the sciences (e.g., the participation of students in mathematics homework and the intrinsic motivation of students majoring in agricultural science) [23,24], and the hospitality industry (e.g., the autonomous motivation of employees in international hotel chains in China) [25]. Some researchers have found that support for autonomy, competence, and relatedness is positively associated with introjected regulation (controlled motivation) among elementary school students in Japan [26], while others have reported that amotivation and autonomy-oriented support are negatively related [27]. Assuming that autonomy-oriented support from project managers affects the autonomous motivation of safety officers, we established Hypothesis 1.

Hypothesis 1a: *Autonomy-oriented support positively affects autonomous motivation.*

Hypothesis 1b: *Autonomy-oriented support positively affects controlled motivation.*

Hypothesis 1c: *Autonomy-oriented support negatively affects amotivation.*

Studies have also uncovered that managers with considerable autonomy exhibit excellent information system planning [28] and that support for autonomy from local authorities directly and positively affects the outcomes of a waste separation programme [29]. A similar situation can occur with respect to safety programmes for construction projects, with improved autonomy-oriented support from project managers potentially engendering superior safety performance. On this basis, Hypothesis 2 was formulated.

Hypothesis 2: *Autonomy-oriented support positively affects safety performance.*

Previous research in various domains has confirmed that autonomous motivation significantly affects engagement levels [30–35]. In [19], for example, the researchers probed into the positive impact of autonomous motivation on the engagement levels of safety personnel in construction projects, and in [34], the authors confirmed that construction workers' autonomous motivation significantly affects their levels of engagement in enhancing construction productivity. In the context of chemical factories, the safety motivation of company crews considerably influences their levels of engagement in safety programmes [35], and research on an international airline verified that perceptions of the importance of a safety programme affect levels of participation in and compliance with safety management [30]. In the industrial sector, the autonomous motivation of employees was found to be positively associated with their safety-related behaviours [32]. Finally, researchers have emphasised that the autonomous motivation to participate in health-related activities significantly affects the intentions and behaviours of individuals [31].

From a conventional view, the literature has reflected a positive association between controlled motivation and intention with regard to other issues, such as the intention of managers in medium-sized manufacturing companies to use computers [36] and the prediction of workaholic behaviours [37]. Conversely, amotivation negatively affects levels of engagement [27,37]. Correspondingly, under the assumption that the motivation of safety officers affects their levels of engagement in cultivating the conditions conducive to the implementation of safety programmes, Hypothesis 3 was established.

Hypothesis 3a: *The autonomous motivation of safety officers positively affects their levels of engagement in cultivating the conditions conducive to safety programme implementation.*

Hypothesis 3b: *The controlled motivation of safety officers positively affects their levels of engagement in cultivating the conditions conducive to safety programme implementation.*

Hypothesis 3c: *Amotivation among safety officers negatively affects their levels of engagement in cultivating the conditions conducive to safety programme implementation.*

Studies have identified a positive link between motivation and outcomes in different fields. An example is research on educational institutions, which has emphasised that enhancing the motivation of employees has become one of the top priorities in ensuring successful institutional performance [38]. Scholars have suggested that autonomous motivation among undergraduate students is essential to achieving a desirable grade point average [39]. Other researchers have discovered that identified regulation (a type of autonomous motivation) significantly affects construction labour productivity [40] and that the greater autonomous motivation of villagers to cooperate in a waste separation programme leads to enhanced waste separation weights [29]. Investigations have also reported that external and introjected types of regulation (types of controlled motivation) drive construction productivity but that considerable amotivation diminishes it [40]. The insights from the studies discussed here imply a positive relationship between safety officers' autonomous and controlled motivation and safety performance and a negative relationship between their amotivation and safety performance.

Meanwhile, limited discussions have been devoted to the direct effects of motivation on outcomes. Of the few studies conducted in this respect, that of Tam et al. [34] documented the effects of workers' autonomous motivation on construction productivity and worker engagement. As discussed in Section 2.3, the activities carried out by safety officers were identified on the basis of a literature review and represented in safety officers' engagement. However, safety officers may engage in tasks other than those identified in the succeeding section, which can uncover the direct impact of motivation on safety performance. This possibility was therefore investigated on the grounds of Hypothesis 4.

Hypothesis 4a: *Autonomous motivation positively affects safety performance.*

Hypothesis 4b: *Controlled motivation positively affects safety performance.*

Hypothesis 4c: *Amotivation negatively affects safety performance.*

2.3. Influencing Factors in Safety Management

A literature review was conducted to identify factors that can affect the implementation of safety programmes in construction projects. Such identification can facilitate safety performance improvement by reducing accidents and promoting a positive safety culture. Researchers have examined five insufficiently pursued measures that affect safety performance in the Chinese construction industry [15]. These measures are safety awareness among top management, sufficient training, safety awareness among project managers, the accumulation of adequate resources, and careful operation. Scholars have also pinpointed four categories of factors for safety programme management at construction sites [7]: worker involvement, safety prevention and control systems, safety arrangement, and management commitment. In the Saudi Arabian context, seven such crucial factors have been explored: management support, clear and reasonable goals, personal attitudes, teamwork, effective enforcement schemes, safety training, and suitable supervision. A study determined safety management commitment, subcontractor and personnel selection, safety supervisors, safety plans, employee involvement, and safety evaluation as significant safety-related factors [9], while another delved into management commitment, worksite analysis, hazard and prevention control, and health and safety training as four important constructs linked to safety programmes [10]. On the grounds of the literature review, we selected 19 potential influencing factors as the foundation for the development of the conceptual paradigm presented in Table 1. Given the scope of these factors, previous studies organised them into four categories [7,8]. Those relevant to and adopted in the current work are (1) worker involvement, (2) safety prevention and control systems, (3) safety arrangement, and (4) safety commitment [7] (Table 1).

The first category, worker involvement, focuses on the encouragement of workers to participate in safety programmes [7]. It comprises continuing participation among construction workers, worker motivation, positive group norms, and personal attitudes [7]. In this category, Al Haadir and Panuwatwanich added safety meetings [8]. The final composition in the present study thus encompassed five factors that represent how well workers are involved in safety endeavours. The next category, safety prevention and control systems, emphasises managing the quality of safety programmes and safety regulations. It covers eight factors: programme evaluation, personal competence, enforcement schemes, safety equipment acquisition and maintenance, effective supervision, safety education and training, safety promotion policy, and safety knowledge [7,41]. The third category, safety arrangement, pertains to the positioning and organisation of resources through effective communication. Three factors are grouped under this category: communication, delegation of authority and responsibility, and sufficient resource allocation [7,8]. The final category, safety commitment, highlights the ambition and determination to achieve safety goals, and it incorporates three factors: management support, teamwork, and clear and realistic goals [7,8].

2.4. Safety Officers' Contributions to Cultivating Conditions Conducive to Safety Programme Implementation

In a construction project, safety officers are responsible for promoting safety-related activities, which include, but are not limited to, establishing safety measures, enforcing safety policies to lessen the risk of accidents, and responding to workers' safety concerns [42,43]. Such professionals are therefore tasked with contributing to the cultivation of the conditions (i.e., the influencing factors) presented in Section 2.3. Assuming that these conditions collectively and exhaustively cover all the necessary tasks of safety officers, Table 1 presents how these practitioners can leverage such conditions to guarantee safety on construction sites.

While safety officers contribute to a variety of aspects of safety programmes, project managers are supposed to encourage safety officers as they manage safety programmes. This entails activities such as supporting safety officers in fostering a decent safety climate at a construction site, allocating sufficient budgets to safety programmes, and supporting safety officers in delivering essential messages to involved individuals.

Table 1. Engagement of safety officers in cultivating conditions conducive to safety programme implementation (influential factors).

Conditions (Factors)		References	Engagement Details
1. Worker involvement	Continuing participation of workers	[5,7,8,41,44,45]	Safety officers obtain feedback from workers to help improve safety programmes. Constructive criticism from construction workers should be taken into consideration to determine ways to enhance safety performance.
	Personal motivation	[7,8,35,45]	Safety officers increase the motivation of workers in different ways, such as by providing them with an opportunity to express their opinions, arranging celebrations after a successful safety programme, and encouraging construction workers to report unsafe behaviours.
	Group norms	[7]	Safety officers focus on the following aspects: positively encouraging open communication as a platform for everyone to criticise the implementation of safety programmes; and educating construction workers on safety. The modification of this statement is expected to fortify the establishment of an exemplary group norm.
	Personal attitudes	[5,7,8]	Safety officers highlight the importance of safety and health issues to increase awareness of safety, encourage workers to obey safety regulations, and motivate them to update their knowledge. This initiative is anticipated to positively influence the cultivation of favourable attitudes among workers concerning safety.
	Safety meetings	[8,15,45]	Safety officers are responsible for arranging regular safety meetings. They also attend these meetings and encourage construction workers to participate during these sessions.

Table 1. *Cont.*

	Conditions (Factors)	References	Engagement Details
2. Safety prevention and control systems	Programme evaluation	[5,7,8,45,46]	During the construction phase, safety officers are responsible mainly for tracking the progress of safety programmes. They arrange regular safety evaluations to identify sensitive aspects, track improvement in relation to these aspects, and analyse safety progress against safety plans.
	Personal competence	[7,8]	Safety officers encourage either site engineers or foremen to support construction workers in several aspects, such as assigning suitable construction activities and ensuring that these activities suit workers.
	Enforcement schemes	[7,8,35,46]	Safety officers regularly establish safety measures for a construction site, such as punishment for violating safety rules and procedures for using safety equipment. The direction of safety programmes should be well established by safety officers as the threshold for implementing safety programmes.
	Safety equipment acquisition and maintenance	[7,8,44–47]	Safety officers provide safety equipment, such as protective helmets, eyewear, ear plugs, dust masks, protective boots, safety gloves, and high-visibility clothing. They ensure the quality of such equipment. This enables construction workers to implement their tasks with a certain level of safety.
	Appropriate supervision	[7,8]	At construction sites, safety officers encourage safety behaviours among either site engineers or foremen. These include encouraging them to obey the same safety regulations, providing site engineers/foremen with beneficial information to support construction workers, and encouraging site engineers/foremen to establish appropriate examples of safety-related practices.
	Appropriate safety education and training	[5,7,8,35,44–46]	Safety officers mainly provide safety training to construction workers. New construction workers should be prepared through adequate training classes to develop their fundamental knowledge of safely implementing construction-related activities.
	Safety promotion policy	[35,44]	Safety officers determine ways to promote safety programmes, such as arranging celebrations of successful safety initiatives or other safety promotional activities to establish awareness of safety and provide construction workers with appropriate rewards when safety programmes are well implemented.
	Safety knowledge	[35]	During site inspections, safety officers can educate workers about some cautionary processes to be noted. They can encourage workers to update regulations, standard work procedures, and emerging instructions on how to use equipment.
3. Safety arrangement	Communication	[5,7,8,44,47]	Safety officers encourage communication at construction sites in several ways, such as conveying major concerns regarding safety-related practices and regularly encouraging open exchange.
	Delegation of authority and responsibility	[7,8,44,45]	Sufficient authority contributes to increased safety performance at construction sites. Thus, safety officers can assist in allocating sufficient safety staff tasked with inspecting construction sites and emphasising sensitive work areas.
	Sufficient resource allocation	[7,8]	Safety officers can encourage site engineers/foremen to provide workers with adequate resources. They can encourage assigning an appropriate number of workers to a specific task, providing sufficient and well-conditioned machines, and allocating sufficient time to practice.

Table 1. *Cont.*

	Conditions (Factors)	References	Engagement Details
4. Management commitment	Clear and realistic goals	[7,8,44]	Safety officers can determine safety goals, which include, but are not limited to, minimising accidents and providing construction workers with a safe work environment.
	Management support	[7,8,44,45,47]	Safety officers can support construction workers in different ways, such as by regularly inspecting construction sites and enthusiastically following the same safety regulations.
	Teamwork	[7,8]	Safety officers are the key contributors to teamwork in safety programmes. They address coping with safety-related issues and communicate that safety programmes require intensive involvement from all parties.

2.5. Influence of Engagement on Safety Performance and Measurement of Performance

2.5.1. Influence of Engagement on Safety Performance

Previous studies demonstrated that greater efforts in implementing safety-related activities can engender better safety performance in construction projects [5,7,8,15,41,44]. Although researchers have identified the effects of such implementation, some have disregarded safety performance indices in their analyses [8,15], while others have used limited samples for data analysis [7,9]. Investigations of the link between safety performance and the effort to engage in safety activities are valuable when both sufficient samples and safety performance indices are covered. Accordingly, we crafted Hypothesis 5.

Hypothesis 5: *The levels of engagement of safety officers in cultivating the conditions conducive to safety programme implementation are positively associated with safety performance.*

2.5.2. Measurement of Safety Performance

Past studies employed diverse indicators to assess safety performance within construction sites, including safety climate, safety orientation, management commitment to safety, near-miss events, job site audits, and accident occurrences. Notably, a perspective revolving around safety climate offers comprehensive coverage of multiple aspects of safety management practices within an organisation [48,49]. Accident occurrence has been widely used worldwide since it is recognised as a fundamental measure of safety performance [7,50,51]. Consistent with these orientations, we used two dimensions of safety performance as indicators: safety climate [45,52–54] and the frequency of construction accidents [45,55,56].

Safety climate is defined as the perception of individuals towards safety policies, safety procedures, and safety practices in the workplace [57]. In other words, it reflects individuals' impressions regarding the safety management efforts exerted by an organisation [45,54]. The safety climate at a construction site has been measured by employing questions based on a Likert scale [49], with the issues addressed including how safety resources are provided in the workplace and the extent to which the company cares about the health and safety of people.

In the present study, the direct employment of safety climate presented a challenge. Safety climate is related to safety officers' engagement in cultivating the conditions conducive to safety programme implementation (Section 2.4): The former stems from the viewpoints of all safety-related parties, whereas the latter derives from the perspectives of safety officers. It is therefore possible to interpret the safety climate as encompassing engagement. Hence, a different performance measure should be identified to prevent an overlap in coverage and potentially inaccurate results.

Construction accidents are the unintentional circumstances that abruptly occur as negative consequences of endeavours to perform construction-related activities. In multiple studies, statistics on construction accidents are used as a safety performance measure. This approach is advantageous in that it lends itself easily to statistical analysis, but it suffers from certain drawbacks. As revealed in an evaluation of statistical records focusing on accidents in construction companies, many cases are undocumented for various reasons [58]. For instance, project managers are afraid that disclosing mishaps might ruin their companies' reputations and call attention to their irresponsibility towards safety. Consultants would enforce stricter on-site safety procedures, which would be excessively extravagant for construction companies. Reporting accidents also disrupts work progress. Construction companies refrain from issuing such reports because of an acceptance among practitioners that construction is a dangerous occupation. Given this reality, safety performance should be cautiously measured because practitioners may be engaged in malpractice, leading to fraudulent information. In some situations, performance indicators (e.g., accident rate, near-miss historical record, and fatality rate) grounded in historical records are difficult to obtain because of confidentiality issues.

In addition to safety climate and construction accidents, satisfaction has been increasingly used as a performance index in the construction industry [59]. The major conception of satisfaction is to measure the difference between how much of something there should be and how much there actually is [59]. The former represents people's expectations, whereas the latter denotes their actual experiences after associated activities have been performed. If an expectation is greater than an actual experience, satisfaction is at an inferior level. Conversely, an actual experience greater than or equal to expectations leads to superior satisfaction. In measuring the success of a construction project, the satisfaction of stakeholders refers to the gratification ensuing from their interactions with other parties [60] as a perceived holistic view of cost, quality, and schedule. Measuring satisfaction has been substantiated as an effective approach to gauging performance in a construction project [59] because it can be used to ascertain the quality of practice and represent inner reality [61].

To improve conventional measurements, we put forward a new method of measuring safety performance. This method has three characteristics: (1) it incorporates the views of three parties: clients, project managers, and safety officers; (2) it entails the evaluation of the parties' views by a safety officer; and (3) it involves the use of satisfaction as a representation of these parties' perspectives.

First, to represent safety performance, the views of clients, project managers, and safety officers were determined. Project success generally means different aspects to different people [60,62]. It is an intangible perceptive feeling that can vary from individual to individual [63]. Similarly, safety performance may be defined in different ways by each stakeholder in a construction project. Contractors may desire only a pleasant historical record, such as few reported accidents and low fatality rates, to maintain their companies' reputations, whereas project owners may aim to secure all possibilities, from minimal accidents and fatalities to a pleasant safety climate. The literature has indicated that clients can impose several requirements in the workplace and should therefore be highlighted as essential parties in a construction project [59]. Project managers affect improvements to a safety programme [16], while safety officers are the main contributors to safety [19,64]. Evaluating the views of each party separately and then synthesising them clears the way for acquiring a holistic picture of the safety situation in construction.

Second, safety officers were asked to evaluate the other two parties' views. It was infeasible to collect data on the satisfaction of clients and project managers, and resorting to the aforementioned evaluation rendered the collection of sufficient data and their statistical analysis feasible. Safety officers are the main actors who bridge the gap among associated parties during the implementation of safety programmes. They interact primarily with project managers and project owners and are therefore assumed to have sufficient information on these parties' perspectives regarding safety performance.

Third, to represent the views of each party, their satisfaction levels were considered. Satisfaction includes components of safety climate and accident occurrence. There are three reasons for incorporating these two components into this study. To begin with, previous research confirmed the significant relationship between these two aspects, indicating that if safety climate is well achieved, then accident occurrences will be low [41,48,49]. This implies that accident records are not necessarily unreliable measures. In addition, since many of the respondents seemed to perceive favourable assurances and expectations regarding this study, they appeared willing to provide accurate information on accidents. This reduced the respondents' anxiety and increased the guarantee that they would disclose accident information. Furthermore, this study represented the feelings of safety officers, which have been minimally illuminated. On this issue, we received encouragement and appreciation from many of the respondents, implying an increase in their positive expectations towards this study and an enhancement of their willingness to provide accurate accident information. Finally, incorporating components of safety climate and accident occurrence enabled the safety officers to more appropriately evaluate the satisfaction levels of all parties, including their own, because each party focused on different aspects of performance. Notwithstanding this justification, however, trade-offs arose. Although integrating the three parties' views and conducting statistical analyses were beneficial, the reliability of the safety officers' evaluation of satisfaction among clients and project managers may have been low. This matter is discussed in Section 5.

Figure 1 illustrates the connections between the developed hypotheses. The conceptual model was used as the basis for developing the survey instrument and data collection, as described in the next section.

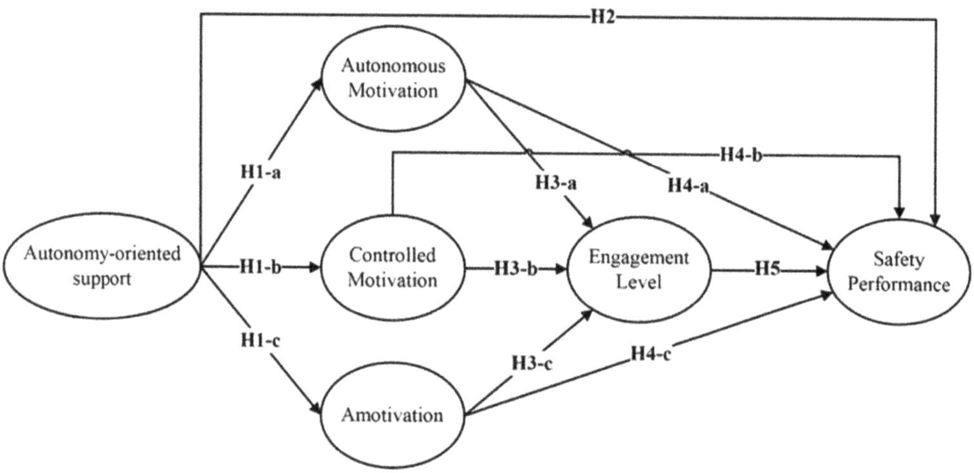

Figure 1. Hypothetical model of the enhancement of safety programmes in construction projects.

3. Survey Instrument

3.1. Questionnaire Design

The questionnaire was created by referencing the literature, and it was divided into five major sections: (1) general information; (2) autonomy-oriented support from project managers; (3) the motivation of safety officers; (4) safety officers' levels of engagement in safety programmes; and (5) safety performance. The first section is intended to obtain demographic information on construction projects, including the types of projects pursued, the educational levels of respondents, years of experience, and gender. The second section (nine items) is meant to determine perceptions regarding the support for autonomy received by safety officers from project managers, with the factors of interest being autonomy, competence, and relatedness. The third section (18 items) is concerned with the degree of motivation held by a safety officer at the time of involvement in a construction project [65,66]. The types of motivation covered are autonomous motivation, controlled motivation, and amotivation. The fourth section (45 items) pertains to potential success factors, which are used to determine the efforts exerted by safety officers in implementing the safety-related tasks categorised into four major groups [7,8]. The last section (three items) is designed to measure safety performance in a construction project on the basis of respondent, contractor, and owner satisfaction with actual safety practices (safety climate and accident occurrence). The questionnaire also includes a blank space for respondents to share additional perspectives on safety programmes.

3.2. Questionnaire Evaluation through Item–Objective Congruence (IOC)

IOC was ascertained to evaluate whether each question was pertinent to a particular objective. This concept was introduced to help researchers confirm the efficacy of a designed questionnaire before commencing with data collection [67]. At this stage, the important task is to evaluate whether questionnaire items and objectives are correspondent [67]. In this study, 75 items were evaluated during the IOC stage. Three experts were selected to assist in the evaluation, with each instructed to rate a question using three scales: +1 for appropriately measuring an objective, 0 for ambiguously measuring an objective, and −1 for not clearly measuring an objective. The IOC index of each objective was then calculated for each question, with the acceptable value being 0.60 [67,68]. Considering the expertise of the professionals involved, the experts selected were those who possessed extensive experience in both academia and the practical application of safety management. In these respects, each of the experts had more than a decade of experience. This careful selection was aimed at ensuring that the experts were qualified to evaluate the devised questionnaire. Two rounds of evaluation were conducted. In the first round, 80% (60 items) of the questionnaire items had IOC indices greater than 0.60, confirming their relevance to the objectives. The remaining 20% (15 items) were adjusted in accordance with the experts' suggestions. The insights provided by the experts covered four principal dimensions: (1) questions for accurately measuring factors of interest; (2) the identification of ambiguous questions that could hinder a comprehensively informed response; (3) lengthy questions that potentially impose a burden on respondents' time; and (4) recommendations for additional questions to enhance the effective measurement of targeted factors. The revised questions were subsequently resubmitted to the experts for further assessment to ensure their appropriate redesign. The evaluation outcome revealed a satisfactory IOC index exceeding 0.60.

3.3. Pilot Survey and Questionnaire Reliability

After the IOC assessment, 15 professional safety officers were invited to voluntarily participate in a pilot survey to ensure both the practicality and reliability of the instrument. They were also asked to provide us with minor recommendations for enhancing the questionnaire's practicality. The reliability of the questions revolving around soft data was tested on the basis of the responses of the volunteers; a Cronbach's alpha greater than 0.70 was regarded as indicative of reliability [67,69]. Accordingly, the four major categories of factors (autonomy-supportive encouragement from project managers, the motivation of safety officers, their degree of participation in implementing safety-related tasks, and safety performance) were subjected to a reliability test. The Cronbach's alpha values of these factors varied from 0.74 to 0.95, confirming that all the items were consistent and pertinent [67,69].

4. Data Collection and Data Profiles

The country selected for survey administration was Thailand, where ministerial regulations require professional safety officers to have a Bachelor of Science degree in occupational health and safety or an equivalent. This requirement guarantees that safety officers are properly trained by the educational system, through which they acquire relevant knowledge, such as that on laws and the ethics governing public health professions, engineering for occupational health and safety, and the management of occupational and environmental health issues. Upcoming safety officers also undergo on-the-job training through internships, ensuring their familiarity with both theory and practice. These qualifications reflect their instrumentality in improving safety in the industry.

The questionnaire was targeted towards professional safety officers who have experience working on construction projects in Thailand. The total population size was estimated using data from the 2013 to 2021 Yearbook of Labour Protection and Welfare Statistics, which documents a survey conducted by the Department of Labour Protection and Welfare [70]. In Thailand, the construction projects required to employ professional-level safety officers are those with a workforce of no fewer than 100 individuals. With the statistical records spanning the past nine years (2013–2021) as bases, the analysis encompassed four primary sizes of construction projects: (1) 684 projects with 100 to 299 employees; (2) 111 projects with 300 to 499 employees; (3) 57 projects with 500 to 999 employees; and (4) 18 projects with 1000 or more employees. Thus, the research sample comprised 870 construction projects. The essential sample size was computed using Equation (1), yielding an approximate requirement of 274 respondents for this research.

$$n = \frac{\mathrm{N}}{1 + \mathrm{N}(e^2)} \quad (1)$$

In the equation above, n denotes the sample size, N indicates the total population, and e refers to precision at a 95% confidence level (e = 0.05), following [71]. We secured permission to conduct the survey from the leaders of an official line group of safety professionals in Thailand. The questionnaire was created in an online form to facilitate the completion of the survey, which was administered to 290 randomly selected safety professionals. Within the questionnaire, three screening questions were embedded to ensure the selection of appropriate respondents: (1) Are you a safety officer at the professional level? (2) Have you served as a contractor for a construction project? (3) Do you hold a full-time safety officer position within a construction project?

In the primary research phase, 224 respondents (77%) actively completed the questionnaire, whereas the remaining 23% were unable to reveal project data because of confidentiality concerns. Among the 224 questionnaires received, 195 were considered valid, constituting the complete project data required for subsequent data analysis. The 195 respondents worked on three categories of construction projects: commercial building initiatives (88 respondents, 45.10%), infrastructural projects (56 respondents, 28.70%), and industrial facilities (51 respondents, 26.20%). Of these respondents, 99 (50.80%) were male and 96 (49.20%) were female. They were also grouped into three categories of experience: less than or equal to five years of experience (112 respondents, 57.44%), more than five but not over 10 years of experience (48 respondents, 24.61%), and more than 10 years of experience (35 respondents, 17.95%). As regards literacy/education, only three levels of education were considered: bachelor's (178 people, 91.3%), master's (16 people, 8.20%), and doctoral (1 individual, 0.50%) degrees.

5. Data Analysis and Results

5.1. Data Screening

This study integrated the responses of the participants as a single view to represent a perspective encompassing the entire construction industry. First, the collected data were screened via a one-way analysis of variance (ANOVA) to ensure the compatibility of the data from all the project types examined [72,73]. In relation to the three groups of construction projects, we tested six major constructs, namely, autonomy-oriented support, autonomous motivation, controlled motivation, amotivation, engagement level, and safety performance. The mean values of these constructs were regarded as representative of the situation in each type of construction project. The one-way ANOVA was conducted to compare the average values of the constructs across the three groups of respondents and determine statistical significance at a 95% confidence interval [72–74]. Table 2 presents the results of the comparison of the mean values of the constructs under the three project types (the extent to which each of the constructs is experienced by the respondents at construction sites).

Table 2. Mean values of the six constructs from the three project types.

Variables	(1) Commercial Buildings		(2) Infrastructure Projects		(3) Industrial Facilities		Sig.
	$\overline{X}$	SD	$\overline{X}$	SD	$\overline{X}$	SD	
Autonomy-oriented support	3.845	0.901	4.026	0.793	4.144	0.720	0.107
Autonomous motivation	4.372	0.524	4.462	0.484	4.326	0.547	0.379
Controlled motivation	3.252	0.872	2.908	0.953	2.944	1.065	0.059
Amotivation	1.913	1.108	1.304	0.529	1.693	1.105	0.002
Engagement level	4.353	0.646	4.520	0.431	4.525	0.496	0.107
Safety performance	3.898	0.777	4.006	0.662	4.078	0.662	0.336

The groups significantly differed only in terms of one construct—amotivation ($p = 0.002$). The p-values generated in the other comparisons were greater than 0.05, indicating no significant difference between the average values of the different types of construction projects [72,73]. This analysis confirms that the collected data on project types can be integrated into a holistic representation of views regarding construction projects in the Thai industry. We therefore integrated the responses revolving around these project types into the analysis of the hypothesised model [74].

5.2. Data Characteristics

To identify the characteristics of the data, the mean values and standard deviations of seven major factorial groups were computed for each type of construction project (Table 3). These groups were demographic information, autonomy-oriented support from project managers (autonomy, competence, and relatedness), autonomous motivation, controlled motivation, amotivation, safety officers' levels of engagement in safety activities (worker involvement, safety prevention system, safety arrangement, and commitment to safety), and safety performance (respondent, contractor, and owner satisfaction). Correlation coefficients were also calculated (Table 4).

Three noteworthy results were obtained. First, the mean value of the respondents' autonomous motivation was significantly higher than that of their controlled motivation. These values were compared using an independent sample t-test ($p = 0.000$). Second, amotivation and controlled motivation were significantly positively correlated (Table 4). Third, the mean values of satisfaction among the three stakeholders mostly followed a descending order—project owner > safety officer > project manager—for each type of project and the overall initiative. The project owners always showed the highest satisfaction levels. The results of the one-way ANOVA indicated that the mean satisfaction of the project owners (clients) and project managers (contractors) with the overall project significantly differed ($p = 0.004$). The implications of the first and second characteristics are discussed in the succeeding chapter. Those of the third are presented here.

The third characteristic may be attributed to the fact that before a client (an owner's representative or consultant) commences site inspection, a general contractor regularly receives an oral or written notification, which helps the contractor prepare a flawless construction site. Thus, clients are possibly exposed only to the best parts of safety management on each site. Few contractors, represented as project managers, focus on construction productivity, particularly when they work under tight schedules and budget constraints. This inclination becomes even stronger when they face considerable uncertainty from developments such as natural disasters or continual rain. In these cases, implementing proactive and reactive safety measures is perceived by project managers to be nothing more than a time-consuming and costly endeavour. This tendency implies that project managers having the lowest satisfaction is caused by the circumstances in which they find themselves. This issue should be analysed and discussed in detail in future research.

Table 3. Analysis of the mean values of associated variables.

Variables	(1) Commercial Buildings n = 88		(2) Infrastructure Projects n = 56		(3) Industrial Facilities n = 51		(4) Overall Project n = 195	
	$\overline{X}$	SD	$\overline{X}$	SD	$\overline{X}$	SD	$\overline{X}$	SD
Demographic information								
Gender	1.534	0.502	1.482	0.504	1.431	0.500	1.492	0.501
Education	1.102	0.340	1.036	0.187	1.137	0.348	1.092	0.307
Experience	1.568	0.828	1.446	0.658	1.843	0.758	1.605	0.775
Autonomy-oriented support	3.845	0.901	4.026	0.793	4.144	0.720	3.975	0.832
Autonomy	3.856	0.881	4.030	0.842	4.176	0.668	3.990	0.826
Competence	3.795	0.953	3.988	0.814	4.105	0.776	3.932	0.876
Relatedness	3.883	0.963	4.060	0.858	4.150	0.847	4.003	0.907
Autonomous motivation	4.372	0.524	4.462	0.484	4.326	0.547	4.386	0.519
Intrinsic regulation	4.159	0.734	4.250	0.766	4.013	0.830	4.147	0.770
Integrated regulation	4.405	0.514	4.482	0.561	4.458	0.508	4.441	0.525
Identified regulation	4.642	0.519	4.750	0.357	4.598	0.640	4.662	0.516
Controlled motivation	3.252	0.872	2.908	0.953	2.944	1.065	3.073	0.958
Introjected regulation	3.322	0.934	3.024	1.029	2.935	1.414	3.135	1.112
External regulation	3.182	1.013	2.792	1.046	2.954	1.037	3.010	1.037
Amotivation	1.913	1.108	1.304	0.529	1.693	1.105	1.680	1.005
Lack of motivation 1	1.909	1.274	1.286	0.706	1.627	1.232	1.656	1.153
Lack of motivation 2	1.716	1.203	1.214	0.706	1.588	1.186	1.538	1.095
Lack of motivation 3	2.114	1.236	1.411	0.733	1.863	1.312	1.846	1.170
Engagement level	4.353	0.646	4.520	0.431	4.525	0.496	4.446	0.557
Worker involvement	4.400	0.626	4.579	0.434	4.545	0.532	4.489	0.556
Safety prevention	4.395	0.651	4.523	0.443	4.516	0.529	4.464	0.567
Safety arrangement	4.201	0.758	4.405	0.595	4.429	0.564	4.319	0.672
Safety commitment	4.417	0.643	4.574	0.434	4.611	0.489	4.513	0.555
Safety performance	3.898	0.777	4.006	0.662	4.078	0.662	3.976	0.717
Safety officer satisfaction	3.898	0.885	4.000	0.688	4.137	0.693	3.990	0.786
Contractor satisfaction	3.761	0.884	3.893	0.779	3.961	0.799	3.851	0.833
Owner satisfaction	4.034	0.837	4.125	0.740	4.137	0.693	4.087	0.772

Notes: Gender served as a dummy variable that was assigned a value of 1 when a subject was male and 2 when female. Education was a dummy variable that took a value of 1 for a bachelor's degree, 2 for a master's degree, and 3 for a doctoral degree. Experience was a dummy variable that was given a value of 1 for respondents with experience less than or equal to five years, 2 for those with experience spanning more than five years but not over 10 years, and 3 for respondents with more than 10 years' experience.

As described in Section 2.5.2, a trade-off exists in the evaluation of satisfaction among clients, project managers, and safety officers. We initially attempted to acquire reliable responses by employing three practitioners in the safety profession from both academic and practical fields to ensure that the respondents understood the intended meaning of the questions and that the designed questions were pertinent to the measurement of safety performance. During the survey, we received positive feedback on the benefits of this research, implying that the respondents were willing to provide pertinent project data, thereby presenting reliable responses. In addition, the results (Table 3) indicated that the satisfaction of the owners and project managers with each project was evaluated consistently, albeit this was assessed by the same safety officer. Therefore, the advantages of integrating the three parties' views and conducting statistical analysis outweigh the disadvantage of the potentially low reliability of the evaluation. Furthermore, the application of confirmatory factor analysis (CFA), statistically confirmed the reliability of the assessment of safety performance by the safety officers. The factor loadings corresponding to the satisfaction of the safety officers, contractors, and owners were 0.871, 0.840, and 0.822, respectively. These findings contribute to the validation of the measurement model and underscore the robustness of the safety performance evaluation by the specified entities.

Table 4. Mean scores, standard deviations, and correlation coefficients of associated variables.

Variables	$\bar{X}$	SD	(1)	(2)	(3)	(4)	(5)	(6)	(7)	(8)	(9)	(10)
(1) Gender	1.492	0.501	1.000									
(2) Education	1.092	0.307	−0.163 *	1.000								
(3) Experience	1.605	0.775	−0.452 **	0.283 **	1.000							
(4) Project type	1.810	0.825	−0.085	0.029	0.124	1.000						
(5) Autonomy-oriented support	3.975	0.832	−0.010	−0.083	−0.034	0.151 *	1.000					
(6) Autonomous motivation	4.386	0.519	−0.048	−0.031	0.000	−0.022	0.424 **	1.000				
(7) Controlled motivation	3.073	0.958	−0.127	0.003	−0.006	−0.146 *	−0.012	0.108	1.000			
(8) Amotivation	1.680	1.005	−0.129	−0.010	−0.026	−0.123	0.007	−0.075	0.522 **	1.000		
(9) Engagement level	4.446	0.557	−0.006	−0.020	−0.033	0.137	0.525 **	0.431 **	0.017	−0.021	1.000	
(10) Safety performance	3.976	0.717	−0.091	0.018	0.005	0.106	0.630 **	0.481 **	0.195 **	0.098	0.540 **	1.000

Notes: Gender served as a dummy variable that was assigned a value of 1 when a subject was male and 2 when female. Education was a dummy variable that took a value of 1 for a bachelor's degree, 2 for a master's degree, and 3 for a doctoral degree. Experience was a dummy variable that was given a value of 1 for respondents with experience less than or equal to five years, 2 for those with experience spanning more than five years but not over 10 years, and 3 for respondents with more than 10 years' experience. Project type is a dummy variable that takes the values of 1 for commercial buildings, 2 for infrastructure construction, and 3 for the construction of industrial facilities; * $p < 0.05$; ** $p < 0.01$.

5.3. *Causal Relationships among the Factors in the Proposed Model*

The proposed model was analysed mainly through a measurement model and structural equation modelling (SEM) [75,76].

5.3.1. Measurement Model for Six Constructs

The measurement model consists of six latent variables, namely, autonomy-oriented support from project managers (detailed questions presented in Appendix A, Table A1), the autonomous motivation of safety officers, their controlled motivation, their amotivation, their levels of engagement in implementing safety programmes, and safety performance. The p-value derived using Bartlett's test of sphericity was 0.000, indicating that the correlation matrix was not an identity matrix and that it was ideal for conducting factor analysis [77]. The Kaiser–Meyer–Olkin (KMO) measure of sampling was used to ensure the plausibility of the collected data. The KMO value was 0.903, reflecting the suitability of the data for factor analysis [77]. As shown in Table 5, CFA was performed to ensure that the observed variables were ideal for each construct, with a factor loading of 0.50 regarded as acceptable [75,76].

The observed variables with standardised factor loadings greater than 0.5 were retained in the measurement model [78,79]. The model was verified with appropriate fit indices (p = 0.397, X^2/DF = 1.017, goodness-of-fit index [GFI] = 0.89, normed fit index [NFI] = 0.923, Tucker–Lewis index = 0.998, comparative fit index [CFI] = 0.999, and root mean square error of approximation [RMSEA] = 0.009). The analysis confirmed that the observed variables were ideal representations of each construct in the conceptual model.

Table 5. CFA results on the six constructs.

Constructs and Observed Variables		Factor Loadings	CR	AVE
Autonomy-oriented support	Autonomous support from project manager 1	0.798	0.961	0.734
	Autonomous support from project manager 2	0.875		
	Autonomous support from project manager 3	0.876		
	Competence support from project manager 1	0.860		
	Competence support from project manager 2	0.820		
	Competence support from project manager 3	0.844		
	Relatedness support from project manager 1	0.864		
	Relatedness support from project manager 2	0.877		
	Relatedness support from project manager 3	0.892		
Autonomous motivation	Intrinsic regulation of safety officers 1	0.591	0.830	0.384
	Intrinsic regulation of safety officers 2	0.600		
	Intrinsic regulation of safety officers 3	0.705		
	Integrated regulation of safety officers 1	0.717		
	Integrated regulation of safety officers 2	0.710		
	Integrated regulation of safety officers 3	0.555		
	Identified regulation of safety officers 1	0.530		
	Identified regulation of safety officers 3	0.505		
Controlled motivation	Introjected regulation of safety officers 1	0.552	0.809	0.416
	Introjected regulation of safety officers 2	0.690		
	Introjected regulation of safety officers 3	0.728		
	External regulation of safety officers 1	0.686		
	External regulation of safety officers 2	0.594		
	External regulation of safety officers 3	0.599		
Amotivation	Lack of motivation of safety officers 1	0.808	0.860	0.673
	Lack of motivation of safety officers 2	0.886		
	Lack of motivation of safety officers 3	0.762		
Engagement level	Worker involvement	0.950	0.963	0.867
	Safety prevention and control systems	0.980		
	Safety arrangement	0.903		
	Safety commitment	0.889		
Safety performance	Safety officer satisfaction	0.871	0.882	0.713
	Contractor satisfaction	0.840		
	Owner satisfaction	0.822		

The composite reliability (CR) values of the constructs (autonomy-oriented support, autonomous motivation, controlled motivation, amotivation, engagement level in safety programmes, and safety performance) were 0.961, 0.830, 0.809, 0.860, 0.963, and 0.882, respectively, which exceeded the acceptable value of 0.60 [80,81]. Their average variance extracted (AVE) values were 0.734, 0.384, 0.416, 0.673, 0.867, and 0.713, respectively. Although most researchers have argued that the AVE should be more than 0.50, other scholars have suggested that a value of approximately 0.40 is still acceptable [82]. Furthermore, if an AVE less than 0.50 is derived but the CR is well above the recommended level, then the internal reliability of measurement items can be considered adequate [80,81]. In the present research, autonomous motivation had an AVE of 0.384.

5.3.2. Structural Model

In the SEM, the final model was constructed by considering the model that was justified using the superior fit indices. The hypotheses regarding the structural model were tested to measure the degree to which each exogenous variable affected the endogenous variable. Figure 2 depicts the hypothetical model with the effect size underlying the relationships of the variables [73,75,76,83].

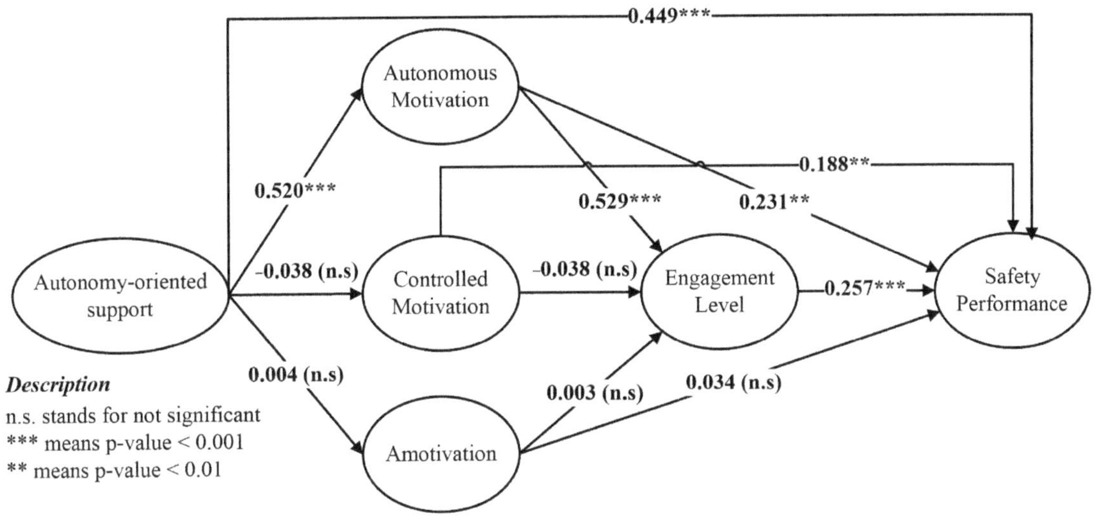

Figure 2. Results of the analysis of the structural model of assumptions.

To begin with, autonomy-oriented support from project managers positively affected both the safety officers' autonomous motivation and safety performance, with the standardised coefficients of these variables being 0.520 and 0.449, respectively. In addition, the safety officers' autonomous motivation positively influenced safety performance ($\beta_{H4\text{-}A} = 0.231$) and their efforts to engage in safety programmes ($\beta_{H3\text{-}A} = 0.529$). The same favourable effect on safety performance was observed with regard to controlled motivation ($\beta_{H4\text{-}B} = 0.188$). Furthermore, the degree of engagement among the safety officers in crucial activities was positively linked to safety performance, with the standardised coefficient being 0.257.

The fit indices were explored to ensure the effectiveness of the proposed model in capturing actual safety situations at construction sites. The fit indices included the p-value, X^2/DF, CFI, GFI, NFI, RMSEA, and the Akaike information criterion (AIC) [73,75,76,83], whose values should satisfy the statistical criteria indicated in Table 6 [84]. The fit indices were at acceptable levels, demonstrating the compatibility of the hypothesised model with the empirical data.

This compatibility implies that the constructed hypotheses are pertinent to real-world practice. This compatibility, together with the advantages presented by the research findings, is discussed in the succeeding section.

In SEM, a consensus regarding appropriate sample sizes is notably absent. Although the literature has proposed a minimum of 200 as a sample size for such analysis, the 195 respondents in the current research closely approach this threshold. A study suggested that an appropriate sample size for SEM should include a minimum of five cases per variable [85]. Considering the parameters dealt with in the present study, which involved 33 observed variables within the proposed model, adhering to this guideline requires that the minimal number of samples be 165. Consequently, our sample was deemed suitable for the execution of SEM.

Table 6. Results on the fit indices of the conceptual model (adapted from [84]).

Fit Indices	Criteria	Initial Model	Final Model
p-value	Greater than 0.05	0.000	0.087
X^2/DF	1 to 2	1.905	1.102
CFI	0 (no fit)–1 (perfect fit)	0.907	0.992
GFI	0 (no fit)–1 (perfect fit)	0.778	0.895
NFI	0 (no fit)–1 (perfect fit)	0.824	0.923
RMSEA	Less than 0.05	0.068	0.023
AIC	Small value	1076.001	791.356

6. Discussion

Table 7 presents each hypothesis, along with its standardised coefficient (β) and p-value. As previously stated, this research delved into four major aspects linked to safety implementation at construction sites: (1) the key role of autonomy-oriented support from project managers (H1a and H2); (2) the essential role of safety officers' autonomous motivation (H3a and H4a); (3) the impact of controlled motivation (H4b, H1b, and H3b); and (4) the engagement of safety officers in safety performance (H5).

Table 7. Details on each hypothesis, with estimated standardised coefficients.

Hypotheses		Descriptions	β	Sig.
H1	**H1a**	autonomy-oriented support → $^{(+)}$ autonomous motivation	0.520	***
	H1 b	autonomy-oriented support → $^{(+)}$ controlled motivation	−0.038	0.635
	H1c	autonomy-oriented support → $^{(-)}$ amotivation	0.004	0.956
H2		autonomy-oriented support → $^{(+)}$ safety performance	0.449	***
H3	**H3a**	autonomous motivation → $^{(+)}$ engagement level	0.529	***
	H3b	controlled motivation → $^{(+)}$ engagement level	−0.038	0.582
	H3c	amotivation → $^{(-)}$ engagement level	0.003	0.961
H4	**H4a**	autonomous motivation → $^{(+)}$ safety performance	0.231	0.007
	H4b	controlled motivation → $^{(+)}$ safety performance	0.188	0.003
	H4c	amotivation → $^{(-)}$ safety performance	0.034	0.561
H5		engagement level → $^{(+)}$ safety performance	0.257	***

Note: *** = $p < 0.001$.

6.1. Essential Role of Autonomy-Oriented Support from Project Managers (H1a and H2)

The results showed that autonomy-oriented support from project managers significantly affected the safety officers' autonomous motivation ($\beta_{H1\text{-}A} = 0.520$). Specifically, it was reinforced and maintained through autonomy-supportive behaviours, as indicated in the comments of multiple respondents. Some of the respondents stated that receiving such support would help them achieve certainty in their roles. Another participant shared an experience with a reputable company where the project manager paid extensive attention to supporting the safety programme, which in turn contributed to the success of such an initiative in the project. This situation may have enhanced the autonomous motivation of the safety officer to engage in safety practices. Note, however, that the high $\beta_{H1\text{-}A}$ means that undesirable situations also existed. One respondent asserted that safety management in Thailand has thus far been ineffective and that this problem stems from project managers' lack of intention to enhance safety programmes. This deficiency can drive safety officers to view safety programmes in the workplace as unimportant. Regardless of the presence of desirable or undesirable situations, autonomy-oriented support from project managers enhanced the autonomous motivation of the safety officers. These findings are similar to those of previous studies [23 25].

Autonomy-oriented project management significantly affected safety performance (β_{H2} = 0.449), which can be attributed to the following reasons: First, every action from project managers as leaders of construction projects can directly influence performance. Second, apart from affecting the autonomous motivation of safety officers, project manager support can be mediated by other relevant parties, such as site engineers and construction labour.

6.2. Essential Role of Safety Officers' Autonomous Motivation (H3a and H4a)

This research acknowledged the key role of autonomous motivation in successful safety programmes. The $\beta_{H3\text{-}A}$ of the effects of the safety officers' autonomous motivation on their degrees of participation in safety initiatives was 0.529; that is, the higher the autonomous motivation, the more intensive the effort exerted towards safety-related activities [19]. Some of the respondents from whom positive survey results were derived declared that they attempt to extensively manage safety programmes, given that they view these as their most important task. A number of the participants added that, as safety professionals, they are obligated to cautiously discuss safety issues with construction practitioners before construction is commenced to ensure their understanding of safety. This empirical evidence confirms that higher autonomous motivation among safety officers translates to greater engagement in safety-related activities. These findings are consistent with the literature on other matters, such as health-related behaviours [31], safety management in a Taiwanese international airline [30], safety management in a coal carbonisation company in China [32], the management of construction worker productivity [34], and the safety practices of a manufacturing factory in Turkey [33].

The results of the analysis pointed to a direct impact of autonomous motivation on safety performance ($\beta_{H4\text{–}A}$ = 0.231). Put differently, the autonomous motivation of the safety officers also significantly influenced safety performance. This result is ascribed to two factors: negotiations with project managers and/or the proposal of recommendations to regulators, and the conduct of health-related activities. With regard to the first factor, the scope of safety officers' engagement in this work is a set of activities involving workers, foremen, and site engineers. Suppose that the safety conditions on a construction site are undesirable and that highly autonomous safety officers propose a new method to project managers or clients or express their honest opinions to government regulators. The considerable contribution of these efforts to improvements in safety programmes counts as a direct influence of autonomous motivation on safety performance. As for the second factor, a respondent shared an experience at the early stage of the COVID-19 pandemic. Given undeveloped regulations at a site, the officer autonomously established additional health regulations, provided test kits and alcohol sprays to workers, and acquired temperature measurement equipment to protect the workers from illness. These two factors may be beyond the job descriptions of safety officers. It is possible that these 'invisible' efforts also support site safety in Thailand. Comprehensive discussions should be devoted to the mechanisms by which the autonomous motivation of safety officers directly elevates safety performance.

The mean autonomous motivation of the safety officers was significantly higher than their mean controlled motivation. Thai safety officers acknowledge the importance of safety programmes over external rewards and egos. As mentioned in the introduction section, they undergo thorough training and are familiar with both theory and practice. They seem to recognise the importance of safety programmes during these training sessions and enter the industry with a strong sense of responsibility. One of the respondents declared that every step of safety programme management should be cautiously performed. Furthermore, they provided support-related information that they had to intensively discuss with construction practitioners regarding safety practices before the initiation of construction-related activities. Thus, such practitioners are likely to value safety management as aligned with their major goals.

6.3. Roles of Controlled Motivation (H1b, H3b, and H4b)

As mentioned previously, the mean value of the controlled motivation of the safety officers was significantly lower than that of their autonomous motivation. The influence of autonomy-oriented support on controlled motivation (H1b) and the influence of controlled motivation on engagement level (H3b) were not validated. However, controlled motivation had a β_{H4-B} of 0.188 with a $p = 0.003$, showing that the controlled motivation of the safety officers also played a role in enhancing safety performance.

Controlled motivation comprises external and introjected regulation, and the extent to which each regulation was exercised in this work was represented by the mean value of three variables. The three variables pertinent to external regulation are the following reasons for engaging in safety programmes: (1) receiving a sizeable salary; (2) receiving external rewards; and (3) avoiding punishment. Introjected regulation covers (1) protecting the ego; (2) avoiding judgement from others; and (3) feeling like a failure under a disregard for safety programme management yet simultaneously dismissing the value of a safety initiative. The coefficient of correlation between safety performance and external regulation was 0.228 ($p < 0.01$). The coefficients of correlation between safety performance and salary and rewards from project managers were 0.250 ($p < 0.01$) and 0.215 ($p < 0.01$), respectively.

Section 6.2 presents a hypothesis that we developed for future study, which states that the direct impact of autonomous motivation on safety performance originates from negotiations with project managers and/or recommendations provided to regulators and the performance of extra health-related activities. These are beyond the job descriptions of safety officers. If this is true, the direct impact of controlled motivation on safety performance can be interpreted as a desire for an appreciation of the additional challenging tasks that safety officers undertake. As previously suggested, then, the mechanisms by which the motivation of safety officers directly affects safety performance should be identified and thoroughly explored in the future.

Another noteworthy result was that the coefficient of correlation between controlled motivation and amotivation was 0.522 and was statistically significant (Table 4). This result, as well as the significant value of β_{H4-B}, implies that enhancing controlled motivation is a double-edged sword. If controlled motivation is inappropriately promoted, safety officers may feel demotivated, consistent with Ryan and Deci's (2000) argument that excessively elevating such motivation may be harmful because it possibly hinders the enhancement of autonomous motivation. Thus, promoting controlled motivation through external rewards should be cautiously strategised to ensure that it does not affect autonomous motivation. In sum, project managers should not neglect the promotion of controlled motivation, but neither should it be a top priority.

6.4. Engagement of Safety Officers and Safety Performance (H5)

As demonstrated by the analytical results, the degree of engagement of the safety officers in safety-related activities positively affected safety performance ($\beta_{H5} = 0.257$). This confirms that increased efforts to conduct safety-related tasks contribute to successful safety performance [20,86]. The results on this aspect also validated the idea that safety officers' levels of engagement are positively associated with safety performance [5]. The findings are similar to those of other research on, for example, the implementation of waste management programmes [29] and health-related behaviours [31].

It is worth noting that the present study did not delineate the motivational levels of safety officers across distinct characteristics. This limitation can be addressed by alternative statistical models, including cluster analysis and latent profile analysis (LPA), which represent person-centred approaches designed to aid researchers in the exploration of potential subpopulations within a sample. As a conventional person-centred technique, cluster analysis exhibits certain limitations: First, it lacks formal criteria for evaluating optimal solution fit, and second, an individual is forcedly classified into only one cluster, which contradicts the reality that individuals can be allocated to various clusters [19,87]. These deficiencies can be addressed in LPA, which provides fit indices that facilitate the evaluation of model adequacy [19] and relies on diverse criteria, including likelihood (i.e., probability). Notably, LPA accommodates the allocation of individuals to multiple clusters. A comprehensive discussion of the potential and limitations of both models was presented by Gabriel et al. [87]. The findings derived from the application of these statistical models can offer a construction company with insightful perspectives, facilitating an exhaustive understanding of the diverse motivational profiles exhibited by safety officers. This understanding, in turn, may inform the formulation of a well-defined strategic framework aimed at supporting the motivational needs of safety officers operating within the construction industry.

7. Conclusions

Securing construction safety remains a considerable challenge, which stems fundamentally from, among other factors, the gap between espoused values and the assumptions that typify the culture of construction organisations [17]. 'Safety first' is a slogan heard on every site, but in reality, project managers focus primarily on other issues, such as construction productivity, and consequently neglect safety programmes at times.

One of the key parties who can bridge the abovementioned gap is safety officers. However, the following issues were not necessarily clarified in previous research: how the motivation of safety officers can be enhanced, whether motivation leads to augmented engagement, and whether engagement among safety officers improves safety performance. To shed light on these matters, we first hypothesised that autonomous motivation, with emphasis on value, responsibility, or interest in work, plays a vital role in safety. Without these characteristics, safety professionals cannot continue working diligently to address the aforementioned gap. In a worst-case scenario, these practitioners might end up resigning and leaving the construction industry, which would lead to a loss of valuable human resources. We then constructed a hypothetical model consisting of four major components: autonomy-oriented support from project managers, the motivation of safety officers, their levels of engagement in safety programmes, and safety performance. A survey was administered to 195 safety officers who had worked on construction projects in Thailand, and the hypothetical model was analysed via SEM.

First, a noteworthy result was that the mean autonomous motivation of the safety officers was significantly higher than their controlled motivation. Thai safety officers acknowledge the importance of safety programmes over external rewards and egos. Second, autonomous support from project managers positively affected the autonomous motivation of the safety officers, indicating that the former was instrumental in enhancing the latter's autonomous motivation. In reality, many project managers follow their companies' policies. Thus, the real key party in providing support to safety officers is company management. Third, the autonomous motivation of the safety officers significantly affected their engagement and safety performance. Fourth, the engagement of the safety officers significantly influenced safety performance. Fifth, controlled motivation significantly affected safety performance. Note, however, that controlled motivation was significantly correlated with amotivation, implying that the former has two sides: it is not a negligible issue but can be harmful if addressed inappropriately. These results mean that in the implementation of safety programmes, the autonomous motivation of safety officers serves as the engine, and autonomy-oriented support from project managers and company management functions as the ignition key.

The findings elucidated the salient interrelationship among the four constructs related to the activities of safety officers in Thailand on the basis of self-determination theory. To the best of our knowledge, this research represents the first systematic definition and investigation of the specific interrelationships concerning safety officers. It has the potential to augment the extant literature on safety management in construction projects.

From a pragmatic perspective, the outcomes provide indispensable insights for both project managers and organisational leadership. Notably, a considerable proportion of project managers adhere to their firms' established policies, positioning organisational management as the pivotal entity in proffering support to safety officers. The defined model can serve as a foundational framework, enabling project managers to devise strategies that augment the efficacy of safety programmes, especially for critical frontline stakeholders such as safety officers. As a result, the responsibility of embracing an autonomy-supportive approach, which encapsulates the incorporation of diverse perspectives, the provision of constructive feedback to safety officers, and the cultivation of an inclusive milieu, lies in the hands of project managers. Additionally, the research outcomes present a strategic blueprint for construction enterprises. Such firms can leverage these insights to either establish training programmes that underscore autonomy-supportive methodologies for project managers or formulate explicit guidelines ensuring the systematic integration of autonomy-centric strategies in construction endeavours.

A notable limitation of this study is its exclusive focus on safety officers operating within Thailand. The observed heightened autonomous motivation among these officers suggests an intrinsic attraction to the sector. This stems from a genuine interest, a commitment to implementing safety protocols, and an acknowledgement of the importance of safety measures. Moreover, even after entering the industry, they continue to maintain a strong sense of autonomous motivation. However, whether these attributes are ubiquitous among safety officers globally or are specific to those in Thailand remains inconclusive. Pursuing comparative analyses across nations would be a worthwhile endeavour. Combining the findings from such cross-national studies could further enrich our understanding of the support, motivation, and contributions inherent to the activities of safety officers. Furthermore, understanding the different characteristics of safety officers may provide a construction company with valuable guidelines on decision-making regarding appropriate strategies for supporting safety officers.

Author Contributions: Conceptualisation, K.K.-a., T.W. and S.T.; methodology, K.K.-a. and T.W.; data curation, K.K.-a.; formal analysis, K.K.-a.; writing—original draft preparation, K.K.-a.; writing—review and editing, T.W.; supervision, T.W. and S.T. All authors have read and agreed to the published version of the manuscript.

Funding: This research received no external funding.

Data Availability Statement: Data that support the findings of this study are available from the corresponding author upon reasonable request. The data are not publicly available due to ethical issues related to data privacy and security.

Acknowledgments: This research was supported by Kochi University of Technology in Japan. Any opinions, findings, or recommendations written here are those of the authors and do not necessarily reflect the perspectives of any agencies.

Conflicts of Interest: The authors declare no conflicts of interest.

Appendix A

The Appendix presents the measurement of autonomy-oriented support from project managers as perceived by the safety officers (Table 6). It covered support for autonomy, competence, and relatedness. Each respondent denoted their perception of the autonomy-oriented support from project managers using a five-point Likert scale (1 = no support at all, 2 = slight support, 3 = moderate support, 4 = high support, and 5 = extensive support).

Table A1. Questions on autonomy-oriented support and examples of evaluation.

Variables	Questions	1	2	3	4	5
Autonomy support 1	To what extent does a project manager encourage safety officers to have alternative choices when managing safety programmes?					
Autonomy support 2	To what extent does a project manager attempt to understand the conditions and responsibilities of safety officers?					
Autonomy support 3	To what extent does a project manager attempt to understand safety officers' perspectives before suggesting an alternative choice?					
Competence support 1	To what extent does a project manager convey their confidence in a safety officer's ability to manage safety?					
Competence support 2	To what extent does a project manager encourage safety officers to ask questions and provide suggestions relating to safety issues?					
Competence support 3	To what extent does a project manager provide safety officers with positive feedback when safety is well performed?					
Relatedness support 1	To what extent does a project manager make safety officers feel accepted by the company?					
Relatedness support 2	To what extent does a project manager encourage safety officers to be an important part of the organisation?					
Relatedness support 3	To what extent does a project manager help safety officers develop trust in project managers?					

References

1. BLS (Bureau of Labor Statistics). *National Census of Fatal Occupational Injuries in 2020*; BLS: Washington, DC, USA, 2021. Available online: https://www.bls.gov/news.release/archives/cfoi_12162021.pdf (accessed on 23 January 2022).
2. CLB (China Labor Bulletin). *Calls to Improve China's Work Safety Go Unheeded in 2015*; CLB: Hong Kong, China, 2016. Available online: https://clb.org.hk/en/content/calls-improve-china%E2%80%99s-work-safety-go-unheeded-2015 (accessed on 10 January 2022).
3. Vongpisal, C.; Yodpijit, N. Construction accidents in Thailand: Statistical data analysis. *Appl. Sci. Eng. Prog.* **2017**, *10*, 7–21. [CrossRef]
4. Social Security Office. *Occupational Injuries and Illnesses between 2016 and 2020*; SSO: Bangkok, Thailand, 2021. Available online: https://www.sso.go.th/wpr/assets/upload/files_storage/sso_th/5ebe42693bf27ca624d2a14a89f99223.pdf (accessed on 20 December 2021).
5. Abudayyeh, O.; Fredericks, T.K.; Butt, S.E.; Shaar, A. An investigation of management's commitment to construction safety. *Int. J. Proj. Manag.* **2006**, *24*, 167–174. [CrossRef]
6. Aksorn, T.; Hadikusumo, B. The unsafe acts and the decision-to-err factors of Thai construction workers. *JCDC* **2007**, *12*, 1–25.
7. Aksorn, T.; Hadikusumo, B.H. Critical success factors influencing safety program performance in Thai construction projects. *Saf. Sci.* **2008**, *46*, 709–727. [CrossRef]
8. Al Haadir, S.; Panuwatwanich, K. Critical success factors for safety program implementation among construction companies in Saudi Arabia. *Procedia Eng.* **2011**, *14*, 148–155. [CrossRef]
9. Bavafa, A.; Mahdiyar, A.; Marsono, A.K. Identifying and assessing the critical factors for effective implementation of safety programs in construction projects. *Saf. Sci.* **2018**, *106*, 47–56. [CrossRef]
10. Buniya, M.K.; Othman, I.; Sunindijo, R.Y.; Karakhan, A.A.; Kineber, A.F.; Durdyev, S. Contributions of safety critical success factors and safety program elements to overall project success. *Int. J. Occup. Saf. Ergon.* **2023**, *29*, 129–140. [CrossRef]
11. Chinda, T. Factors Influencing Construction Safety Equipment Selection. *Int. J. Struct. Civ. Eng. Res.* **2016**, *5*, 333–336. [CrossRef]
12. Findley, M.; Smith, S.; Kress, T.; Petty, G.; Enoch, K. Safety program elements in construction. *Prof. Saf.* **2004**, *49*, 14.
13. Häkkinen, K. A learning-by-doing strategy to improve top management involvement in safety. *Saf. Sci.* **1995**, *20*, 299–304. [CrossRef]
14. Mohammadi, A.; Tavakolan, M.; Khosravi, Y. Factors influencing safety performance on construction projects: A review. *Saf. Sci.* **2018**, *109*, 382–397. [CrossRef]
15. Tam, C.M.; Zeng, S.X.; Deng, Z. Identifying elements of poor construction safety management in China. *Saf. Sci.* **2004**, *42*, 569–586. [CrossRef]
16. Teo, E.A.L.; Ling, F.Y.Y.; Chong, A.F.W. Framework for project managers to manage construction safety. *Int. J. Proj. Manag.* **2005**, *23*, 329–341. [CrossRef]
17. Schein, E.H. *Organizational Culture and Leadership*; John Wiley & Sons: Hoboken, NJ, USA, 2010; Volume 2.

18. Toole, T.M. Construction site safety roles. *J. Constr. Eng. Manag.* **2002**, *128*, 203–210. [CrossRef]
19. Ju, C. Work motivation of safety professionals: A person-centred approach. *Saf. Sci.* **2020**, *127*, 104697. [CrossRef]
20. Ryan, R.M.; Deci, E.L. Intrinsic and extrinsic motivations: Classic definitions and new directions. *Contemp. Educ. Psychol.* **2000**, *25*, 54–67. [CrossRef] [PubMed]
21. Ryan, R.M.; Deci, E.L. *Self-Determination Theory: Basic Psychological Needs in Motivation, Development, and Wellness*; Guilford Publications: New York, NY, USA, 2017.
22. Katz, I.; Kaplan, A.; Buzukashvily, T. The role of parents' motivation in students' autonomous motivation for doing homework. *Learn. Individ. Differ.* **2011**, *21*, 376–386. [CrossRef]
23. Hagger, M.S.; Sultan, S.; Hardcastle, S.J.; Chatzisarantis, N.L.D. Perceived autonomy support and autonomous motivation toward mathematics activities in educational and out-of-school contexts is related to mathematics homework behavior and attainment. *Contemp. Educ. Psychol.* **2015**, *41*, 111–123. [CrossRef]
24. Karimi, S.; Sotoodeh, B. The mediating role of intrinsic motivation in the relationship between basic psychological needs satisfaction and academic engagement in agriculture students. *Teach. High. Educ.* **2019**, *25*, 959–975. [CrossRef]
25. Zhang, J.; Huang, R.; Chen, Q.; Zhao, G. The relationships between supervisor-subordinate guanxi, perceived supervisor autonomy support, autonomous motivation, and employee job satisfaction: Evidence from international hotel chains in China. *Int. J. Hosp. Manag.* **2023**, *108*, 103354. [CrossRef]
26. Carreira, J.M. Motivational orienations and psychological needs in EFL learning among elementary school students in Japan. *System* **2012**, *40*, 191–202. [CrossRef]
27. Van Tuin, L.; Schaufeli, W.B.; Van Rhenen, W. The satisfaction and frustration of basic psychological needs in engaging leadership. *J. Leadersh.* **2020**, *14*, 6–23. [CrossRef]
28. Mirchandani, D.A.; Lederer, A.L. The impact of autonomy on information systems planning effectiveness. *Omega* **2008**, *36*, 789–807. [CrossRef]
29. Nguyen, T.; Watanabe, T. Autonomous motivation for the successful implementation of waste management policy: An examination using an adapted institutional analysis and development framework in Thua Thien Hue, Vietnam. *Sustainability* **2020**, *12*, 2724. [CrossRef]
30. Chen, C.F.; Chen, S.C. Measuring the effects of Safety Management System practices, morality leadership and self-efficacy on pilots' safety behaviors: Safety motivation as a mediator. *Saf. Sci.* **2014**, *62*, 376–385. [CrossRef]
31. Hagger, M.S.; Hardcastle, S.; Chater, A.; Mallett, C.; Pal, S.; Chatzisarantis, N. Autonomous and controlled motivational regulations for multiple health-related behaviors: Between-and within-participants analyses. *Health Psychol. Behav.* **2014**, *2*, 565–601. [CrossRef]
32. Jiang, L.; Tetrick, L.E. Mapping the nomological network of employee self-determined safety motivation: A preliminary measure in China. *Accid. Anal. Prev.* **2016**, *94*, 1–7. [CrossRef] [PubMed]
33. Peker, M.; Doğru, O.C.; Meşe, G. Role of supervisor behavioral integrity for safety in the relationship between top-management safety climate, safety motivation, and safety performance. *Saf. Health Work* **2022**, *13*, 192–200. [CrossRef]
34. Tam, N.V.; Watanabe, T.; Hai, N.L. Importance of Autonomous Motivation in Construction Labor Productivity Improvement in Vietnam: A Self-Determination Theory Perspective. *Buildings* **2022**, *12*, 763. [CrossRef]
35. Vinodkumar, M.; Bhasi, M. Safety management practices and safety behaviour: Assessing the mediating role of safety knowledge and motivation. *Accid. Anal. Prev.* **2010**, *42*, 2082–2093. [CrossRef]
36. Fagan, M.H.; Neill, S.; Wooldridge, B.R. Exploring the intention to use computers: An empirical investigation of the role of intrinsic motivation, extrinsic motivation, and perceived ease of use. *J. Comput. Inf. Syst.* **2008**, *48*, 31–37.
37. Stoeber, J.; Davis, C.R.; Townley, J. Perfectionism and workaholism in employees: The role of work motivation. *Pers. Individ. Differ.* **2013**, *55*, 733–738. [CrossRef]
38. Zlate, S.; Cucui, G. Motivation and performance in higher education. *Procedia Soc. Behav. Sci.* **2015**, *180*, 468–476. [CrossRef]
39. Di Domenico, S.I.; Fournier, M.A. Able, ready, and willing: Examining the additive and interactive effects of intelligence, conscientiousness, and autonomous motivation on undergraduate academic performance. *Learn. Individ. Differ.* **2015**, *40*, 156–162. [CrossRef]
40. Johari, S.; Jha, K.N. Impact of work motivation on construction labor productivity. *J. Manag. Eng.* **2020**, *36*, 04020052. [CrossRef]
41. Vinodkumar, M.; Bhasi, M. Safety climate factors and its relationship with accidents and personal attributes in the chemical industry. *Saf. Sci.* **2009**, *47*, 659–667. [CrossRef]
42. Li, H.; Lu, M.; Hsu, S.C.; Gray, M.; Huang, T. Proactive behavior-based safety management for construction safety improvement. *Saf. Sci.* **2015**, *75*, 107–117. [CrossRef]
43. Tang, S.; Poon, S.; Ahmed, S.M.; Wong, F.K. *Modern Construction Project Management*; Hong Kong University Press: Hong Kong, China, 2003; Volume 2.
44. Choudhry, R.M.; Fang, D.; Ahmed, S.M. Safety management in construction: Best practices in Hong Kong. *J. Prof. Iss. Eng. Ed. Pr.* **2008**, *134*, 20–32. [CrossRef]
45. Alruqi, W.M.; Hallowell, M.R. Critical success factors for construction safety: Review and meta-analysis of safety leading indicators. *J. Constr. Eng. Manag.* **2019**, *145*, 04019005. [CrossRef]
46. Khalid, U.; Sagoo, A.; Benachir, M. Safety Management System (SMS) framework development–Mitigating the critical safety factors affecting Health and Safety performance in construction projects. *Saf. Sci.* **2021**, *143*, 105402. [CrossRef]

47. Yap, J.B.H.; Lee, W.K. Analysing the underlying factors affecting safety performance in building construction. *Prod. Plan. Control* **2020**, *31*, 1061–1076. [CrossRef]
48. Alruqi, W.M.; Hallowell, M.R.; Techera, U. Safety climate dimensions and their relationship to construction safety performance: A meta-analytic review. *Saf. Sci.* **2018**, *109*, 165–173. [CrossRef]
49. Kim, N.K.; Rahim, N.F.A.; Iranmanesh, M.; Foroughi, B. The role of the safety climate in the successful implementation of safety management systems. *Saf. Sci.* **2019**, *118*, 48–56. [CrossRef]
50. Hwang, B.G.; Shan, M.; Phuah, S.L. Safety in green building construction projects in Singapore: Performance, critical issues, and improvement solutions. *KSCE J. Civ. Eng.* **2018**, *22*, 447–458. [CrossRef]
51. Wachter, J.K.; Yorio, P.L. A system of safety management practices and worker engagement for reducing and preventing accidents: An empirical and theoretical investigation. *Accid. Anal. Prev.* **2014**, *68*, 117–130. [CrossRef] [PubMed]
52. Dedobbeleer, N.; Béland, F. A safety climate measure for construction sites. *J. Saf. Res.* **1991**, *22*, 97–103. [CrossRef]
53. Hallowell, M.R.; Hinze, J.W.; Baud, K.C.; Wehle, A. Proactive construction safety control: Measuring, monitoring, and responding to safety leading indicators. *J. Constr. Eng. Manag.* **2013**, *139*, 04013010. [CrossRef]
54. Schwatka, N.V.; Hecker, S.; Goldenhar, L.M. Defining and measuring safety climate: A review of the construction industry literature. *Ann. Occup. Hyg.* **2016**, *60*, 537–550. [CrossRef]
55. Awolusi, I.G.; Marks, E.D. Safety activity analysis framework to evaluate safety performance in construction. *J. Constr. Eng. Manag.* **2017**, *143*, 05016022. [CrossRef]
56. Chinda, T.; Mohamed, S. Structural equation model of construction safety culture. *Eng. Constr. Archit. Manag.* **2008**, *15*, 114–131. [CrossRef]
57. Neal, A.; Griffin, M.A. A study of the lagged relationships among safety climate, safety motivation, safety behavior, and accidents at the individual and group levels. *J. Appl. Psychol.* **2006**, *91*, 946. [CrossRef]
58. Kartam, N.A.; Bouz, R.G. Fatalities and injuries in the Kuwaiti construction industry. *Accid. Anal. Prev.* **1998**, *30*, 805–814. [CrossRef]
59. Nguyen, L.H. Relationships between critical factors related to team behaviors and client satisfaction in construction project organizations. *J. Constr. Eng. Manag.* **2019**, *145*, 04019002. [CrossRef]
60. Davis, K. Different stakeholder groups and their perceptions of project success. *Int. J. Proj. Manag.* **2014**, *32*, 189–201. [CrossRef]
61. Layard, R. Measuring subjective well-being. *Science* **2010**, *327*, 534–535. [CrossRef] [PubMed]
62. Chan, A.P.C.; Chan, A.P.L. Key performance indicators for measuring construction success. *Benchmarking Int. J.* **2004**, *11*, 203–221. [CrossRef]
63. Parfitt, M.K.; Sanvido, V.E. Checklist of critical success factors for building projects. *J. Manag. Eng.* **1993**, *9*, 243–249. [CrossRef]
64. Beaumont, P.B.; Leopold, J.W.; Coyle, J.R. The safety officer: An emerging management role? *Pers. Rev.* **1982**, *11*, 35–38. [CrossRef]
65. Murcia, J.; Gimeno, E.C.; Camacho, A.M. Measuring self-determination motivation in a physical fitness setting: Validation of the Behavioural Regulation in Exercise Questionnaire-2 (BREQ-2) in a Spanish sample. *J. Sports Med. Phys. Fit.* **2007**, *47*, 366–378.
66. Vallerand, R.J. Toward a hierarchical model of intrinsic and extrinsic motivation. *Adv. Exp. Soc. Psychol.* **1997**, *29*, 271–360.
67. Rovinelli, R.J.; Hambleton, R.K. On the use of content specialists in the assessment of criterion-referenced test item validity. *ERIC* **1976**, *2*, 49–60.
68. Fouzul Kareema, M.I.; Bt Zubairi, A.M. Item objective congruence analysis for multidimensional items content validation of a reading test in Sri Lankan university. *ERIC* **2021**, *15*, 106–117.
69. Taber, K.S. The use of Cronbach's alpha when developing and reporting research instruments in science education. *Res. Sci. Educ.* **2018**, *48*, 1273–1296. [CrossRef]
70. Department of Labor Protection and Welfare. *Statistics on Labor Protection and Welfare Fiscal Year 2013 to 2021*; DLPW: Bangkok, Thailand, 2021. Available online: https://www.labour.go.th/index.php/service-statistic/service-statistic-m/category/22-report2 (accessed on 25 December 2021).
71. Yamane, T. *Elementary Sampling Theory*; Prentice-Hall: Englewood Cliffs, NJ, USA, 1967.
72. Kim, T.K. Understanding one-way ANOVA using conceptual figures. *Korean J. Anesthesiol.* **2017**, *70*, 22–26. [CrossRef] [PubMed]
73. Norman, G. Likert scales, levels of measurement and the "laws" of statistics. *Adv. Health Sci. Educ.* **2010**, *15*, 625–632. [CrossRef]
74. Carifio, J.; Perla, R. *Resolving the 50-Year Debate around Using and Misusing Likert Scales*; Wiley-Blackwell Commerce Place: Hoboken, NJ, USA, 2008; Volume 42, pp. 1150–1152.
75. Maruyama, G. *Basics of Structural Equation Modeling*; Sage: New York, NY, USA, 1997.
76. Ullman, J.B.; Bentler, P.M. *Structural Equation Modeling*; Wiley: Hoboken, NJ, USA, 2012; Volume 2.
77. Williams, B.; Onsman, A.; Brown, T. Exploratory factor analysis: A five-step guide for novices. *Australas. J. Paramed.* **2010**, *8*, 1–13. [CrossRef]
78. Awang, Z. *SEM Made Simple: A Gentle Approach to Learning Structural Equation Modeling*; MPWS Rich Publication: Bangi, Malaysia, 2015.
79. Lopez, R.; Ahmad, J.; Zarim, N.Z.A. The psychometric properties of authentic leadership measurement in school context: Confirmatory factor analysis approach. *Turk. J. Physiother. Rehabil.* **2021**, *32*, 6975–6986.
80. Fornell, C.; Larcker, D.F. Evaluating structural equation models with unobservable variables and measurement error. *J. Mark. Res.* **1981**, *18*, 39–50. [CrossRef]
81. Lam, L.W. Impact of competitiveness on salespeople's commitment and performance. *J. Bus. Res.* **2012**, *65*, 1328–1334. [CrossRef]

82. Huang, C.C.; Wang, Y.M.; Wu, T.W.; Wang, P.A. An empirical analysis of the antecedents and performance consequences of using the moodle platform. *Int. J. Inf. Educ.* **2013**, *3*, 217. [CrossRef]
83. Iacobucci, D. Structural equations modeling: Fit indices, sample size, and advanced topics. *J. Consum. Psychol.* **2010**, *20*, 90–98. [CrossRef]
84. Doloi, H.; Sawhney, A.; Iyer, K.C. Structural equation model for investigating factors affecting delay in Indian construction projects. *Constr. Manag. Econ.* **2012**, *30*, 869–884. [CrossRef]
85. Bentler, P.M.; Chou, C.P. Practical issues in structural modeling. *Sociol. Methods Res.* **1987**, *16*, 78–117. [CrossRef]
86. Deci, E.L.; Flaste, R. *Why We Do What We Do: The Dynamics of Personal Autonomy*; GP Putnam's Sons: New York, NY, USA, 1995.
87. Gabriel, A.S.; Campbell, J.T.; Djurdjevic, E.; Johnson, R.E.; Rosen, C.C. Fuzzy profiles: Comparing and contrasting latent profile analysis and fuzzy set qualitative comparative analysis for person-centered research. *Organ. Res. Methods* **2018**, *21*, 877–904. [CrossRef]

buildings

Article

Psychological Distress and Work Engagement of Construction Workers during the COVID-19 Pandemic: A Differential Study by Sex

Carlos Gómez-Salgado [1], Juan Carlos Camacho-Vega [2], Regina Allande-Cussó [3], Carlos Ruiz-Frutos [4,5], Mónica Ortega-Moreno [6], Marta Linares-Manrique [7], Juan Jesús García-Iglesias [4], Javier Fagundo-Rivera [8,†], Luciano Rodríguez-Díaz [9], Juana María Vázquez-Lara [9] and Juan Gómez-Salgado [4,5,*,†]

1 School of Doctorate, University of Huelva, 21007 Huelva, Spain
2 Department of Building Construction II, Higher Technical School of Building Engineering, University of Seville, 41012 Sevilla, Spain
3 Department of Nursing, Faculty of Nursing, Physiotherapy and Podiatry, University of Seville, 41009 Sevilla, Spain
4 Department of Sociology, Social Work and Public Health, Faculty of Labour Sciences, University of Huelva, 21007 Huelva, Spain
5 Safety and Health Postgraduate Programme, Universidad Espíritu Santo, Guayaquil 092301, Ecuador
6 Department of Economy, Faculty of Labour Sciences, University of Huelva, 21007 Huelva, Spain
7 Department of Nursing, Faculty of Health Sciences of Melilla, University of Granada, 52005 Melilla, Spain
8 Centro Universitario de Enfermería Cruz Roja, Universidad de Sevilla, 41009 Sevilla, Spain
9 Department of Nursing, Faculty of Health Sciences of Ceuta, University of Granada, 51005 Ceuta, Spain
* Correspondence: salgado@uhu.es; Tel.: +34-959-219-700
† These authors have contributed equally to this work and share senior authorship.

Citation: Gómez-Salgado, C.; Camacho-Vega, J.C.; Allande-Cussó, R.; Ruiz-Frutos, C.; Ortega-Moreno, M.; Linares-Manrique, M.; García-Iglesias, J.J.; Fagundo-Rivera, J.; Rodríguez-Díaz, L.; Vázquez-Lara, J.M.; et al. Psychological Distress and Work Engagement of Construction Workers during the COVID-19 Pandemic: A Differential Study by Sex. *Buildings* **2024**, *14*, 773. https://doi.org/10.3390/buildings14030773

Academic Editors: David Arditi and Mariusz Szóstak

Received: 15 January 2024
Revised: 28 February 2024
Accepted: 8 March 2024
Published: 13 March 2024

Abstract: Since the beginning of the COVID-19 pandemic, a major impact on the mental health of the population has been observed, with women being one of the most affected groups. From the lockdown to "de-escalation" phases, sex differences have been recognised as significant determinants of mental health. Thus, equally ensuring physical and mental protection at work remains one of the challenges faced by industrial companies, especially in the construction sector, where the percentage of employed women has increased in recent years. This study aims to examine the impact of sex differences on psychological distress and work engagement in the productive construction sector, as well as related variables. For this, a cross-sectional descriptive study was performed. Descriptive statistical analyses were completed, and non-parametric Mann-Whitney U and Chi-squared tests were used to identify differences between men and women. This was followed by logistic regression analysis by sex. Psychological distress is more prevalent among women, even after controlling for most variables. Both sexes receive equal preventive measures and training from the companies, yet women still experience higher levels of psychological distress. At the beginning of the pandemic, women reported higher levels of anxiety and fear of COVID-19 and of perceived danger associated with the pandemic than men. However, these differences were not present by 2023. For men, work engagement appeared to be a determining factor for a stable mental health, while for women, health and physical status seemed to be more influential. In both sexes, psychological distress was found to be conditioned by mental and emotional well-being. In a sector where women are increasingly present, the differences observed in terms of how physical and mental health are affected across the two sexes justify the need to promote data analysis that acknowledges this reality.

Keywords: mental health; construction workers; construction industry; sex; work conditions; anxiety; stress; fear; COVID-19; public health

1. Introduction

The COVID-19 health crisis has caused a severe global economic downturn, leading to psychological insecurity [1]. This has had a disproportionate impact on the most vulnerable groups [2], and has required significant technical and lifestyle adjustments [3]. Since its onset, the World Health Organization (WHO) has classified the COVID-19 pandemic as a public health concern due to its association with occupational exposure and close contact between co-workers [4]. Consequently, many governments authorised a gradual return to work in various sectors, including construction, as part of the 'de-escalation' process after lockdown restrictions [5]. The construction industry, known for its high number of health hazards and one of the highest occupational accident rates globally [6,7], faces occupational health and safety challenges [8]. Certain working conditions in the construction sector, such as stress, anxiety, and fear, have been linked to mental health issues among workers, leading to a higher incidence of workplace accidents [9]. Additionally, prolonged working hours have been associated with certain physical ailments, like musculoskeletal disorders [10]. Technical staff may experience pressure to expedite building projects, resulting in decreased productivity and increased absenteeism [11,12]. Moreover, small companies may perceive COVID-19 risk-reduction measures as less effective compared to their effect in larger companies [13]. Highly educated workers have shown greater satisfaction with their organisation's response to the pandemic, yet they experienced heavier workloads and increased anxiety and depression, particularly among labourers [14,15]. Similarly, stress related to workplace safety measures can affect participation, with psychological factors acting as moderators [16].

The construction sector is predominantly male-dominated worldwide, although in Spain the percentage of women has increased since 2016 and reached 11.1% in 2022. Women are more likely to be found in administrative positions (46.8%) and professions requiring higher education, especially architecture, engineering, and urban planning (12.2%), as well as in cleaning activities. Women are predominantly employed in large and medium-sized companies, with the majority working in building construction (30.5%) and to a lesser extent in civil engineering (4.1%), with the majority aged between 30 and 54 and 74.3% working full-time [9]. In 2020, the percentage of women in building construction in Andalusia, southern Spain, was 11.75%, while the sector as a whole had a female share of 8.03% in 2022 [17], with a greater growth rate over the last 13 years (3.95%) than for men (2.43%) [18].

As stated before, from the onset of the COVID-19 pandemic, a major impact on the mental health of the population was observed [19], with women being one of the most affected groups [20]. Sex differences are recognised as significant determinants of mental health inequalities. However, recent studies indicate that public policies are still lacking, resources are scarce, bureaucratisation is high, and women's participation in decision-making is still limited, thus showing that the implementation of sex-sensitive policies, which are essential for achieving sex equality in health, is poor [21]. These differences in mental health status by sex have been noted by WHO [22], with women having experienced higher levels of mental distress than men during the pandemic [23]. Additionally, variations exist throughout the different stages of the pandemic, particularly when preventive measures, social distancing, or isolation became commonplace among the workforce [24,25]. It has been already observed that psychological distress was significantly and negatively correlated with work engagement [26], and that work engagement moderated the direct and indirect effects of the stress related to job insatisfaction [27].

These facts, together with the increase in the percentage of female workers in the construction sector [28], led to the research question of this study, which was aimed to know what differences in the effects of COVID-19 on psychological distress and work engagement exist between male and female construction workers.

In this context, this study aimed to investigate disparities in the impact that the COVID-19 pandemic has had on the psychological distress and work engagement of construction workers of both sexes in southern Spain and to correlate these impacts with socio-demographic and health-related factors. These findings may help identify the risk factors that influence the mental well-being of women and men in the construction industry during pandemics. This will in turn facilitate the establishment of preventive measures, including safeguards for mental health, by establishing the respective pathways.

2. Materials and Methods

2.1. Study Procedure

A cross-sectional descriptive study using online questionnaires was carried out. The study population comprised workers in the construction sector in the autonomous region of Andalusia, Spain. The government was interested in understanding the situation of the construction sector and any differences in its workers based on sex. Thus, the research team worked in collaboration with the Andalusian Observatory of Occupational Diseases of the Andalusian Institute for the Prevention of Occupational Risks [29].

Considering a population size, whatever their occupation within the sector, for the year 2023 of 204,100 subjects, of which 8.03% were women [17], a total sample of 384 individuals was estimated with a confidence level of 95%, precision of 3.5%, and adjustment for losses of 10%. However, using non-probability convenience sampling, the final sample size obtained was 857 participants, of whom 154 were women (17.9%). The study was conducted from March to May 2022.

A pilot study was conducted on a sample of 50 participants to identify comprehension errors, test the data collection tool, and estimate the mean total completion time. Following the pilot study, no errors were identified, and the mean completion time was 9 min.

The questionnaire was distributed online by sending emails to companies in the sector located in Andalusia (southern Spain), as well as to workers' unions, sector associations, and construction material distribution companies. To this end, an application was made to the Spanish Government's Register of Accredited Companies and Workers Associations [30], specifically for companies in the construction sector in Andalusia. After obtaining a list of 21,000 companies, the research team contacted all of them by e-mail presenting the project and the objectives of the study, and requesting their collaboration in disseminating the link to the data collection tool among their employees. Thus, the companies that accepted the collaboration disseminated by e-mail the objectives of the project and the link to the data collection tool among those employees who had agreed to be contacted for training and research purposes in their work affiliation with that company.

A fact sheet outlining the research project, including its objectives, and a leaflet with the link to the data collection tool and a QR code were included. All participants were duly informed of the voluntary nature of their participation, the confidentiality of their data, and that they were free to withdraw from the project at any time. To participate, participants had to be 18 years or older, an active worker in the construction sector in Andalusia, and give their informed consent (Figure 1).

2.2. Instruments

The Emotional Impact Questionnaire COVID-19 (EIQ COVID-19), previously validated by Gómez-Salgado J. and collaborators [28] and which includes questions adapted from previous studies [31], was used. This questionnaire had been previously used in numerous studies in Spain, and in Brazil to assess the emotional well-being of its population about COVID-19 [32].

This instrument was supplemented with other variables specific to the construction sector. Thus, the questionnaire included socio-demographic data such as age; sex; province; employment status (self-employed, full-time employee, part-time employee, temporary redundancy procedure, unemployed); professional category (manager, skilled worker, intermediate manager, manual worker, administrative or cleaning staff); type of work; type of construction project (building, industrial, or civil engineering); place of work (outdoors or indoors); economic income (sufficient to make ends meet or not, and euros earned per month in the household); persons living with them; the size of the dwelling in square metres; and use of company canteens (yes, no, or only when there were few people, they did not exist, or they were closed during the pandemic).

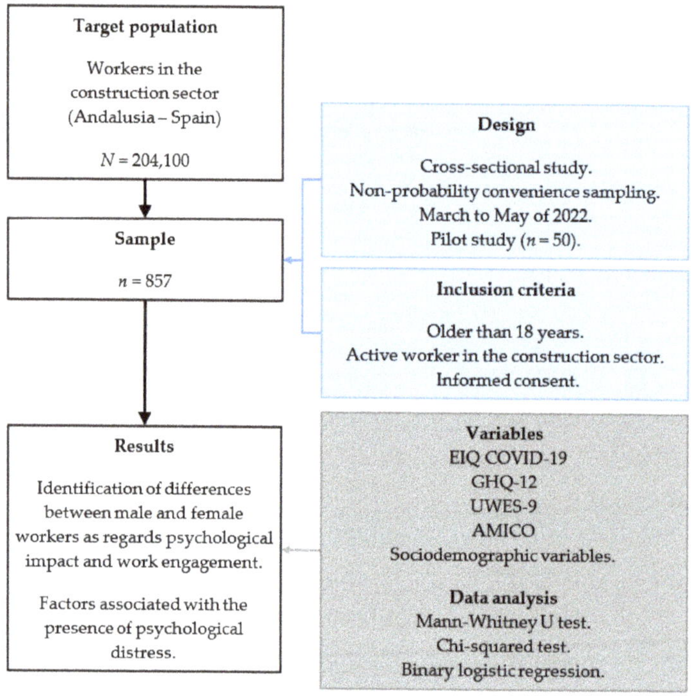

Figure 1. Research methodology flowchart.

Data on COVID-19 disease were also included: diagnosis; isolation; severity; hospitalisation; vaccination and side effects; availability and use of preventive measures; and training received. Perceptions of being protected at work and whether their work had been affected by the pandemic were also included.

As in the case of the EIQ COVID-19, two more questionnaires were selected and validated in Spain, which had already been used to assess psychological distress concerning COVID-19 in Spain in previous studies [23,33]. To measure psychological distress, the Goldberg General Health Questionnaire (GHQ-12) was used, which consists of 12 items with 4 response options and a total score ranging from 0 to 12 points [34]. This is a self-administered instrument that evaluates psychological well-being and identifies non-psychotic disorders. The cut-off point used was >3, following the cut-off point used in national surveys in Spain and adhering to the authors Rocha et al. [35]. The GHQ-12 has shown good reliability in the different studies, with Cronbach's alphas varying between 0.82 and 0.86 [34]. The internal consistency index obtained was $\alpha = 0.905$, being higher than 0.7 and demonstrating good reliability.

Work engagement (WE) was measured using the short version of the Utrecht Work Engagement Scale (UWES-9) [36]. The UWES-9 scale, based on the Job Demands-Resources model, has been empirically validated in various work contexts including industry, supporting its usefulness. It is an instrument designed to be self-administered and consists of nine items, with Likert-type response options ranging from 0 (Never) to 6 (Always). It consists of three dimensions, namely vigour, dedication, and absorption. The score for each of the different dimensions is obtained by adding the items of each dimension and dividing the result by the number of items that compose each dimension. The Spanish version of the instrument achieved the following Cronbach's internal consistency indices:

vigour (α = 0.82), dedication (α = 0.86), and absorption (α = 0.8), being higher than 0.7 and demonstrating good reliability concerning the construct measured by each scale [37].

To measure anxiety and fear of COVID-19, four ad hoc items were designed to assess perceived level of anxiety and fear and level of danger on a scale ranging from 0 to 10 points at the onset of the pandemic and at the time of data collection for the present study. In addition, the Anxiety and Fear of COVID-19 (AMICO) scale, designed and validated in previous studies by Gómez-Salgado et al. in 2021 [38], was also included. The scale obtained an internal consistency index of α = 0.92, and the established cut-off point was 6.4 points [34]. The response options of the AMICO scale range from 1 to 10 points, where 1 corresponds to strongly disagree and 10 to strongly agree. The total score of the scale is obtained by calculating the mean score of the self-reported responses.

2.3. Data Analysis

The statistical analyses were carried out using SPSS version 27 [39].

A descriptive analysis of the variables used in the study was performed, providing frequencies, means, and standard deviations according to the type of variable. Subsequently, the normality of the data distribution was analysed using the Kolmogorov-Smirnov test, which showed the non-normality of the sample. To determine the existence or not of significant differences depending on the presence of distress and between men and women, non-parametric tests such as the Mann-Whitney U test or the Chi-squared test were used. Specifically, the chi-squared test has been used for categorical variables, while the Mann-Whitney U test was used to compare two independent groups of an ordinal variable [40].

A binary logistic regression analysis was also carried out to assess sex differences [41] on the presence or absence of psychological distress and to identify those variables among the ones studied that played a relevant role. The variables were included based on statistical significance tests. In this sense, UWES score, AMICO score, current level of anxiety and fear of COVID-19, effect of the pandemic on mental/emotional well-being, and the score for health and physical status were entered. ORs were estimated and confidence intervals were provided for this measure of association. To verify the adequacy of the model, different goodness-of-fit measures were used, such as the Hosmer-Lemeshow test, the percentage of correctly classified values, or the sensitivity and specificity.

2.4. Ethical Considerations

The 2013 Declaration of Helsinki was considered [42]. Participants were required to explicitly provide informed consent before their participation, which was voluntary and confidential, and the data collected were processed following the current laws on personal data protection and digital rights of 2018 [43] and the Spanish Biomedical Research Act [44]. The project was approved by the Research Ethics Committee of Huelva, part of the Andalusian Ministry of Health (PI 036/20).

3. Results

3.1. Socio-Demographic Characteristics by Sex

The percentage of women in the study (17.9%) was lower than that of men (81.7%), in line with the number of workers in the sector in Andalusia in 2023, which was 8.03%, according to data from the Andalusian Institute of Statistics and Cartography [17].

The mean age of women, 38.96 (SD = 9.98), was lower than that of men, 43.56 years (SD = 10.4), $p < 0.001$, as was the percentage of women with a partner (42.2%) lower than that of men (63.3%), $p < 0.001$. Regarding socio-economic level, the percentage of women who stated that their household income was over 1200 euros per month (63.6%) was lower than that of men (78.0%), $p < 0.001$. No statistically significant differences were found regarding the question of whether they considered that they had sufficient income to make ends meet nor regarding the size of the dwelling (Table 1).

Table 1. Socio-demographic characteristics by sex.

Variable	Total	Men	Women	X^2 Test	GFI
	n (%)	*n* (%)	*n* (%)	*p*-Value	
Sex	857	703 (81.7%)	154 (17.9%)		
Age [mean (SD)]	42.7 (10.4)	43.56 (10.3)	38.96 (9.98)	<0.001 *	
Marital status					
Married or cohabiting	524 (61.1%)	459 (63.3%)	65 (42.2)	<0.001	0.967
Other situations	333 (38.9%)	244 (34.7%)	89 (57.8)		
0–50 m^2	28 (3.3%)	21 (3.0%)	7 (4.5%)	0.734	0.999
51–75 m^2	154 (18.0%)	123 (17.5%)	31 (20.1%)		
76–100 m^2	310 (36.2%)	258 (36.7%)	52 (33.8%)		
101–125 m^2	169 (19.7%)	143 (20.3%)	26 (16.9%)		
126–150 m^2	105 (12.3%)	85 (12.1%)	20 (13.0%)		
More than 151 m^2	91 (10.6%)	73 (10.4%)	18 (11.7%)		
Yes	373 (43.5%)	299 (42.5%)	74 (48.1%)	0.211	0.998
No or depending on the month	484 (56.5%)	404 (57.5%)	80 (51.9%)		
Between 0 and 1200 euros	211 (24.6%)	155 (22.0%)	56 (36.4%)	<0.001	0.984
More than 1200 euros	646 (75.4%)	548 (78.0%)	98 (63.6%)		
Self-employed	104 (12.1%)	87 (12.4%)	17 (11.0%)	<0.001	0.994
Full-time employee	673 (78.5%)	563 (80.1%)	110 (71.4%)		
Part-time employee	51 (6.0%)	30 (4.3%)	21 (13.6%)		
Temporary Redundancy Procedure	5 (0.6%)	4 (0.6%)	1 (0.6%)		
Unemployed	24 (2.8%)	19 (2.7%)	5 (3.2%)		
Degree of responsibility					
Managers and skilled workers	207 (24.2%)	147 (20.9%)	60 (40.0%)	<0.001	0.939
Intermediate management	135 (15.8%)	114 (16.2%)	21 (13.6%)		
Manual workers	409 (47.7%)	392 (55.8%)	17 (11.0%)		
Others (administration staff, cleaning…)	106 (12.4%)	50 (7.1%)	56 (35.4%)		
Type of project					
Building work	489 (57.1%)	411 (58.5%)	78 (50.6%)	0.033	0.997
Civil Works	158 (18.4%)	117 (16.6%)	41 (26.6%)		
Industrial Works	123 (14.4%)	102 (14.5%)	21 (13.6%)		
More than one type of work	86 (10.0%)	73 (10.4%)	13 (8.4%)		
Place of work					
Outdoors	357 (41.7%)	319 (45.4%)	38 (24.7%)	<0.001	0.974
Indoors (of buildings, facilities….)	500 (58.3%)	384 (54.6%)	116 (75.3%)		
Use of canteens					
Yes	122 (14.2%)	111 (15.8%)	11 (7.1%)	0.003	0.996
Yes, only when there were not many workers	85 (9.9%)	74 (10.5%)	11 (7.1%)		
No, they have been closed during the pandemic	61 (7.1%)	42 (6.0%)	19 (12.3%)		
No, there are no canteens in the places I work in	184 (21.5%)	145 (20.6%)	39 (25.3%)		
No, I never use them	366 (42.7%)	302 (43.0%)	64 (41.6%)		
Other cases	39 (4.6%)	29 (4.1%)	10 (6.5%)		

*: Mann-Whitney U test; GFI: Goodness-of-Fit Index.

In terms of employment status, there were significant differences between men and women ($p < 0.001$), with a higher percentage of part-time workers among women (13.6%) than among men (4.3%) and higher percentages of full-time workers among men (80.1%) than among women (70.4%). Similarly, there were differences concerning the degree of responsibility at work $p < 0.001$, with 40.0% of the women occupying managerial or skilled positions, compared to 20.9% of the men. On the other hand, while 55.8% of men held manual labour positions, the percentage was 11.0% for women. It should also be noted that women occupied 35.4% of the 'Other' section (administration, cleaning, etc.) and men, 7.1% (Table 1).

Regarding the type of work, differences were found between women and men $p = 0.033$, with a higher proportion of both sexes working in 'Building works', whereas a higher proportion of women were engaged in 'Civil engineering' (26.6%) than men (16.6%). Regarding the place of work, the proportion of women working outdoors (24.7%) was lower than that of men (45.4%). It was also observed that women were less likely to use canteens ($p = 0.003$) (Table 1).

3.2. Psychological Distress by Socio-Demographic Characteristics and Differentiated by Sex

Overall, 29.2% of respondents had psychological distress (GHQ > 3), with statistically significant differences by sex, i.e., higher among women (37.7%) than among men (27.3%) ($p = 0.37$). In terms of marital status, among those without a partner, women (43.8%) were more likely to have psychological distress than men (31.1%) ($p = 0.031$) (Table 2).

Table 2. Psychological distress by socio-demographic characteristics and differentiated by sex.

	Men				Women					
	No GHQ < 3		Yes GHQ > 3		No GHQ < 3		GFI		χ^2 Test	GFI
	Cases	%	Cases		Cases	%	Cases	%	*p*-Value	
Total	511	72.7%	192	27.3%		62.3%	58	37.7%	0.037 *	
Marital status										
Married or cohabiting	343	74.7%	116	0.999	46	70.8%	19	29.2%	0.495	0.999
Other situations	168	68.9%	76	0.986	50	56.2%	39	43.8%	0.031	0.986
Approximately how many square metres (m^2) does your dwelling have?										
0–50 m^2	11	52.4%	10		4	57.1%	3	42.9%	N/A	
51–75 m^2	89	72.4%	34	0.984	18	58.1%	13	41.9%	0.122	0.984
76–100 m^2	198	76.7%	60	0.979	31	59.6%	21	40.4%	0.010	0.979
101–125 m^2	97	67.8%	46	0.982	13	50.0%	13	50.0%	0.079	0.982
126–150 m^2	66	77.6%	19	0.998	17	85.0%	3	15.0%	0.467	0.998
More than 151 m^2	50	68.5%	23	0.999	13	72.2%	5	27.8%	0.759	0.999
Do you consider that your income is sufficient to make ends meet?										
Yes	232	77.6%	67	0.991	50	67.6%	24	32.4%	0.072	0.991
No, or depending on the month	279	69.1%	125	0.992	46	57.5%	34	42.5%	0.044	0.992
How much income do you receive in your household every month?										
Between 0 and 1200 euros	103	66.5%	52	0.997	34	60.7%	22	39.3%	0.441	0.997
More than 1200 euros	408	74.5%	140	0.992	62	63.3%	36	36.7%	0.022	0.992
Employment situation										
Self-employed	60	69.0%	27	0.996	13	76.5%	4	23.5%	0.536	0.996
Full-time employee	419	74.4%	144	0.982	64	58.2%	46	41.8%	<0.001	0.982
Part-time employee	15	50.0%	15	0.929	16	76.2%	5	23.8%	0.059	0.929
Temporary Redundancy Procedure	2	50.0%	2			0.0%	1	100.0%	N/A	
Unemployed	15	78.9%	4		3	60.0%	2	40.0%	N/A	

Table 2. *Cont.*

	Men				Women					
	No GHQ < 3		Yes GHQ > 3		No GHQ < 3		GFI		χ^2 Test	GFI
	Cases	%	Cases		Cases	%	Cases	%	*p*-Value	
Degree of responsibility										
Managers and skilled workers	104	70.7%	43	0.982	34	56.7%	26	43.3%	0.051	0.982
Intermediate management	86	75.4%	28	1.000	13	61.9%	8	38.1%	0.197	1.000
Manual workers	287	73.2%	105	1.000	13	76.5%	4	23.5%	0.766	1.000
Others (administration staff, cleaning...)	34	68.0%	16	0.998	36	64.3%	20	35.7%	0.687	0.998
Type of project										
Building work	296	72.0%	115	0.997	51	65.4%	27	34.6%	0.237	0.997
Civil Works	98	83.8%	19	0.942	25	61.0%	16	39.0%	0.003	0.942
Industrial Works	70	68.6%	32	0.992	12	57.1%	9	42.9%	0.309	0.992
More than one type of work	47	64.4%	26	1.000	8	61.5%	5	38.5%	0.844	1.000
Place of work										
Outdoors	233	73.0%	86	0.976	19	50.0%	19	50.0%	0.003	0.976
Indoors (of buildings, facilities....)	278	72.4%	106	0.997	77	66.4%	39	33.6%	0.211	0.997
Use of canteens										
Yes	139	75.1%	46	0.993	14	63.6%	8	36.4%	0.246	0.993
No	351	71.8%	138	0.992	75	61.5%	47	38.5%	0.027	0.992

*: Chi-squared test; GFI: Goodness-of-Fit Index; N/A: not available.

At the socio-economic level, among the respondents who reported that their income was not enough to make ends meet, or was sometimes not enough to make ends meet, the percentage with psychological distress was higher among women (42.5%) than among men (30.1%) ($p = 0.044$). Similarly, among participants who reported a household income of more than 1200 euros per month, psychological distress was higher in women (36.7%) than among men (25.5%) ($p = 0.022$), with no statistically significant differences observed among those with an income below 1200 euros (Table 2).

With regard to the work situation, the percentage of female full-time employees with psychological distress was higher (41.8%) than that of male employees under the same conditions (25.6%) ($p < 0.001$). There were no differences between men and women regarding the degree of responsibility. In the group of managers and skilled workers, the higher percentage of psychological distress among women was not found to be statistically significant ($p = 0.051$). The higher proportion of women reporting psychological distress in relation to the type of work they did was only statistically significant in the case of 'civil work', where it was 39.0% for women and 16.2% for men ($p = 0.003$). Women working outdoors were found to be more likely to experience psychological distress (50.0%) than men under the same conditions (27.0%) ($p = 0.003$). Women who did not use the canteen were more likely to have psychological distress (38.5%) than male workers who did not use it either (28.2%) ($p = 0.027$) (Table 2).

Psychological Distress, Working Conditions and COVID-19 Pandemic by Sex

The higher percentage of overall psychological distress among women (37.7%) than among men (27.3%) ($p = 0.037$) was also observed when specifically analysing cases where it was stated that the companies provided means of protection to avoid infection: 39.4% among women versus 24.0% among men ($p < 0.001$); when they had received specific training from the company on ways of infection, routes of transmission, prevention measures, or warning signs of COVID-19: 40.1% among women and 23.1% among men ($p = 0.001$); and when they felt safe and protected from infection during the performance of their work duties: 33.3% of women with psychological distress and 19.3% of men ($p = 0.006$); also when they had been vaccinated against COVID-19: 37.3% of women with psychological distress and 27.2% of men ($p = 0.013$), as well as when they reported side effects after vaccination: 44.2% of women with psychological distress and 31.7% of men ($p = 0.041$) (Table 3).

Table 3. Psychological distress, working conditions, and COVID-19 pandemic by sex.

	Men				Women					
	No	GHQ < 3	Yes	GFI	Yes	GHQ < 3	Yes	GHQ > 3	χ^2 Test	GFI
	Cases	%	Cases		Cases	%	Cases	%	*p*-Value	
Total	511	72.7%	192	27.3%		62.3%	58	37.7%	0.037 *	
Have you been diagnosed with COVID-19?										
Yes	199	69.1%	89	0.999	47	66.2%	24	33.8%	0.637	0.999
No	312	75.2%	103	0.982	49	59.0%	34	41.0%	0.003	0.982
Has anyone in your circle been diagnosed with COVID-19?										
Yes	446	72.9%	166	0.995	88	64.2%	49	35.8%	0.043	0.995
No	65	71.4%	26	0.964	8	47.1%	9	52.9%	0.049	0.964
Has anyone in your circle died from COVID-19?										
Yes	57	69.5%	25	1.000	17	68.0%	8	32.0%	0.886	1.000
No	454	73.1%	167	0.980	79	61.2%	50	38.8%	0.007	0.980
Have you been isolated for having the disease or been in contact with a positive person?										
Yes	268	70.5%	112	0.977	65	62.5%	39	37.5%	0.117	0.977
No	243	75.2%	80	0.995	31	62.0%	19	38.0%	0.049	0.995
Have you been hospitalised for COVID-19?										
Yes	4	33.3%	8		1	100.0%		0.0%	N/A	
No, but I had mild symptoms	180	69.2%	80	0.998	45	64.3%	25	35.7%	0.430	0.998
No	327	75.9%	104	0.983	50	60.2%	33	39.8%	0.003	0.983
Have your working conditions been affected by the pandemic?										
Yes	222	62.2%	135	0.995	50	53.2%	44	46.8%	0.113	0.995
No	289	83.5%	57	0.996	46	76.7%	14	23.3%	0.197	0.996
Have your managers or your company provided and do they provide you with the means of protection to avoid contagion (masks, gloves, gels, eye protection)?										
Yes	389	76.0%	123	0.982	66	60.6%	43	39.4%	<0.001	0.982
No	117	63.2%	68	1.000	26	63.4%	15	36.6%	0.984	1.000
Other	5	83.3%	1		4	100.0%		0.0%	N/A	

Table 3. *Cont.*

	Men				Women					
	No	GHQ < 3	Yes	GFI	Yes	GHQ < 3	Yes	GHQ > 3	χ^2 Test	GFI
	Cases	%	Cases		Cases	%	Cases	%	p-Value	
Did you receive or have you ever received specific training on COVID-19 disease (transmission routes, self-protection measures, warning signs) organised by your managers or your company?										
Yes	299	76.9%	90	0.981	42	60.0%	28	40.0%	0.003	0.981
No	207	68.1%	97	0.998	48	62.3%	29	37.7%	0.338	0.998
Other (self-employed, other means of training…)	5	50.0%	5		6	85.7%	1	14.3%	N/A	
In general, do you feel safe and protected from infection in the performance of your job duties?										
Yes, totally safe	305	80.7%	73	0.983	54	66.7%	27	33.3%	0.006	0.983
Somewhat safe	195	67.0%	96	0.992	39	56.5%	30	43.5%	0.101	0.992
No, not safe at all	11	32.4%	23		3	75.0%	1	25.0%	N/A	
Have you been vaccinated against COVID-19?										
Yes	504	72.8%	188	0.993	94	62.7%	56	37.3%	0.013	0.993
No	7	63.6%	4		2	50.0%	2	50.0%	N/A	
Have you had any side effects after vaccination?										
Yes	200	68.3%	93	0.989	43	55.8%	34	44.2%	0.041	0.989
No	311	75.9%	99	0.997	53	68.8%	24	31.2%	0.193	0.997
Do you think that the situation experienced during the COVID-19 pandemic has negatively affected your mental/emotional well-being?										
Yes	215	56.9%	163	0.998	56	52.3%	51	47.7%	0.404	0.998
No	296	91.1%	29	0.995	40	85.1%	7	14.9%	0.196	0.995

*: Chi-squared test; GFI: Goodness-of-Fit Index; N/A: not available.

As can be seen in Table 3, this greater incidence of psychological distress in women was seen in cases where they had not been diagnosed with COVID-19, nor had people around them been diagnosed with COVID-19, had not had to isolate themselves because of illness or contact with a positive person, and had not been hospitalised because of COVID-19. In contrast, there was no difference by sex in terms of greater occurrence of psychological distress when asked whether the situation experienced during the COVID-19 pandemic had negatively affected their mental/emotional well-being.

3.3. Psychological Distress Related to Anxiety and Fear of COVID-19 by Sex

As can be seen in Table 4, the mean age of women with psychological distress was lower (mean 36.6 years old) than that of women without it (mean 40.3 years old), and no such difference was observed among men (43.9 years old vs 42.6 years old; $p = 0.125$). The results also showed that women in the sample were younger than men (38.9 years old vs. 43.56 years old; $p < 0.001$). Health and physical status were reported to be worse in both women ($p = 0.004$) and men with PD ($p < 0.001$), but the difference by sex was not statistically significant ($p = 0.075$).

The level of anxiety and fear of COVID-19 at the start of the pandemic ($p < 0.001$) and perceived anxiety and fear at the time of answering the questionnaire were higher among participants with PD, for both women (4.03 vs. 2.79; $p < 0.001$) and men (4.31 vs 2.86; $p < 0.001$). Again, there was a difference by sex, with more women reporting anxiety and fear at the start of the pandemic than men ($p < 0.001$), but this difference by sex was not statistically significant in terms of perceived anxiety and fear of COVID-19 at the time of answering the questionnaire.

Table 4. Psychological distress related to anxiety and fear of COVID-19 by sex.

	Men (n = 703)				Women (n = 154)				Men-Women
	Total	No GHQ < 3	Yes GHQ > 3	Mann-Whitney U Test	Total	No GHQ < 3	Yes GHQ > 3	Mann-Whitney U Test	Mann-Whitney U Test
	Mean (SD)	Mean (SD)	Mean (SD)	p-Value	Mean (SD)	Mean (SD)	Mean (SD)	p-Value	p-Value
Age	43.56 (10.3)	43.91 (10.33)	42.64 (10.15)	0.125	38.96 (9.98)	40.35 (10.07)	36.66 (9.56)	0.027	<0.001
General health and physical status *:	7.85 (1.59)	7.98 (1.53)	7.49 (1.70)	<0.001	7.69 (1.22)	7.93 (1.19)	7.31 (1.15)	0.004	0.075
What level of anxiety and fear of COVID-19 did you perceive at the start of the pandemic? *	7.05 (2.67)	6.73 (2.69)	7.89 (2.43)	<0.001	8.41 (2.19)	8.07 (2.42)	8.97 (1.58)	0.019	<0.001
What level of anxiety and fear of COVID-19 do you currently perceive? *	3.25 (2.10)	2.86 (1.90)	4.31 (2.25)	<0.001	3.26 (1.94)	2.79 (1.68)	4.03 (2.07)	<0.001	0.631
How dangerous did you consider COVID-19 to be at the beginning of the pandemic? *	7.81 (2.46)	7.68 (2.47)	8.17 (2.41)	<0.001	8.69 (2.12)	8.40 (2.34)	9.19 (1.53)	0.026	<0.001
How dangerous do you consider COVID-19 to be at present? *	3.92 (2.37)	3.62 (2.27)	4.72 (2.42)	<0.001	3.82 (2.21)	3.38 (2.05)	4.55 (2.25)	0.001	0.786
AMICO mean score *	4.10 (1.65)	3.76 (1.47)	5.02 (1.77)	<0.001	4.30 (1.71)	3.87 (1.58)	5.00 (1.68)	<0.001	0.157

* Score from 1 to 10.

The mean score of the AMICO questionnaire (anxiety and fear of COVID-19) was higher among those who perceived PD (5.02 in men; 5.00 in women) than among those who did not (3.76 in men; 3.87 in women), for both women and men ($p < 0.001$), with no statistically significant differences by sex ($p = 0.157$) (Table 4).

The perceived level of danger of COVID-19 at the start of the pandemic was higher among those who had PD than among those who did not have it, for both women and men, and this difference was also found to be greater among women ($p < 0.001$). The perceived level of danger of COVID-19 at the time of answering the questionnaire was also higher among those with PD for both women and men ($p < 0.001$), but there was no statistically significant difference by sex ($p = 0.786$) (Table 4).

Psychological Distress Related to the Level of Work Engagement (UWES) by Sex

It was found that men with psychological distress had lower levels of work engagement than those who did not report psychological distress, both in the total score (UWES) and in its three dimensions (vigour, dedication, and absorption), a difference that was not observed for women. Differences by sex were indeed found for the global level of work engagement (UWES), which was lower for women (M = 38.60; SD = 13.12) than for men (M = 41.3; SD = 12.32) $p = 0.010$; these differences were maintained across the three dimensions of the scale (Table 5).

Table 5. Psychological distress related to the level of work engagement (UWES) by sex.

	Men (n = 703)				Women (n = 154)				Men-Women
	TOTAL	NO GHQ < 3	Yes GHQ > 3	Mann-Whitney U Test	Total	No GHQ < 3	Yes GHQ > 3	Mann-Whitney U Test	Mann-Whitney U Test
	Mean (SD)	Mean (SD)	Mean (SD)	p-Value	Mean (SD)	Mean (SD)	Mean (SD)	p-Value	p-Value
Vigour	13.56 (4.23)	14.18 (3.87)	11.91 (4.66)	<0.001	12.64 (4.39)	12.79 (4.40)	12.40 (4.32)	0.667	0.011
Dedication	13.88 (4.39)	14.42 (4.09)	12.45 (4.81)	<0.001	12.92 (4.60)	12.99 (4.64)	12.81 (4.48)	0.706	0.007
Absorption	13.86 (4.26)	14.24 (3.96)	12.84 (4.81)	<0.001	13.03 (4.56)	13.19 (4.53)	12.78 (4.56)	0.524	0.025
UWES mean score	41.30 (12.32)	42.84 (11.36)	37.20 (13.73)	<0.001	38.60 (13.12)	38.97 (13.14)	37.98 (12.94)	0.574	0.010

3.4. Logistic Regression of Psychological Distress Differentiated by Sex

Psychological distress in women was identified in 72.1% of cases based on 'effect of the pandemic on mental/emotional well-being' (OR = 5.457; 95% CI = 2.101–14.178), 'current level of anxiety and fear of COVID-19' (OR = 1.390; 95% CI = 1.137–1700) and 'health and physical status' (OR = 0.630; 95% CI = 0.453–0.876), with a specificity of 83.3% and a sensitivity of 53.4%. In men, psychological distress was identified in 78.7% of cases based on the variables 'effect of the pandemic on mental/emotional well-being' (OR = 5.942; 95% CI = 3.766–9.373), 'current level of anxiety and fear of COVID-19' (OR = 1.137; 95% CI = 1.167–1.271), 'the AMICO model' (OR = 1.347; 95% CI = 1.167–1.551) and 'level of work engagement (UWES)' (OR = 0.966; 95% CI = 0.952–0.981), with a specificity of 91.0% and a sensitivity of 45.8% (Table 6).

Table 6. Logistic regression of psychological distress (GHQ) by sex.

Psychological Distress	Men	Women
	Odds Ratio (Confidence Interval at the 95% Level)	**Odds Ratio (Confidence Interval at the 95% Level)**
UWES	0.966 ** (0.952; 0.981)	
AMICO	1.347 ** (1.167; 1.556)	
Current level of anxiety and fear of COVID-19	1.137 * (1.167; 1.271)	1.390 ** (1.137; 1.700)
Effect of the pandemic on mental/emotional well-being (Ref. NO)	5.941 ** (3.766; 9.373)	5.457 ** (2.101; 14.178)
Health and physical status		0.630 ** (0.453; 0.876)
Sensitivity (%)/Specificity (%)	45.8/91.0	53.4/83.3
Correctly classified percentage	78.7%	72.1%
Nagelkerke's R^2	0.336	0.294
Hosmer-Lemoshov test	$\chi^2 = 11.029$ (p = 0.200)	$\chi^2 = 9.409$ (p = 0.309)
Omnibus test	$\chi^2 = 185.670$ (p < 0.001)	$\chi^2 = 37.411$ (p < 0.001)

* p < 0.005; ** p < 0.001.

4. Discussion

4.1. Contributions of This Research

The findings have allowed to meet the objectives of the study and to identify sex differences in mental health derived from the COVID-19 pandemic, in particular psychological distress and work engagement. By analysing differences by sex, this study fulfils the ethical commitment to investigate the behaviour of women in a sector where they are a minority; failure to do so would constitute discrimination. It is widely known that there are differences in health between women and men and, as the World Health Organization states, it is necessary to investigate in this area to move towards a more gender-sensitive model [22].

4.2. Differences Regarding Sociodemographic Variables

In 2023, the percentage of women in the construction sector in Andalusia, southern Spain, was 8.03%, according to official data [17]. However, the Labour Foundation for Construction reports a higher proportion of 11.1%, indicating a substantial increase since 2016 [45]. The increasing number of women being hired in the construction sector highlights the importance of including them in any study aimed at analysing their working conditions, as it is well known that there are differences in health between men and women, and for ethical and efficiency reasons, in the preventive measures that companies should adopt. It has been previously suggested that there is a need to have studies of this kind in place during health crises [46], and that one of the possible explanations for the lack of such studies is the low proportion of female researchers in certain scientific fields, even though women now outnumber men in terms of graduates in these fields [47].

The ageing of the population is an internationally identified concern and one of the most important socio-economic challenges faced by developed countries in Europe and Asia [48] which may have an impact on the productivity of companies [49]. The recent increase in the number of women in the construction sector may justify the statistically significant differences found in terms of age, with women being younger in this study. This may contribute to reducing the ageing of the sector and increasing its productivity. Besides, the methodology used in the study required internet access, which may have resulted in a higher participation rate of young people of both sexes.

The higher percentage of women not living with a partner in this study may also be explained by the younger age of the women in this study, which has been associated with the presence of PD [50,51] and with reduced well-being [46], although other studies have not found such an association [31].

4.3. Working Conditions

According to the data, the proportion of men in full-time employment (80.1%) was higher than that of women (71.4%), yet full-time employment was predominant in both sexes. Moreover, among full-time employees, women were found to have a higher level of PD than men. The higher percentage of women working part-time was not unexpected, as women globally account for 75% of all part-time contracts, according to the Spanish National Statistics Institute's Labour Force Survey. It is worth noting that in this study, 53% of women reported that they had been unable to find a full-time job, while only 14% had chosen such a part-time contract to be able to look after children or elderly relatives [52].

The study found that psychological distress was more prevalent in women than in men, regardless of their family income level. This is consistent with the well-established link between financial situation and psychological distress [53]. This is in line with the global economic crisis generated by the pandemic, which has had a greater impact on women [54].

Similarly, the degree of responsibility in the job could be a reason for the lower salary, as women were more likely to be engaged in 'Other' occupations (administration, cleaning, etc.) and less likely to be employed as 'Manual workers'. However, the proportion of women in the 'Managers and skilled workers' group (40.0%) was higher than that of men

(20.9%). This study found no statistically significant difference by sex regarding the degree of responsibility at work and the development of PD, but greater PD was found in women working in civil works or outdoors. Yet, no hypotheses have been found to explain this finding. In Spain, the pay gap in the sector was reduced in 2023 compared to previous years, with a mean salary of 19,122 euros. This was higher than the salary in the commerce, repairs and transport sector, but lower than that in the industry, social services, and finance and insurance sectors, the latter being the sector that pays women the most [55]. It may be surprising that, despite the fact that women in general seem to receive lower salaries than men [56], no sex differences were found when asked whether they had problems making ends meet, which may be due to their younger age compared to that of men, or to the fact that they have obtained a job in a sector that was almost inaccessible to women until a few years ago.

4.4. Preventive Measures against COVID-19

Although no differences have been observed between male and female workers in terms of the preventive measures or specific training provided by companies to prevent contagion, the percentage of PD was higher among women than among men when companies did offer these preventive measures, showing that the impact on mental health can differ according to sex when faced with similar preventive measures. Similarly, the proportion of women with PD was higher than among men, especially amid those who reported feeling safe and protected from infection while performing their job duties, having been vaccinated, having experienced side effects from vaccination, or not having had to isolate themselves.

Even though no discernible disparities were noted between male and female employees in terms of the preventative measures or specific training provided by companies to mitigate contagion, the incidence of PD was notably higher among women compared to men in instances where companies did implement such preventive measures. This highlights the possibility of a difference in the impact on mental health based on sex, even in the face of similar preventive measures. In addition, the prevalence of PD was higher among women than men. This was especially true for women who felt secure and protected from infection while performing their job duties, those who had been vaccinated, those who experienced side effects from the vaccine, or those who had not required isolation.

A plausible explanation for the increased psychological distress experienced by women during the most severe phases of the pandemic lies in the fact that during the most severe phases of the pandemic, women may have experienced increased psychological distress as a result of the double burden of balancing work responsibilities with domestic and/or family responsibilities. Despite societal progress, women still retain the burden of housework, childcare and caring for family members [18]. Managing time becomes increasingly challenging for women employed part-time, which exacerbates their burden. Therefore, factors such as family protection, income levels, and concerns regarding family finances have emerged as mediators of women's psychological distress during the COVID-19 pandemic. These disparities in social protection policies for women during the pandemic underscore the urgent need to examine the impact of role overload on women's access to personal protective equipment, safety at work and at home, and personalised health care [57].

4.5. Psychological Distress, Anxiety and Fear

The greater level of psychological distress among women is consistent with what was observed in a systematic review [9], as previous studies had found a positive association between family members' fear of COVID-19 and psychological distress, fully mediated by individual fear, and a negative association between family well-being and psychological distress, moderated by income level [49]. These results may contradict the absence of sex differences when people were asked whether the COVID-19 pandemic had had a negative impact on their mental/emotional well-being. However, studies have shown that during critical situations, such as the COVID-19 pandemic, during which strict movement restrictions were imposed, a large proportion of people were able to adjust to the situation and maintain their lifestyle favorably [58,59].

The effects of teleworking are an interesting example of the impact that the pandemic has had, with known negative effects on sleep, reported to be greater for women and young people [60]. Yet, there are also positive effects such as men taking on more responsibilities in the home, which leads to a more equal distribution of roles in the household [61]. It has been reported that the effects of developing PD from teleworking are less significant than those from exposure to infection by essential workers, because of the risk of transmitting the infection to the family when they return home [23].

Finally, it is worth noting that women initially perceived higher levels of anxiety and fear of COVID-19 at the beginning of the pandemic, as well as higher levels of perceived danger associated with the pandemic, but these levels declined significantly, and these sex differences did not persist in the later phases of the pandemic. This contrasts with previous studies conducted in Spain, where the differences were more significant during the 'new normal' phase than during the initial phase of maximum restriction [25].

The study identified several factors that appear to have influenced PD to a greater extent in both women and men, namely 'the impact of the pandemic on mental/emotional wellbeing' and 'the level of anxiety and fear of COVID-19 at the time of completing the questionnaire'. In women, these factors were influenced by 'health and physical condition', whereas in men they were influenced by 'level of work engagement', giving rise to hypotheses that should be tested in future studies.

5. Conclusions

In recent years, the proportion of women in the construction sector has increased and in the case of this study in the south of Spain, they are younger than men, which may help to reduce the ageing of the sector. Women are mainly engaged in high-skilled activities such as administration or cleaning and, to a lesser extent, manual labour.

After controlling for most variables, the level of psychological distress remained higher for women. The preventive measures taken by the companies and the training received did not differ across sexes, but there was a greater PD among women with equal preventive measures. For men, work engagement appeared to be a determining factor, while for women, health and physical status seemed to be more influential. For both sexes, the impact of the pandemic on mental and emotional well-being and the level of anxiety and fear of COVID-19 at the time of completing the questionnaire played a role in the development of PD.

At the start of the pandemic, women exhibited higher levels of fear and anxiety of COVID-19 and perceived danger of the pandemic, a difference that was not maintained in 2023.

There is a growing need for studies that assess women and men separately, given the scientific evidence of health variations across sexes and the increasing involvement of women in the construction sector. This study identified remarkable differences that would make it possible to define public policies to facilitate progress towards sex equality in working environments.

Limitations

One limitation of this study was that the proportion of women in the sample was much lower than that of men (a 17.9% of the final sample were women), which is in line with current employment data in the construction sector [18]; this may be the reason for the lack of significant differences in some variables. Nonetheless, this study is focused on the population of construction workers of Andalusia, one of the biggest and inhabited regions in Spain, and so, 204,100 subjects have been considered, leading to an over-estimated sample size of 857 participants. As the study focused on this geographical region, the results do not allow the conclusions to be generalised to the Spanish population, although it is true that the interpretations could support the hypothesis that the rest of the Spanish population would have behaved in the same way. Likewise, they could also support the approximation of the results to the international population, given that the adoption of preventive measures by companies and the levels of fear and anxiety of COVID-19 in workers have been fairly homogeneous among the working population worldwide [9]. Yet, these data should be handled with caution due to the design of this study, as discussed below.

Another limitation is the potential over-representation of women with higher educational level and younger age due to the methodology used to collect the information, which required access to the internet. Thirdly, this methodology (a cross-sectional study) does not establish causal associations and, thus, requires the application of other types of quantitative methods in future studies. However, the present study presents data from the construction sector in the south of Spain, analysed on the basis of the sex variable, showing for the first time in quantitative terms the differences between the two sexes.

Author Contributions: Conceptualization, C.G.-S., J.C.C.-V., R.A.-C., C.R.-F., M.O.-M., M.L.-M., J.J.G.-I., J.F.-R., L.R.-D., J.M.V.-L. and J.G.-S.; Data curation, C.G.-S., J.C.C.-V., R.A.-C., C.R.-F., M.O.-M., M.L.-M., J.J.G.-I., J.F.-R. and J.G.-S.; Formal analysis, C.G.-S., J.C.C.-V., R.A.-C., C.R.-F., M.O.-M., M.L.-M., J.J.G.-I., J.F.-R., L.R.-D., J.M.V.-L. and J.G.-S.; Funding acquisition, J.G.-S.; Investigation, C.G.-S., J.C.C.-V., R.A.-C., C.R.-F., M.O.-M., M.L.-M., J.J.G.-I., J.F.-R., L.R.-D., J.M.V.-L. and J.G.-S.; Methodology, C.G.-S., J.C.C.-V., R.A.-C., C.R.-F., M.O.-M., M.L.-M., J.J.G.-I., J.F.-R., L.R.-D., J.M.V.-L. and J.G.-S.; Project administration, C.G.-S., J.C.C.-V., R.A.-C., J.F.-R. and J.G.-S.; Resources, C.G.-S., J.C.C.-V., R.A.-C., C.R.-F., M.O.-M., M.L.-M., J.J.G.-I., J.F.-R., L.R.-D., J.M.V.-L. and J.G.-S.; Software, C.G.-S., J.C.C.-V., R.A.-C., C.R.-F., M.O.-M. and J.G.-S.; Supervision, C.G.-S., J.C.C.-V., R.A.-C., C.R.-F., M.O.-M., M.L.-M., J.J.G.-I., J.F.-R., L.R.-D., J.M.V.-L. and J.G.-S.; Validation, C.G.-S., J.C.C.-V., R.A.-C., C.R.-F., M.O.-M., M.L.-M., J.J.G.-I., J.F.-R., L.R.-D., J.M.V.-L. and J.G.-S.; Visualization, C.G.-S., J.C.C.-V., R.A.-C., C.R.-F., M.O.-M., M.L.-M., J.J.G.-I., J.F.-R., L.R.-D., J.M.V.-L. and J.G.-S.; Writing—original draft, C.G.-S., J.C.C.-V., R.A.-C., C.R.-F., M.O.-M., M.L.-M., J.J.G.-I., J.F.-R., L.R.-D., J.M.V.-L. and J.G.-S.; Writing—review & editing, C.G.-S., J.C.C.-V., R.A.-C., C.R.-F., M.O.-M., M.L.-M., J.J.G.-I., J.F.-R., L.R.-D., J.M.V.-L. and J.G.-S. All authors have read and agreed to the published version of the manuscript.

Funding: This research is part of the project "Impact of the COVID-19 pandemic on the emotional and occupational health of construction workers: Evaluation and intervention proposals" which was funded by the Andalusian Institute for the Prevention of Occupational Hazards (IAPRL), part of the Ministry of Employment, Training and Self-Employment of Andalusia (CTC-2022142721).

Data Availability Statement: Data available within reasonable request to the corresponding author.

Acknowledgments: All authorizations for the use of the scales in this study, especially the GHQ-12, were obtained by the authors.

Conflicts of Interest: The authors declare no conflicts of interest.

References

1. Ren, T.; Cao, L.; Chin, T. Crafting Jobs for Occupational Satisfaction and Innovation among Manufacturing Workers Facing the COVID-19 Crisis. *Int. J. Environ. Res. Public Health* **2020**, *17*, 3953. [CrossRef] [PubMed]
2. International Food Policy Research Institute. How Much will Poverty Increase Because of COVID-19? 2020. Available online: https://www.ifpri.org/blog/how-much-will-global-poverty-increase-because-covid-19 (accessed on 31 July 2023).
3. Conversano, C.; Marchi, L.; Miniati, M. Psychological Distress Among Healthcare Professionals Involved in the COVID-19 Emergency: Vulnerability and Resilience Factors. *Clin. Neuropsychiatry* **2020**, *17*, 94–96. [CrossRef] [PubMed]
4. Martí-Amengual, G.; Sanz-Gallen, P.; Arimany-Manso, J. Valoración medicolegal de la infección por COVID-19 en el ámbito laboral. *Rev. Esp. Med. Leg.* **2020**, *46*, 146–152. [CrossRef]
5. Moreno-Sueskun, I.; Díaz-González, A.; Acuña Juanbeltz, A.; Pérez-Murillo, A.; Garasa Jiménez, A.; García-Osés, V.; Extramiana-Cameno, E. Return to work in the context of the COVID-19 pandemic in the industrial and construction sectors in Navarre (Spain). *Arch. Prev. Riesgos Labor.* **2020**, *23*, 443–457. [CrossRef] [PubMed]
6. Kamas, W.M.; Hasan, A.A.; Fadel, A.H. Economic benefits for the application of standards of sustainability in construction projects. *J. Eng.* **2019**, *25*, 117–126. [CrossRef]
7. Meng, X.; Chan, A.H.S. Cross-Regional Research in Demographic Impact on Safety Consciousness and Safety Citizenship Behavior of Construction Workers: A Comparative Study between Mainland China and Hong Kong. *Int. J. Environ. Res. Public Health* **2022**, *19*, 12799. [CrossRef]

8. Sierra, F. COVID-19: Main challenges during construction stage. *Eng. Constr. Archit. Manag.* **2021**, *29*, 1817–1834. [CrossRef]
9. Gómez-Salgado, C.; Camacho-Vega, J.C.; Gómez-Salgado, J.; García-Iglesias, J.J.; Fagundo Rivera, J.; Allande-Cussó, R.; Martín-Pereira, J.; Ruiz-Frutos, C. Stress, fear and anxiety among construction workers: A systematic review. *Front. Public Health* **2023**, *11*, 1226914. [CrossRef]
10. Seo, H.C.; Lee, Y.S.; Kim, J.J.; Jee, N.Y. Analyzing safety behaviors of temporary construction workers using structural equation modeling. *Saf. Sci.* **2015**, *77*, 160–168. [CrossRef]
11. Pheng, L.S.; Hou, L.S. The Economy and the Construction Industry. In *Construction Quality and the Economy*; Springer Singapore: Singapore, 2019; pp. 21–54. [CrossRef]
12. European Agency for Safety and Health at Work (EU-OSHA). *Managing Psychosocial Risks in European Micro and Small Enterprises: Qualitative Evidence from the Third European Survey of Enterprises on New and Emerging Risks (ESENER 2019)*; Publications Office of the European Union: Luxembourg, 2022.
13. Nnaji, C.; Jin, Z.; Karakhan, A. Safety and Health Management Response to COVID-19 in the Construction Industry: A Perspective of Fieldworkers. *Process Saf. Environ. Prot.* **2022**, *159*, 477–488. [CrossRef]
14. Liang, H.; Yang, W.; Liu, T.; Xia, F. Demogrphic Influences on Perceived Stressors of Construction Workers during the COVID-19 Pandemic. *Int. J. Environ. Res. Public Health* **2022**, *19*, 4192. [CrossRef]
15. Olatoye, O.E.; Jang, Y.; Song, K.; Ahn, J. Impact of COVID-19 on Social and Psychological Well-Being of Workers in Construction Projects: A Comparative Analysis of Managers and Laborers. *Buildings* **2023**, *13*, 1902. [CrossRef]
16. Wang, D.; Wang, X.; Xia, N. How safety-related stress affects workers' safety behaviour: The moderating role of psychological capital. *Saf. Sci.* **2018**, *103*, 247–249. [CrossRef]
17. Instituto de Estadística y Cartografía de Andalucía. Encuesta de Población Activa-Visualización del Número Ocupados por Sector de Actividad. 2022. Available online: https://www.juntadeandalucia.es/institutodeestadisticaycartografia/epa/visualizacion-actividad.htm (accessed on 31 July 2023).
18. Observatorio Argos. La mujer en el mercado de trabajo andaluz 2020. Sistema de Prospección Permanente del Mercado de Trabajo de Andalucía. Servicio Andaluz de Empleo. Consejería de Empleo, Formación y Trabajo Autónomo. Junta de Andalucía, Sevilla, Spain. 2021. Available online: https://www.juntadeandalucia.es/servicioandaluzdeempleo/web/argos/web/es/ARGOS/Publicaciones/pdf/20230308_Mujer_2022.pdf (accessed on 14 August 2023).
19. Esteban-Gonzalo, S.; González-Pascual, J.L.; Caballero-Galilea, M.; Esteban-Gonzalo, L. Psychosocial Correlates of Mental Health and Well-Being During the COVID-19: The Spanish Case. *Front. Psychol.* **2020**, *11*, 609815. [CrossRef] [PubMed]
20. Ahmed, F.; Ahmed, N.; Pissarides, C.; Stiglitz, J. Why inequality could spread COVID-19. *Lancet Public Health* **2020**, *5*, e240. [CrossRef] [PubMed]
21. Crespí-Lloréns, N.; Hernández-Aguado, I.; Chilet-Rosell, E. Have Policies Tackled Gender Inequalities in Health? A Scoping Review. *Int. J. Environ. Res. Public Health* **2021**, *18*, 327. [CrossRef] [PubMed]
22. World Health Organization. *Gender in Mental Health Research*; WHO: Ginebra, Switzerland, 2005; Available online: http://www.who.int/gender/documents/MentalHealthlast2.pdf (accessed on 31 July 2023).
23. Ruiz-Frutos, C.; Ortega-Moreno, M.; Allande-Cussó, R.; Domínguez-Salas, S.; Dias, A.; Gómez-Salgado, J. Health-related factors of psychological distress during the COVID-19 pandemic among non-health workers in Spain. *Saf. Sci.* **2021**, *133*, 104996. [CrossRef] [PubMed]
24. Mediavilla, R.; Fernández-Jiménez, E.; Martínez-Alés, G.; Moreno-Küstner, B.; Martínez-Morata, I.; Jaramillo, F.; Morán-Sánchez, I.; Minué, S.; Torres-Cantero, A.; Alvarado, R.; et al. Role of access to personal protective equipment, treatment prioritization decisions, and changes in job functions on health workers' mental health outcomes during the initial outbreak of the COVID-19 pandemic. *J. Affect. Disord.* **2021**, *295*, 405–409. [CrossRef]
25. Matud, M.P.; Zueco, J.; Díaz, A.; Del Pino, M.J.; Fortes, D. Gender differences in mental distress and affect balance during the first wave of COVID-19 pandemic in Spain. *Curr. Psychol.* **2023**, *42*, 21790–21804. [CrossRef]
26. Unda-López, A.; Paz, C.; Hidalgo-Andrade, P.; Hermosa-Bosano, C. Variations of work engagement and psychological distress based on three working modalities during the COVID-19 pandemic. *Front. Public Health* **2023**, *11*, 1191314. [CrossRef]
27. Mukaihata, T.; Greiner, C.; Fujimoto, H. Testing the relationship between patient-related stressor, psychological distress, work engagement, job satisfaction and recovery attitude among psychiatric nurses in Japan. *J. Adv. Nurs.* **2022**, *78*, 1348–1365. [CrossRef] [PubMed]
28. Gómez-Salgado, J.; Andrés-Villas, M.; Domínguez-Salas, S.; Díaz-Milanés, D.; Ruiz-Frutos, C. Related Health Factors of Psychological Distress During the COVID-19 Pandemic in Spain. *Int. J. Environ. Res. Public Health* **2020**, *17*, 3947. [CrossRef] [PubMed]
29. Instituto Andaluz de Prevención de Riesgos Laborales. Observatorio de Enfermedades Profesionales de Andalucía. Consejería de Empleo, Empresa y Trabajo Autónomo de la Junta de Andalucía. 2023. Available online: https://www.juntadeandalucia.es/organismos/iaprl/areas/investigacion/ladep.html (accessed on 14 August 2023).
30. Ministerio del Trabajo y Economía Social. Register of Accredited Companies. Government of Spain. 2023. Available online: https://expinterweb.mites.gob.es/rea/ (accessed on 1 February 2023).
31. Wang, C.; Pan, R.; Wan, X.; Tan, Y.; Xu, L.; Ho, C.S.; Ho, R.C. Immediate Psychological Responses and Associated Factors during the Initial Stage of the 2019 Coronavirus Disease (COVID-19) Epidemic among the General Population in China. *Int. J. Environ. Res. Public Health* **2020**, *17*, 1729. [CrossRef] [PubMed]

32. Alonso, M.S.; Lima, M.C.P.; Dias, A.; Camacho-Vega, J.C.; García-Iglesias, J.J.; Ruiz-Frutos, C.; Bernardes, J.M.; Gómez-Salgado, J. Psychological distress among Brazilian workers during the initial stage of the COVID-19 pandemic: A descriptive study. *Front. Public Health* **2024**, *12*, 1283310. [CrossRef] [PubMed]
33. Ruiz-Frutos, C.; Ortega-Moreno, M.; Allande-Cussó, R.; Ayuso-Murillo, D.; Domínguez-Salas, S.; Gómez-Salgado, J. Sense of coherence, engagement, and work environment as precursors of psychological distress among non-health workers during the COVID-19 pandemic in Spain. *Saf. Sci.* **2021**, *133*, 105033. [CrossRef]
34. Goldberg, D.P.; Gater, R.; Sartorius, N.; Ustun, T.B.; Piccinelli, M.; Gureje, O.; Rutter, C. The validity of two versions of the GHQ in the WHO study of mental illness in general health care. *Psychol. Med.* **1997**, *27*, 191–197. [CrossRef]
35. Rocha, K.B.; Pérez, K.; Rodríguez-Sanz, M.; Borrell, C.; Obiols, J.E. Propiedades psicométricas y valores normativos del general health questionnaire (GHQ-12) en población general española. *Int. J. Clinl. Health Psychol.* **2011**, *11*, 125–139.
36. Schaufeli, W.; Salanova, M.; González-Romá, V.; Bakker, A. The Measurement of Engagement and Burnout: A Two Sample Confirmatory Factor Analytic Approach. *J. Happiness Stud.* **2002**, *3*, 71–92. [CrossRef]
37. Domínguez-Salas, S.; Rodríguez-Domínguez, C.; Arcos-Romero, A.I.; Allande-Cussó, R.; García-Iglesias, J.J.; Gómez-Salgado, J. Psychometric Properties of the Utrecht Work Engagement Scale (UWES-9) in a Sample of Active Health Care Professionals in Spain. *Psychol. Res. Behav. Manag.* **2022**, *15*, 3461–3472. [CrossRef]
38. Gómez-Salgado, J.; Allande-Cussó, R.; Domínguez-Salas, S.; García-Iglesias, J.J.; Coronado-Vázquez, V.; Ruiz-Frutos, C. Design of Fear and Anxiety of COVID-19 Assessment Tool in Spanish Adult Population. *Brain. Sci.* **2021**, *11*, 328. [CrossRef]
39. IBM Corp. *IBM SPSS Statistics for Windows*, version 26.0; IBM Corp: Armonk, NY, USA, 2019.
40. Schlittgen, R.; Young, F.W.; Valero-Mora, P.M.; Friendly, M. Visual statistics: Seeing data with dynamic interactive graphics. *AstA* **2008**, *92*, 115. [CrossRef]
41. Domínguez-Almendros, S.; Benítez-Parejo, N.; Gonzalez-Ramirez, A.R. Logistic regression models. *Allergol. Immunopathol.* **2011**, *39*, 295–305. [CrossRef]
42. Barrios, I.; Anido, V.; Morera, M. Declaración de Helsinki: Cambios y exégesis. *Rev. Cuba. Salud Pública* **2016**, *42*, 1–11. Available online: http://www.revsaludpublica.sld.cu/index.php/spu/article/view/597 (accessed on 7 March 2024).
43. Ley Orgánica 3/2018, de 5 de Diciembre, de Protección de Datos Personales y Garantía de los Derechos Digitales. Boletín Oficial del Estado, nº 294, (6-12-2018). Available online: https://www.boe.es/eli/es/lo/2018/12/05/3 (accessed on 14 August 2023).
44. Ley 14/2007, de 3 de Julio, de Investigación Biomédica. Boletín Oficial del Estado, nº159, (4-7-2007). Available online: https://www.boe.es/eli/es/l/2007/07/03/14 (accessed on 14 August 2023).
45. Observatorio de la Fundación Laboral de la Construcción. Mujeres en el Sector de la Construcción 2022. Available online: https://www.observatoriodelaconstruccion.com/informes/detalle/mujeres-en-el-sector-de-la-construccion-2022 (accessed on 31 July 2023).
46. Smith, J. Overcoming the 'tyranny of the urgent': Integrating gender into disease outbreak preparedness and response. *Gend. Dev.* **2019**, *27*, 355–369. [CrossRef]
47. Kotozaki, Y.; Kozasa, E.H. Editorial: Women in science: Public mental health 2022. *Front. Public Health* **2023**, *10*, 1123506. [CrossRef] [PubMed]
48. Rožman, M.; Tominc, P.; Milfelner, B.A. Comparative study using two SEM techniques on different samples sizes for determining factors of older employee's motivation and satisfaction. *Sustainability* **2020**, *12*, 2189. [CrossRef]
49. Hashiguchi, N.; Sengoku, S.; Kubota, Y.; Kitahara, S.; Lim, Y.; Kodama, K. Age-Dependent Influence of Intrinsic and Extrinsic Motivations on Construction Worker Performance. *Int. J. Environ. Res. Public Health* **2020**, *18*, 111. [CrossRef]
50. Rodríguez-Rey, R.; Garrido-Hernansaiz, H.; Collado, S. Psychological impact and associated factors during the initial stage of the coronavirus (COVID-19) pandemic among the general population in Spain. *Front. Psychol.* **2020**, *11*, 1540. [CrossRef]
51. McCallum, S.M.; Calear, A.L.; Cherbuin, N.; Farrer, L.M.; Gulliver, A.; Shou, Y.; Dawel, A.; Batterham, P.J. Associations of loneliness, belongingness and health behaviors with psychological distress and wellbeing during COVID-19. *J. Affect Disord. Rep.* **2021**, *6*, 100214. [CrossRef]
52. Asenjo-Domínguez, A. Las Mujeres Ocupan ya el 75% de los Contratos a Tiempo Parcial. Available online: https://www.elindependiente.com/economia/2022/02/14/las-mujeres-ocupan-ya-el-75-de-los-contratos-a-tiempo-parcial/ (accessed on 14 August 2023).
53. Chen, B.; Gong, W.; Lai, A.Y.K.; Sit, S.M.M.; Ho, S.Y.; Yu, N.X.; Wang, M.P.; Lam, T.H. Family context as a double-edged sword for psychological distress amid the COVID-19 pandemic with the mediating effect of individual fear and the moderating effect of household income. *Front. Public Health* **2023**, *11*, 1109446. [CrossRef]
54. Tejani, S.; Fukuda-Parr, S. Gender and COVID-19: Workers in global value chains. *Int. Labour Rev.* **2021**, *160*, 649–667. [CrossRef] [PubMed]
55. CSV Empresa Constructora. Available online: https://csvconstruccion.com/es/la-construccion-reduce-la-brecha-salarial-a-solo-1-000-euros-y-bate-su-record-de-presencia-femenina/ (accessed on 31 July 2023).
56. Blau, F.D.; Kahn, L.M. The Gender Wage Gap: Extent, Trends, and Explanations. *J. Econ. Lit.* **2017**, *55*, 789–865. [CrossRef]
57. Lokot, M.; Bhatia, A. Unequal and Invisible: A Feminist Political Economy Approach to Valuing Women's Care Labor in the COVID-19 Response. *Front. Sociol.* **2020**, *5*, 588279. [CrossRef]
58. Patel, V.; Burns, J.K.; Dhingra, M.; Tarver, L.; Kohrt, B.A.; Lund, C. Income inequality and depression: A systematic review and meta-analysis of the association and a scoping review of mechanisms. *World Psychiatry* **2018**, *17*, 76–89. [CrossRef]

59. Cancello, R.; Soranna, D.; Zambra, G.; Zambon, A.; Invitti, C. Determinants of the Lifestyle Changes during COVID-19 Pandemic in the Residents of Northern Italy. *Int. J. Environ. Res. Public Health* **2020**, *17*, 6287. [CrossRef]
60. Figueiredo, S.; João, R.; Alho, L.; Hipólito, J. Psychological Research on Sleep Problems and Adjustment of Working Hours during Teleworking in the COVID-19 Pandemic: An Exploratory Study. *Int. J. Environ. Res. Public Health* **2022**, *19*, 14305. [CrossRef]
61. González Ramos, A.M.; García-de-Diego, J.M. Work-Life Balance and Teleworking: Lessons Learned during the Pandemic on Gender Role Transformation and Self-Reported Well-Being. *Int. J. Environ. Res. Public Health* **2022**, *19*, 8468. [CrossRef]

buildings

MDPI

Article

Near-Miss Detection Metrics: An Approach to Enable Sensing Technologies for Proactive Construction Safety Management

Filzah Hashmi [1], Muhammad Usman Hassan [1,*], Muhammad Umer Zubair [2], Khursheed Ahmed [1] Taha Aziz [1] and Rafiq M. Choudhry [3]

[1] School of Civil and Environmental Engineering (SCEE), National University of Sciences and Technology (NUST), Islamabad 44000, Pakistan; filzah.cem19@student.nust.edu.pk (F.H.); khursheed@nice.nust.edu.pk (K.A.); taziz.bece19nice@student.nust.edu.pk (T.A.)

[2] School of Civil & Environmental Engineering, College of Engineering, King Faisal University, Al Hofuf 31982, Saudi Arabia; mzubair@kfu.edu.sa

[3] College of Engineering, Imam Mohammad Ibn Saud Islamic University (IMSIU), Riyadh 11432, Saudi Arabia; rchoudhry@imamu.edu.sa

* Correspondence: usman.hassan@nice.nust.edu.pk

Abstract: One in every five occupational deaths occurs in the construction sector. A proactive approach for improving on-site safety is identifying and analyzing accident precursors, such as near-misses, that provide early warnings of accidents. Despite the importance of near-misses, they are frequently left unreported and unrecorded in the construction sector. The adoption of modern technologies can prevent accidents by automated data collection and analysis. This study aims to develop near-miss detection metrics to facilitate the automated detection of near-misses through sensors. The study adopted a mixed method approach including both qualitative and quantitative approaches. First, a quantifiable definition of near-misses was developed from the literature. Hazards, accidents, and the causes of accidents were identified. Through empirical and statistical analyses of accidents from the OSHA repository, combinations of unsafe acts and conditions responsible for a near-miss were identified. The identified factors were analyzed using a frequency analysis, correlation, and a lambda analysis. The results revealed twelve significant near-misses, such as A1—approach to restricted areas and C2—unguarded floor/roof openings, A5—equipment and tool inspection was incomplete and C8—unsafely positioned ladders and scaffolds, A2—no or improper use of PPE and C2—unguarded floor or roof openings, etc. Lastly, measurable data required by sensors for autonomous detection of near-misses were determined. The developed metric set the basis for automating near-miss reporting and documentation using modern sensing technology to improve construction safety. This study contributes to improving construction safety by addressing the underreporting of near-miss events. Overall, the developed metrics lay the groundwork for enhancing construction safety through automated near-miss reporting and documentation. Furthermore, it helps for the establishment of safety management schemes in the construction industry, specifically in identifying near-misses. This research offers valuable insight into developing guidelines for safety managers to improve near-miss reporting and detection on construction sites. In sum, the findings can be valuable for other industries also looking to establish or assess their own safety management systems.

Keywords: near-miss; sensing technology; proactive safety management; accident prevention; early warnings

Citation: Hashmi, F.; Hassan, M.U.; Zubair, M.U.; Ahmed, K.; Aziz, T.; Choudhry, R.M. Near-Miss Detection Metrics: An Approach to Enable Sensing Technologies for Proactive Construction Safety Management. *Buildings* **2024**, *14*, 1005. https://doi.org/10.3390/buildings14041005

Academic Editors: Marek Sawicki and Mariusz Szóstak

Received: 24 February 2024
Revised: 27 March 2024
Accepted: 2 April 2024
Published: 4 April 2024

1. Introduction

The construction industry has the highest fatality rate due to its temporary, dynamic, decentralized, and complex nature [1]. The construction site's ever-changing environment may result in unlimited hazards—for example, poor layouts, confined spaces, work in restricted areas, and many others [2]. Due to the high accident rate, researchers and professionals constantly seek ways to improve construction safety. One way to enhance construction safety is to collect and analyze safety indicators or precursors. These indicators are used by safety personnel to assess if they are taking suitable precautions and controlling risks in the workplace [3]. Lagging and leading indicators are the most common types of safety indicators [4]. Lagging indicators commonly used in the construction industry are after-the-damage safety measures, e.g., injury, fatality, or illness [3]. The disadvantage of using lagging indicators is that it takes an accident to discover and address an unsafe act or condition [5]. To address this shortcoming, attention has recently shifted to leading indicators linked with proactive measures that do not require lagging indicators to predict future results [6]. Several researchers have proposed programs for gathering, monitoring, and controlling leading indicators [7–9].

Previous research has highlighted the necessity of investigating accidents to reveal leading signs or precursors because their numbers are typically higher than the accidents [10]. One of the important leading indicators nearest to the accident's final state is near-misses [11]. Near-misses are defined as an unexpected event that did not cause an accident but may have done so under certain conditions [12]. According to the literature, increasing the reporting of near-misses significantly reduces the number of accidents; it helps managers to analyze the data, identify risk factors, and take action to reduce risks, preventing severe accidents [13,14]. For instance, 75% of lost-time injuries were avoided by implementing near-miss reporting programs [15]. These reports can help organizations spot new risks, gain insight into system safety, and spark learning opportunities [16]. Therefore, a near-miss must be detected, recorded, and reviewed to decide the best course of action for preventing accidents [17].

Various initiatives have been undertaken in the construction sector to increase near-miss reporting due to the evident importance of near-misses. For example, Cambraia et al. [18] performed a case study to identify, analyze, and disseminate near-miss information in the construction industry. Construction industry institute (CII) members created and implemented near-miss reporting programs based on workers' feedback [13]. Other strategies include safety audits and governmental initiatives [19,20]. Despite these efforts, the construction industry still lacks the efficient use of near-miss information because near-misses are often left unreported and undocumented on construction sites [21]. Since most of the reporting systems are mainly reliant on manual reports [22], the reporting of near-misses directly depends on workers' feedback [23]. The construction workers may fail to report near-misses due to a lack of awareness regarding near-misses [24], the reporting process, or how organizations use near-misses [25]. Even if near-misses are collected, the existing reporting mechanism fails to assure that the data are accurate and reliable or that they offer valid event information [26] due to their dependence on workers, which brings subjectivity into the reporting process. Despite technological advancements, very few organizations presently use systematized near-miss management systems [27]. More specifically, the conditions in developing countries are the worst in term of recognizing and reporting near-misses in the construction industry [28]. However, a few studies have shown improvements in safety in some sectors such as manufacturing, services, and chemical ones, following the implementation of incident learning systems [29]. These improvements include increased engagement in safety improvement, heightened awareness of safety issues, reduced fear of corrective measures, and greater confidence that reported incidents are utilized to enhance the system.

Real-time methods that can consistently and reliably collect and monitor near-misses from construction sites are needed. Some of the technologies present in the literature are promising but have limitations in terms of advancement and applicability; for example, Shen and Marks [30] contributed to the visualization of near-misses by developing a BIM-based near-miss visualization tool. The limitation of the developed tool was its dependence on manual reporting. Automated monitoring systems can collect, process, and deliver meaningful data to users in real-time. However, progress in the automated detection and real-time monitoring of near-misses has not been fully explored [21]. Despite the growing attention to incident reporting, there is limited understanding of its effectiveness in enhancing construction safety within the construction industry. Although there are numerous reports in the near-miss incident construction literature, it has been challenging to quantify its impact on automatic detection. Therefore, this research study put forward a mechanism to develop a near-miss detection metric that could serve as a foundation for deploying sensors to predict future accidents by real-time monitoring of near-misses.

The rest of the paper is organized as follows: First, an overview of the near-miss definitions found in the literature is presented, followed by the research approach used in this study. Then, according to the technique, a near-miss detection metric is established. The Discussion and Conclusion are presented at the end based on the developed metric.

2. An Overview of Near-Misses

2.1. Near-Miss Definition: Worldwide Perspective

The idea of near-misses is vastly expanding from numerous industries, including aviation, medical, retail, chemical, construction, and manufacturing industries. This widespread belief underscores the importance of reporting near-misses for enhancing safety. According to the WHO, near-misses are defined as a severe mistake with the potential to inflict harm but not due to luck or interception [31]. The OSHA defines a near-miss as an incident which could have led to severe injury or sickness but did not [32]. Another study defines a near-miss as "an unintended incident which, under different circumstances, could have become an accident" [33]. Jones et al. [33] suggested two concepts for near-misses based on the incident's consequences: major near-miss and near-miss. The first concept defines near-misses as incidents that might lead to a significant accident with long-term implications. In the second concept, near-misses are risky situations that lead to accidents. According to Phimister et al. [27], a near-miss is an occurrence that indicates a system flaw that, if not addressed, could have severe effects in the future [27]. From the chemical and hospital sector perspective, Vanderschaaf [34] and Caspi et al. [23] described a near-miss as a safety effect that does not result in accidents, but there is a risk of injury. In the railway industry, Ritwik [35] defined a near-miss as an unsafe condition with the potential for damage. All these worldwide definitions combined reveal that a near-miss results in a successful outcome, where no harmful result takes place. Accidents are widely acknowledged to be the tip of the iceberg. An estimate shows that 300 near-miss incidents exist before a workplace accident [36].

2.2. Near-Miss Incidents in the Construction Sector

Researchers in the construction sector have used near-misses, close calls, near hits, and other terms to describe near-misses relevant to their fields. These definitions center on the observer's perceptions, risk tolerance, experiences, and how organizations see near-miss incidents. For ease of comprehension, some have defined a near-miss as an unsafe act or condition with the potential for injury or property damage [37], and others have defined it as an unsafe act linked to some action (e.g., the release of energy) [18]. Near-misses are often considered accident precursors [38], warnings of prospective accidents when out of luck [39], or close signals of accidents [11]. Cambraia et al. [18] defined a near-miss as an unanticipated event that requires a rapid burst of energy and may end in an accident. Williamsen [40] stated that near-misses occur when there are no injuries, property damage, or other proof that they had taken place.

There is no standardized definition of a "near-miss". The existing definitions focus on using near-miss information to improve safety management rather than reporting near-misses. Previous definitions have viewed a near-miss as merely an event-driven occurrence requiring an energy burst, with identification relying on subjective judgment. In these definitions, there is much reliance on the observer's perspective. Workers on the construction sites are the primary source of discovery of near-misses, but some factors may prevent workers from reporting near-misses: (a) an apprehension of administrative action; (b) risk acceptance; (c) a lack of knowledge about how data reports are used, and (d) data collecting is complex and time-consuming [25]. Since the management efficiency of near-misses depends on the accurate identification of near-misses, a complete and easy-to-understand definition of near-misses is required.

2.3. Measures for Detecting Near-Miss Incidents in the Construction Industry

Prominent indicators serve as a proactive approach to safety metrics, focusing on evaluating the processes, events, and conditions that are indicative of safety performance and have the potential to forecast future outcomes [3]. A prime example of such an indicator is the reporting of near-misses. A significant benefit of tracking near-misses lies in the ability to collect and scrutinize data that can inform safety improvements without the occurrence of actual injuries [33,36]. Understanding the causation of accidents may facilitate quantifying near-miss definitions, as near-misses and accidents have the same causation model [41–44]. Based on this recognition, analyzing the underlying causes of accidents can assist in identifying previous near-misses, reducing uncertainty regarding the possibility of an accident. For example, falls are the predominant cause of worker deaths in the construction industry [45]. The detection of near-misses is vital for halting and averting the events that lead to falls. However, traditional methods for recognizing near-miss events rely on the self-reporting of workers, which can lead to inconsistent data. The adoption of modern technologies can prevent accidents by automated data collection and analysis. Therefore, this study aimed to develop near-miss detection metrics to facilitate the automated detection of near-misses through sensors.

3. Research Approach

This research intended to establish a technique for automating near-miss data collection from construction sites by implementing sensing technologies. Figure 1 depicts the layout of the research methodology adopted in this study.

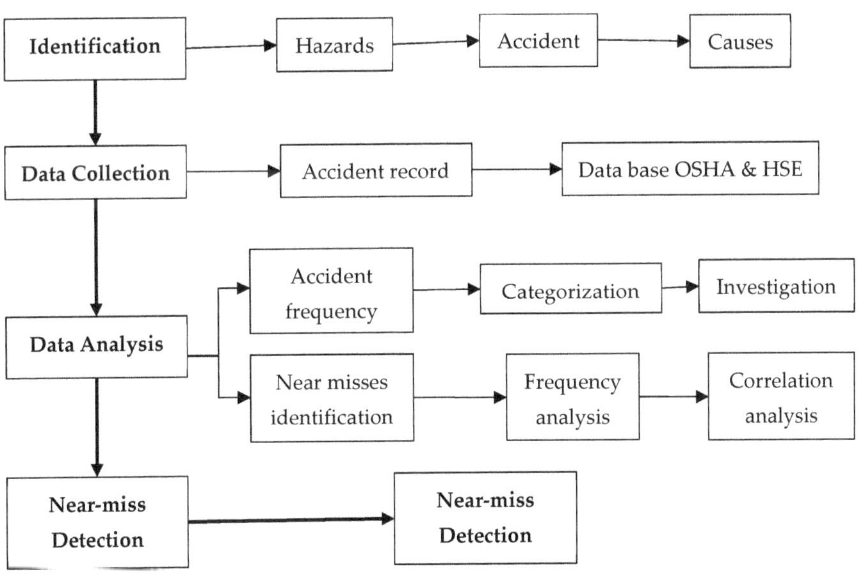

Figure 1. Layout of research methodology.

The research methodology included four major stages, including identification (i.e., hazards, accidents, and causes), data collection (accident and hazard records from the OSHA and HSE), data analysis (frequency analysis and correlation), and near-miss detection (establishing a near-miss metric). In the identification process, first, the authors identified significant hazards and accidents associated with these hazards. Then, they investigated the summary of the accidents to identify the underlying causes and performed a correlation analysis to find the significant causes responsible for these accidents. The authors subsequently created a well-designed near-miss data repository in terms of underlying causes identified from the summary of historical accidents. Lastly, they developed near-miss detection metrics consisting of quantifiable data required to autonomously detect near-misses. The detection metrics offered a strategic approach for selecting and deploying relevant sensors to collect and evaluate near-miss data. In the end, the key findings of this study are presented in the Discussion, and a conclusion is derived. The data required for near-miss identification were acquired based on the specified definition, and the near-miss detection metrics were established.

3.1. Developing Near-Miss Detection Metrics

Near-Miss Definition

Sensor-based safety management systems can be established more efficiently if each accident precursor is well investigated and quantitative metrics are defined [21]. Therefore, a quantifiable definition of near-miss is required to select and employ sensing technologies for autonomous near-miss data collection. Understanding the causation of accidents may facilitate quantifying near-miss definitions, as near-misses and accidents have the same causation model [33,36,43,44]. Near-misses share common causes with accidents, but unlike the latter, the effects of a near-miss are negligible because the opportunity factor is absent (Equation (1)) [37]. Ritwik [35] defined an opportunity factor as an uncertainty factor beyond control that decides the event's consequences. Based on the above recognition, analyzing the underlying causes of accidents can assist in identifying their previous near-misses, reducing the uncertainty regarding the possibility of an accident.

$$\text{Nearmiss} = \text{Accident} - \text{Uncertainty factor} \quad (1)$$

Heinrich [36] found that workers' unsafe activities combined with unsafe conditions were responsible for 88% of construction accidents. A recent systematic study of accidents also stated that the causes frequently responsible for accidents consist of combinations of unsafe behaviors and risky situations [43]. An unsafe working space (unsafe conditions) and weakness in safety practices result in an inadequate safety performance (unsafe act) [46]. For example, risky behavior such as not wearing personal protective equipment (PPE) in an unsafe situation, like working near unprotected machinery, may increase the risk of being struck by sharp objects.

Since near-misses have the same causation model as accidents, the preceding facts conclude that, similar to accidents, the frequency of near-misses caused by an unsafe act and conditions is much higher than near-misses caused by either an unsafe act or unsafe conditions. Defining near-misses through a full breakdown of each accident will facilitate identifying quantitative metrics for developing sensor-based near-miss monitoring systems. A near-miss can be defined as an occurrence that lacks an opportunity factor and mainly consists of unsafe acts and unsafe conditions that did not result in an injury. Therefore, identifying and tracking the interaction between unsafe acts and conditions aids in near-miss autonomous detection.

3.2. Data Collection

As mentioned above, near-misses mainly consist of unsafe behaviors and unsafe working conditions. To identify significant near-misses that occur on construction sites, the underlying causes of these near-misses need to be identified and collected in a database. Three types of data, including hazards, accidents, and their causes, were collected and analyzed. Figure 2 illustrates the overall flow of data collection.

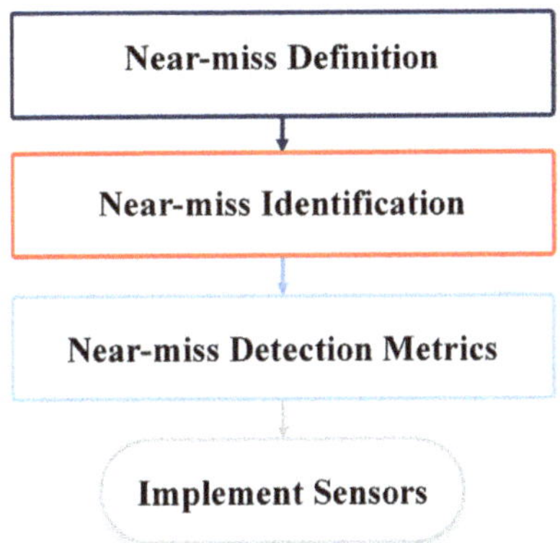

Figure 2. Overall flow of data collection.

3.2.1. Hazards

As the primary source of accidents, hazards are considered the basis for collecting relevant data. Organizations such as the Occupational Safety and Health Administration (OSHA) and the Health and Safety Executive (HSE) preserve accident data and analyze them to find their sources. According to the OSHA, the following hazards frequently contribute to fatal accidents: fall, electrocution, struck-by or -against, and caught-in or -between. In 2014, there were 782 fatal incidents reported in Europe. Among these accidents, falls (26%), breakage/fall/collapse (20%), and loss of control (19%) were the most common sources [44]. Falls and struck-by were two significant hazards responsible for 45% of fatal injuries in the United Kingdom in the last five years [47]. According to a survey, the most frequent causes of fatalities among construction workers in North Carolina between 1988 and 1994 were vehicles (21%), followed by falls (20%), machines (15%), electrocution (14%), and falling objects (14%) [48]. Another study on construction accidents discovered that substantial accidents occurred due to falls and contact with a fixed machine's moving parts [49]. Swuste et al. [45] stated that the most prevalent hazards were falls, struck-by, electric shock, and caught-in or -between. Naveen Kumar et al. [50] carried out a study in Bangalore stating that between 2014 and 2016, 41.1% of fall fatalities were related to construction activities. There were 1102 construction worker deaths in the United States in 2019, including 401 deaths from falls, 170 deaths from struck-by, 79 deaths from electrocution, 59 deaths from caught-in, and 393 deaths from other hazards [51]. Considering the available records and literature analysis, the leading hazards this study addressed were fall, struck-by, electrocution, and caught-in or -between.

3.2.2. Accidents

An accident demonstrates the effects of hazards or how a hazard will affect a worker's safety. The leading hazards identified in the preceding step were responsible for significant construction accidents. Accidents like falling from height [52], being hit by falling objects [53], slip and trip [54], machine-related accidents [55], exposure to electricity [56], and others may frequently occur on construction sites. Much research and various reports have identified construction accidents. Ale et al. [49] conducted a study to determine accidents. The findings revealed that non-fatal accidents involved tools and machinery, working at height, and being struck by flying objects, while fatal accidents were due to enormous falling objects, explosions, and heavy vehicles.

Fall is the leading cause of fatality on construction sites [57]. An increasingly common fall-related accident is falling from the roof, structures, and other falls [57]. The leading fall-related accident in the Indonesian construction industry is falling from height [58]. Other fall-related accidents include falls from openings, trenches, service pits, and falling at the same level. In the construction industry, struck-by accidents are the second most common cause of fatalities [59]. Accidents due to struck-by-equipment are 58% and falling objects are 42% of all accidents [59]. Wang et al. [60] stated that working with heavy machines (17.1%), working under elevated weights (15.6%), working on foot (13.5%), and moving equipment (13.5%) are risky elements that play a role in struck-by accidents. Haslam et al. [61] stated that 17% of construction accidents are caused by being struck by a falling or moving object. Caught-in or -between has been the leading cause of permanent injuries [62]. Most of the caught-in or -between accidents occur in activities that involve machinery, vehicles, cranes, and elevators [62]. Following falling from heights, vehicle accidents, and being struck-by, contact with electricity is fourth among the top ten accident scenarios that lead to construction fatalities [63]. The five electrocution accident patterns are worker or equipment contact with a power line, vehicle collision with an electrical power line, and incorrectly installed equipment [64].

Studies have identified a variety of accidents but have not grouped them into broad categories for simple recognition and comprehension. Correct and massive amounts of accident data are required to accurately identify near-misses. Accident data from various industries, including construction, are collected and stored by the OSHA. This study utilized an OSHA-provided data pool of historical accidents to identify major accidents on construction sites. The authors collected 8598 accidents over the past ten years between 2010 and 2020. The accident data were only related to the construction of residential buildings by screening the standard industrial classification (SIC) provided by the OSHA.

The final data set comprised 6663 accidents, among which 4532 were fatal and 2131 were non-fatal. The accidents included 4489 (67.4%) fall-related accidents, 933 (14.1%) struck-by-related accidents, 574 (8.61%) electrocution-related accidents, and 667 (10.2%) caught-in- or -between-related accidents. Figure 3 summarizes the statistical results of the OSHA accident records [65]. According to the occupational codes provided by the OSHA, the victims involved in these accidents were masons, carpenters, electricians, painters, plasterers, plumbers, roofers, duct installers, welders, cutters, and machines operators, and other workers.

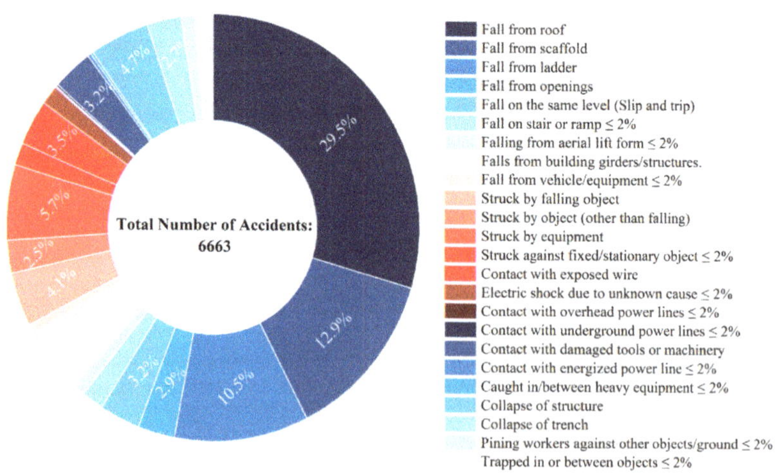

Figure 3. Categories of accidents.

From the detailed investigation, accidents with high occurrence rates were identified. The investigation was called off when no new accident was discovered. Accidents were categorized into the following four sources: (1) fall, (2) struck-by, (3) caught-in or -between, and (4) electrocution (see Table 1).

Table 1. Frequent accidents on construction sites.

S. No	Fall	Struck-by or -against	Electrocution	Caught-in or -between
1	From roof	By equipment	Contact with exposed wires	Collapse of structure
2	From scaffold	By falling or flying object	Contact with a damaged tool or machinery	Collapse of a trench (cave in)
3	From ladder	By object (other than falling, e.g., moving)	Electric shock by unknown cause	Trapped in or between objects
4	From building girders or other structures	Against a fixed or stationary object	Contact with overhead power lines	Pinned workers against other objects or the ground
5	On the same level (slip and trip)		Contact with underground, buried power lines	Contact of hand tools with an electrified wire
6	From openings (e.g., trench)			By heavy equipment

3.2.3. Unsafe Acts and Conditions

Identifying near-misses and implementing preventative actions before they become a problem requires determining their underlying cause. Primary causes were divided into workers' unsafe behavior and exposure to unsafe conditions. Some of the workers' unsafe acts found in previous research included non-compliance with work and safety procedures [66], inadequate or no use of PPE [67], unauthorized use of equipment [68], deliberately risking one's life [66], inadequate knowledge of hazardous situations [67], unsafe working posture [69], failure to use equipment safely [67], and a lack of experience [66]. Similarly, the primary unsafe conditions included poor site layout [70], excessive and loud noise [71], unsafe working procedures [72], defective PPE or equipment, insufficient supports or guards, poor warning systems, clothing hazards [72,73], task complexity [74], and poor lighting and weather [75]. The OSHA's original accident summaries and investigation reports [76] were reviewed to accurately view the unsafe acts and conditions responsible for the previously identified accidents. Summaries provided by the OSHA were entered into a spreadsheet along with accidents, inspection IDs, date, and occupation code, and the probable cause was determined by personally examining the summaries. Table 2 illustrates the methodology used to identify unsafe acts and conditions from the OSHA's accident summary.

Table 2. Investigation of accidents.

ID	Date	Occupation	Accident
316268895	**15 February 2013**	**Construction Laborers**	**Fall from Scaffold**
Summary			**Causes**
The employee was working on a two-story scaffold. The scaffold was not fully planked. The employee was not wearing any fall protection. The employee fell from the scaffold.			Working at height/Not wearing PPE/Unsafely positioned scaffold

The accidents where the causes were ambiguous were neglected. Out of 6663 accidents, 5725 accidents were carefully analyzed. Twelve (12) unsafe acts and twelve (12) unsafe conditions were identified. Table 3 summarizes the unsafe acts and conditions identified.

Table 3. Summary of identified unsafe acts and conditions.

Unsafe Acts	Unsafe Conditions
A1—Approach to restricted areas	C1—Poor site layout
A2—No or improper use of PPE	C2—Unguarded floor or roof opening
A3—Inappropriate equipment usage	C3—Defective PPE
A4—Unsafe posture and position	C4—Unprotected excavations and trenches
A5—Equipment and tool inspection was incomplete or non-existent	C5—Unsuitable weather conditions
A6—Operating at an unusually high altitude	C6—Missing or defective warning sign
A7—Operating close to moving vehicles and equipment	C7—Places under unfixed materials/tools towards the edges
A8—Ignoring alarms and warning signages	C8—Unsafely positioned ladders and scaffolds
A9—Working close to overhead power lines	C9—Lack of training and poor experience
A10—Running heavy equipment near edges	C10—Lack of machine guards
A11—Worker carrying a heavy load	C11—Defective or damaged equipment
A12—Unauthorized use of equipment	C12—Unsuitable working conditions

4. Results and Discussion

4.1. Near-Miss Identification

Accident causation studies have concluded that risky behavior, combined with unsafe conditions, is the primary cause of accidents and near-misses. This study intended to find the relationships between unsafe acts and conditions that contribute to construction near-miss incidents through statistical analyses. Figure 4 illustrates the near-miss identification process. The process aimed to use a correlation analysis to identify the influential unsafe act and condition combinations responsible for significant near-misses on construction sites.

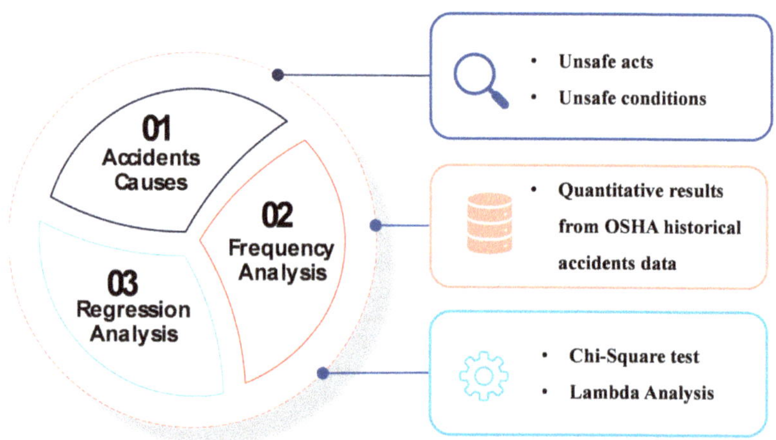

Figure 4. Near-miss identification process.

Because most accidents occur when an unsafe act and condition are combined, the authors performed a frequency analysis to identify the number of unsafe acts and conditions responsible for construction accidents among the unsafe acts and conditions previously identified. The 5725 construction accident cases were analyzed. Figure 5 shows the frequency of accidents caused by a particular combination.

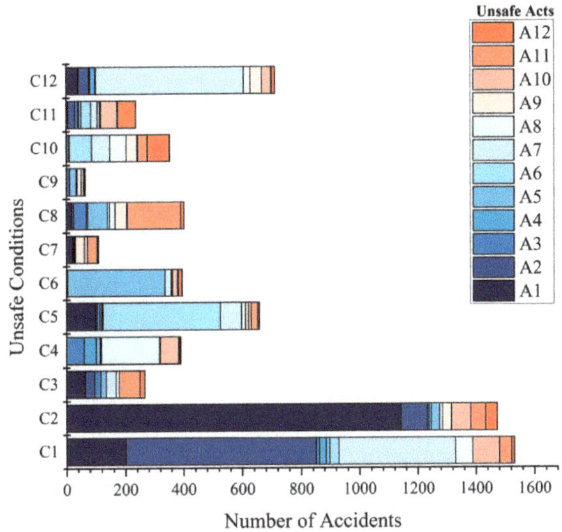

Figure 5. Results of the frequency analysis.

When determining major near-misses, dealing with highly correlated combinations would be more efficient than relying on frequencies, as combinations can have high frequencies and weak correlations. A correlation analysis was performed to identify the most effective combination that causes near-misses. The correlation analysis included the chi-square test and a lambda analysis. The chi-square analysis was used to determine whether there was a relationship between unsafe acts and conditions. On the other hand, a lambda analysis was used to determine how strong of a relationship existed between unsafe acts and conditions.

The SPSS 29 (IBM, Armonk, NY, USA) analysis software was used to perform this statistical analysis. No cell had an expected frequency of zero, and only 20% of the cells had an expected frequency of less than five. As a result, the data were eligible for the chi-square test. The chi-square test was performed at a significance level (p-value) of 0.01 (1%). The null hypothesis that "there is no association between the unsafe act (A) and unsafe condition (C) under test" was rejected when the estimated p-value was less than the accepted significance level, implying a strong correlation between the two variables. Only combinations that passed the chi-square test were considered for the lambda test. Figure 6 summarizes the results of the correlation analysis.

A lambda analysis was performed on the 24 combinations that passed the chi-square significance level of 0.01. The value of lambda is usually between 0 and 1, and the closer it is to 1, the more cohesive the variables are thought to be. The criteria to consider a combination as an effective combination is that the lambda value must be greater than 0.20. After deduction, twelve (12) significant near-misses were identified (see Table 4).

Table 4. Major near-misses on construction sites.

S. No	Near-Misses
1	A1—Approach to restricted areas and C2—Unguarded floor or roof openings
2	A5—Equipment and tools inspection was incomplete or non-existent and C8-Unsafely positioned ladders and scaffolds
3	A2—No or improper use of PPE and C2—Unguarded floor or roof openings
4	A2—No or improper use of PPE and C11—Inappropriately installed or defective tools and equipment
5	A6—Operating at an unusually high altitude and C11—Inappropriately installed or defective tools and equipment

Table 4. *Cont.*

S. No	Near-Misses
6	A7—Operating close to moving vehicles and equipment and C1—Poor site layout (congestion and overcrowding)
7	A8—Ignoring alarms and warning signages and C12—Unsuitable working conditions (e.g., limited visibility and excessive noise)
8	A7—Operating close to moving vehicles and equipment and C12—Unsuitable working conditions (e.g., limited visibility)
9	A12—Unauthorized use of equipment and C10—Lack of machine guards
10	A9—Working close to overhead power lines and C1—Poor site layout (congestion and overcrowding)
11	A9—Working close to overhead power lines and C5—Unsuitable weather conditions (heavy rain, poor lightning, high temperature, etc.)
12	A10—Running heavy equipment near edges and C4—Unprotected excavations and trenches

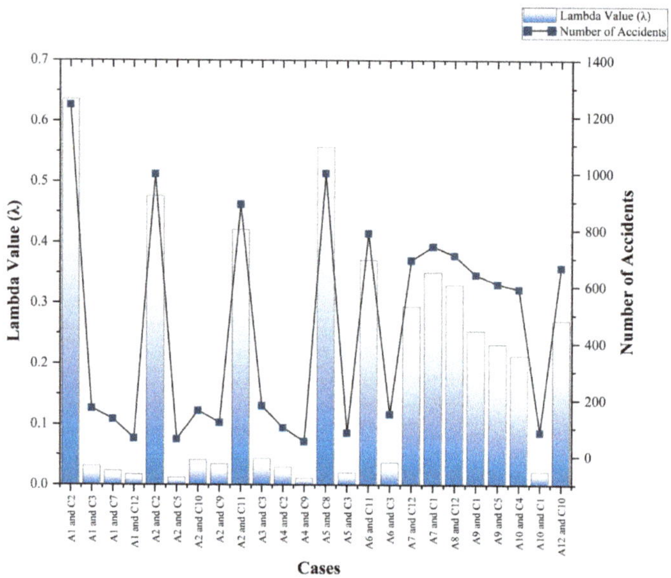

Figure 6. Correlation between unsafe acts and conditions.

4.2. Near-Miss Detection Metrics

The development of efficient real-time monitoring systems relies on a detailed understanding of specific accident precursors and quantitative measures for evaluating risky behaviors or situations [77]. The construction site risks, accidents, and causes discussed in the previous section were used to identify quantitative parameters that sensing technologies can collect and analyze to measure and monitor near-misses. Analyzing each near-miss in Table 4, the data required for autonomous near-miss detection were determined (see Table 5).

While analyzing the near-miss cases, an early warning system for employees operating in unsafe conditions can be constructed if the real-time location is estimated. As a result, worker position was required in N1, N5, and N3; relative positions of workers and equipment were required in N6 and N8; relative positions of workers and fixed powerlines were required in N10 and N11; and real-time location of equipment and vehicles was required in N12. Along with the location, these near-misses also required hazardous locations to spot workers' interactions for near-miss detection. N1 and N3 required the location of unsecured edges, N6 and N10 required updated site layouts, including equipment placement, and N12 required the location of unprotected excavation and trenches; these conditions were pre-programmed into the system. Similarly, N7 and N8 required environment information that included lightning, noise, visibility, etc. Real-time PPE information that included PPE status and a brief description of how to utilize PPE correctly was required in N3 and N4. N9 required real-time worker information that included the workers' experience, operators' working certificates, workers' training details, and others. Weather information was required by N11. Real-time information on equipment or tools included a summary of the most recent inspection reports, other information regarding their placement, and a correct procedure guide was required in near-misses N2, N4, N5, and N9.

Table 5. Data required for near-miss detection.

S. No	Near-Misses	Data Required
N1	A1—Approach to restricted areas and C2—Unguarded floor or roof openings	Real-time worker location and location of the mentioned unsafe condition
N2	A5—Equipment and tool inspection was incomplete or non-existent and C8—Unsafely positioned ladders and scaffolds	Real-time information on equipment/tools
N3	A2—No or improper use of PPE and C2—Unguarded floor or roof openings	Real-time worker location, real-time information on PPE status, and location of the mentioned unsafe condition
N4	A2—No or improper use of PPE and C11—Inappropriately installed or defective tools and equipment	Real-time information on PPE status and real-time information on equipment/tools
N5	A6—Operating at an unusually high altitude and A5—Equipment and tool inspection was incomplete or non-existent.	Real-time worker location and real-time information on equipment/tool inspections
N6	A7—Operating close to moving vehicles and equipment and C1—Poor site layout (congestion and overcrowding)	Real-time worker and vehicle location and construction site layout
N7	A8—Ignoring alarms and warning signages and C12—Unsuitable working conditions	Real-time environment information
N8	A7—Operating close to moving vehicles and equipment and C12—Unsuitable working conditions	Real-time worker and vehicle location and real-time environment information
N9	A12—Unauthorized use of equipment and C10—Lack of machine guards	Real-time information on equipment/tools and real-time information on workers
N10	A9—Working close to overhead power lines and C1—Poor site layout (congestion and overcrowding)	Real-time worker location, location of the mentioned unsafe condition, and construction site layout
N11	A9—Working close to overhead power lines and C5—Unsuitable weather conditions	Real-time worker location and location of the mentioned unsafe condition and real-time information about weather
N12	A10—Running heavy equipment/vehicles near edges and C4—Unprotected excavations and trenches	Real-time location of equipment/vehicles and location of the mentioned unsafe condition

The data required in Table 5 revealed four divisions of near-miss detection metrics: location (i.e., worker, vehicle, or equipment location), environment (e.g., temperature, noise level, light intensity, rain, and wind), real-time identity information (e.g., workers' information), and proximity (e.g., for locating the interaction to unsafe conditions, e.g., unguarded edges) (See Table 6). These detection metrics quantized all the previously reviewed unsafe acts and conditions required for near-miss detection. These measurable metrics provided a foundation for the proper deployment of sensing technologies (i.e., GPS, IMU, light sensors, anemometers, proximity sensors, RFID, etc.) to collect and monitor these metrics for real-time monitoring of near-misses.

Table 6. Near-miss detection metrics.

Detection	Metrics	Parameters	Device/Sensor	Remarks
Location	Worker, vehicle, or equipment location	Geometric coordinates	GPS, IMU	Sensor-based system for monitoring environmental conditions in confined workspaces
Environment	Temperature	Celsius	Temperature sensor	Sensor-based system for monitoring environmental conditions in confined workspaces
	Noise level	Decibel	Sound meter	
	Light intensity	Lumens	Light sensor	
	Rain	Mm	Rain gauge	
	Wind	Windspeed	Anemometer	
Proximity	Distance from unsafe conditions such as from unguarded edge	Distance	Proximity sensor, distance sensor, cameras	Near-miss/proximity analysis; proximity monitoring for struck-by hazard identification
Identity	Worker information	Name, gender, age, experience, trade	RFID cards, database	Assessing workers' perceived risk through monitoring workers' physiological and emotional response; awkward posture recognition, work-related musculoskeletal disorders (WMSDs), ergonomics

5. Practical Implications and Future Research

Effective accident prevention necessitates a proactive approach to monitoring hazards by focusing on leading indicators [4]. This study focused on near-misses, an important leading indicator often left unreported and undocumented. Regularly collecting and processing leading indicators can help improve safety decision-making [78]. Similarly, if the near-miss data are rapidly collected and updated, in that case, it will allow safety personnel to make quick decisions. This can be done by adopting modern sensing technologies that continuously collect and analyze data. This study developed a technique for facilitating the deployment of sensors for the autonomous monitoring of near-misses. This study's main findings included the near-miss detection metrics. The near-miss detection metrics consisted of measurable parameters that could be sensed and analyzed by sensing technology.

This research study was only the beginning of future research into using sensors for autonomous collection and analysis of near-misses and other precursors in preventing future accidents by providing early warnings. Although advanced sensing technologies have the potential to reveal construction safety improvement opportunities [79], compared to other industries, the use of sensors is limited in the construction industry. The advancement of promising digital technologies, such as cloud computing, RFID, wireless sensor networks (WSNs), and the Internet of Things (IoT), has sparked interest in researching their use in construction. Simultaneously, significant attempts have begun to automate the construction safety management process through sensors. Sensor technology offers safety managers measurements of the subjects' status on construction sites, allowing them to make more informed decisions on the efficacy of ongoing treatments and, if necessary, to make quick changes to the approach. A few applications of sensors include proximity detection and generating alerts when workers are present in unsafe zones [60,80,81]. Despite various research studies, the construction sector has been extremely slow to adopt sensors; there is much room for sensing technology for tailored construction safety management.

The developed detection metrics provided the data required by sensors for near-miss identification and data collection without relying on workers' feedback. Once the data were collected, they could be analyzed, processed, and used to predict accidents and provide other on-demand services. There are numerous commercially available sensor systems, each with their strengths and weaknesses, that can be effectively managed by combining two or more sensing devices to generate complimentary benefits. The findings of this study will enable future researchers to select suitable sensors as well as integrate several sensors and systems per the required metrics to detect near-misses on construction sites. Further, this research suggested that technology developers should concentrate their efforts on obtaining valuable data from many sensors merged into a single device that is simple to deploy and can measure all the required metrics.

In addition, sensor-based technologies hold the promise of continuously tracking exposures to safety risks, which can be used as a proactive indicator for near-misses. To ensure that these proposed systems are embraced by industry professionals, they must be seamlessly integrated into current practices, requiring minimal expertise and maintenance. Moreover, such systems should be merged with Building Information Modeling (BIM) to create comprehensive solutions. Presently, the focus of many studies has been on creating monitoring systems that gather and analyze specific types of data from different sources on construction sites, such as workers' movements, health statuses, activities, environmental conditions, and more. An integrated framework that consolidates this diverse information would be a valuable addition. Future studies should determine the most effective ways to convert this collected data into practical, actionable insights. Further research could also look into the development of data-driven platforms that support safety decisions for site managers, which could significantly encourage the industry to adopt sensor-based safety management practices.

6. Conclusions

Near-misses in the construction sector have long been seen as a great way to improve safety performance. Existing near-miss data collection practices are manual and face significant accuracy, interpretation, and efficiency challenges. Modern sensor technologies offer a non-intrusive solution for gathering and delivering real-time data that can be used to make proactive and efficient decisions. It is time for construction stakeholders and experts to fully embrace these rapidly evolving technological advancements in order to considerably improve safety performance.

The study proposed a quantifiable near-miss definition that defined a near-miss as an interaction of unsafe behaviors and conditions. Twelve significant near-misses were identified in terms of unsafe acts and conditions responsible for severe construction accidents through a correlation analysis. These near-misses were investigated to develop a detection metric that consisted of quantitative parameters for automatic detection and the documentation of near-misses. The detection metric was further divided into location, proximity, environment, and identity information. Based on the metric requirements, the best suitable sensors from a wide range of sensors could be chosen to collect the metrics by going over each division in detail. This research intended to shift the interest of researchers towards deploying advanced sensors to predict future accidents and generate early warnings by the real-time monitoring of near-misses.

This study contributes to improving construction safety by addressing the under-reporting of near-miss events. It developed near-miss detection metrics for automated detection using sensors. The study established a quantifiable definition of near-misses and identified combinations of unsafe acts and conditions leading to near-misses through an empirical analysis. Additionally, it determined the measurable data needed for autonomous near-miss detection. Overall, the developed metrics lay the groundwork for enhancing construction safety through automated near-miss reporting and documentation. Furthermore, it helped for the establishment of safety management schemes in the construction industry, specifically identifying near-misses. This research offers valuable insight into developing safety guidelines for managers to improve near-miss reporting and detection on construction sites.

Author Contributions: Conceptualization, F.H., M.U.H., T.A. and R.M.C.; methodology, F.H., M.U.H., R.M.C., M.U.Z. and K.A.; software, F.H., T.A. and M.U.H.; validation, M.U.Z., F.H., R.M.C. and K.A.; formal analysis, F.H., T.A. and M.U.H.; investigation, M.U.Z., F.H. and K.A.; resources, M.U.H., M.U.Z. and R.M.C.; data curation, F.H., T.A. and M.U.H.; writing—original draft preparation, F.H., M.U.H. and R.M.C.; writing—review and editing, M.U.Z., T.A. and K.A.; visualization, M.U.Z., F.H., M.U.H. and R.M.C.; supervision, M.U.H. and R.M.C.; project administration, M.U.H., M.U.Z., R.M.C. and K.A. All authors have read and agreed to the published version of the manuscript.

Funding: This research received no external funding.

Data Availability Statement: The data that support the findings of this study are openly available in Data and Statistic, Fatality and Catastrophe Summaries at https://www.osha.gov/data, https://www.osha.gov/pls/imis/accidentsearch.html, reference number [65,76].

Conflicts of Interest: The authors declare that they have no conflicts of interest.

References

1. Lozano-Díez, R.V.; López-Zaldívar, O.; del Cura, S.H.; Verdú-Vázquez, A. Analysis of the impact of health and safety coordinator on construction site accidents: The case of Spain. *J. Saf. Res.* **2019**, *68*, 149–156. [CrossRef]
2. Sarasanty, D. Safety hazards identification of construction site layout based on geographic information system (GIS). *Int. J. Adv. Sci. Eng. Inf. Technol.* **2020**, *10*, 2021–2027. [CrossRef]
3. Hallowell, M.R.; Bhandari, S.; Alruqi, W. Methods of safety prediction: Analysis and integration of risk assessment, leading indicators, precursor analysis, and safety climate. *Constr. Manag. Econ.* **2020**, *38*, 308–321. [CrossRef]
4. Versteeg, K.; Bigelow, P.; Dale, A.M.; Chaurasia, A. Utilizing construction safety leading and lagging indicators to measure project safety performance: A case study. *Saf. Sci.* **2019**, *120*, 411–421. [CrossRef]
5. Hinze, J.; Thurman, S.; Wehle, A. Leading indicators of construction safety performance. *Saf. Sci.* **2013**, *51*, 23–28. [CrossRef]
6. Lingard, H.; Hallowell, M.; Salas, R.; Pirzadeh, P. Leading or lagging? Temporal analysis of safety indicators on a large infrastructure construction project. *Saf. Sci.* **2017**, *91*, 206–220. [CrossRef]
7. Choudhry, R.M.; Fang, D. Why operatives engage in unsafe work behavior: Investigating factors on construction sites. *Saf. Sci.* **2008**, *46*, 566–584. [CrossRef]
8. Hallowell, M.R.; Hinze, J.W.; Baud, K.C.; Wehle, A. Proactive construction safety control: Measuring, monitoring, and responding to safety leading indicators. *J. Constr. Eng. Manag.* **2013**, *139*, 04013010. [CrossRef]
9. Costin, A.; Wehle, A.; Adibfar, A. Leading indicators—A conceptual IoT-based framework to produce active leading indicators for construction safety. *Safety* **2019**, *5*, 86. [CrossRef]
10. Andriulo, S.; Gnoni, M.G. Measuring the effectiveness of a near-miss management system: An application in an automotive firm supplier. *Reliab. Eng. Syst. Saf.* **2014**, *132*, 154–162. [CrossRef]
11. Saleh, J.H.; Saltmarsh, E.A.; Favarò, F.M.; Brevault, L. Accident precursors, near misses, and warning signs: Critical review and formal definitions within the framework of Discrete Event Systems. *Reliab. Eng. Syst. Saf.* **2013**, *114*, 148–154. [CrossRef]
12. Ritwik, U. Risk-based approach to near miss. *Hydrocarb. Process.* **2002**, *81*, 93–96.
13. Marks, E.; Teizer, J.; Hinze, J. Near-miss reporting program to enhance construction worker safety performance. In Proceedings of the Construction Research Congress, Atlanta, GA, USA, 19–21 May 2014; pp. 2315–2324. [CrossRef]
14. Nyflot, M.J.; Zeng, J.; Kusano, A.S.; Novak, A.; Mullen, T.D.; Gao, W.; Jordan, L.; Sponseller, P.A.; Carlson, J.C.; Kane, G.; et al. Metrics of success: Measuring impact of a departmental near-miss incident learning system. *Pract. Radiat. Oncol.* **2015**, *5*, e409–e416. [CrossRef]
15. Macrae, C. The problem with incident reporting. *BMJ Qual. Saf.* **2016**, *25*, 71–75. [CrossRef] [PubMed]
16. Gnoni, M.G.; Saleh, J.H. Near-miss management systems and observability-in-depth: Handling safety incidents and accident precursors in light of safety principles. *Saf. Sci.* **2017**, *91*, 154–167. [CrossRef]
17. Cambraia, F.B.; Saurin, T.A.; Formoso, C.T. Identification, analysis and dissemination of information on near misses: A case study in the construction industry. *Saf. Sci.* **2010**, *48*, 91–99. [CrossRef]
18. HSE. Types of Reportable Incidents. Available online: https://www.hse.gov.uk/riddor/reportable-incidents.htm (accessed on 12 October 2021).
19. OSHA. Near-Miss Reporting Policy. Available online: https://www.osha.gov/sites/default/files/2021-07/Template%20for%20Near%20Miss%20Reporting%20Policy.pdf (accessed on 28 December 2021).
20. Asadzadeh, A.; Arashpour, M.; Li, H.; Ngo, T.; Bab-Hadiashar, A.; Rashidi, A. Sensor-based safety management. *Autom. Constr.* **2020**, *113*, 103128. [CrossRef]
21. Hislop, R.D. *Construction Site Safety*; Taylor & Francis Ltd.: London, UK, 1999; pp. 203–210. [CrossRef]
22. Caspi, H.; Perlman, Y.; Westreich, S. Managing near-miss reporting in hospitals: The dynamics between staff members' willingness to report and management's handling of near-miss events. *Saf. Sci.* **2023**, *164*, 106147. [CrossRef]
23. Bugalia, N.; Maemura, Y.; Ozawa, K. A system dynamics model for near-miss reporting in complex systems. *Saf. Sci.* **2021**, *142*, 105368. [CrossRef]

24. Winkler, M.; Perlman, Y.; Westreich, S. Reporting near-miss safety events: Impacts and decision-making analysis. *Saf. Sci.* **2019**, *117*, 365–374. [CrossRef]
25. Thoroman, B.; Salmon, P.; Goode, N. Evaluation of construct and criterion-referenced validity of a systems-thinking based near miss reporting form. *Ergonomics* **2020**, *63*, 210–224. [CrossRef]
26. Shen, X.; Marks, E. Near-miss information visualization tool in BIM for construction safety. *J. Constr. Eng. Manag.* **2016**, *142*, 04015100. [CrossRef]
27. Gnoni, M.; Tornese, F.; Guglielmi, A.; Pellicci, M.; Campo, G.; De Merich, D. Near miss management systems in the industrial sector: A literature review. *Saf. Sci.* **2022**, *150*, 105704. [CrossRef]
28. WHO. *WHO Draft Guidelines for Adverse Event Reporting and Learning Systems: From Information to Action*; WHO: Geneva, Switzerland, 2005.
29. OSHA. *Incident Investigations: A Guide for Employers*; Occupational Safety and Health Administration: Washington, DC, USA, 2015; p. 24.
30. Zhou, Z.; Li, C.; Mi, C.; Qian, L. Exploring the potential use of near-miss information to improve construction safety performance. *Sustainability* **2019**, *11*, 1264. [CrossRef]
31. Awolusi, I.; Marks, E.; Hainen, A.; Alzarrad, A. Incident Analysis and Prediction of Safety Performance on Construction Sites. *Civileng* **2022**, *3*, 669–686. [CrossRef]
32. Lee, S.; Koo, B.; Yang, S.; Kim, J.; Nam, Y.; Kim, Y. Fall-from-height detection using deep learning based on IMU sensor data for accident prevention at construction sites. *Sensors* **2022**, *22*, 6107. [CrossRef]
33. Williamsen, M. Near-miss reporting: A missing link in safety culture. *Prof. Saf.* **2013**, *58*, 46–50.
34. Jones, S.; Kirchsteiger, C.; Bjerke, W. The importance of near miss reporting to further improve safety performance. *J. Loss Prev. Process Ind.* **1999**, *12*, 59–67. [CrossRef]
35. Wright, L.; Vanderschaaf, T. Accident versus near miss causation: A critical review of the literature, an empirical test in the UK railway domain, and their implications for other sectors. *J. Hazard. Mater.* **2004**, *111*, 105–110. [CrossRef] [PubMed]
36. Chen, Q.; Wu, W.; Zhang, X. The differentiation and decision matrix risk assessment of accident precursors and near-misses on construction sites. *Int. J. Eng. Technol.* **2012**, *12*, 38–53.
37. Heinrich, H.W. *Industrial Accident Prevention: A Scientific Approach*, 2nd ed.; McGraw-Hill: New York, NY, USA; London, UK, 1941.
38. Hollnagel, E. *Barriers and Accident Prevention*; Routledge: London, UK, 2016.
39. Vanderschaaf, T.W. Near miss reporting in the chemical process industry: An overview. *Microelectron Reliab.* **1995**, *35*, 1233–1243. [CrossRef]
40. Cui, J.; Rosoff, H.; John, R.S. Public response to a near-miss nuclear accident scenario varying in causal attributions and outcome uncertainty. *Risk Anal.* **2018**, *38*, 947–961. [CrossRef] [PubMed]
41. Park, I.; Kim, J.; Han, S.; Hyun, C. Analysis of fatal accidents and their causes in the Korean construction industry. *Sustainability* **2020**, *12*, 3120. [CrossRef]
42. Eurostat. European Statistics on Accidents at Work (ESAW). 2014. Available online: https://ec.europa.eu/eurostat/web/health/health-safety-work/data/database (accessed on 22 December 2021).
43. HSE. Kinds of Accident Statistics in Great Britain. 2020. Available online: www.hse.gov.uk/statistics/causinj/kinds-of-accident.pdf (accessed on 12 December 2021).
44. Lipscomb, H.J.; Dement, J.M.; Rodriguez-Acosta, R. Deaths from external causes of injury among construction workers in North Carolina, 1988–1994. *Appl. Occup. Environ. Hyg.* **2000**, *15*, 569–580. [CrossRef] [PubMed]
45. Brown, S.; Harris, W.; Brooks, R.D.; Dong, X.S. Fatal injury trends in the construction industry. *CPWR Data Bull.* 2021, pp. 1–6. Available online: https://www.cpwr.com/wp-content/uploads/DataBulletin-February-2021.pdf (accessed on 1 April 2024).
46. Musasa, T.; Jerie, S. Challenges of near miss incidents reporting at a mine in the Midlands Province of Zimbabwe. *East. Afr. Soc. Sci. Res. Rev.* **2020**, *36*, 95–114. [CrossRef]
47. Ale, B.J.M.; Bellamy, L.J.; Baksteen, H.; Damen, M.; Goossens, L.H.J.; Hale, A.R.; Mud, M.; Oh, J.; Papazoglou, I.A.; Whiston, J.Y. Accidents in the construction industry in the Netherlands: An analysis of accident reports using story builder. *Reliab. Eng. Syst. Saf.* **2008**, *93*, 1523–1533. [CrossRef]
48. Swuste, P.; Frijters, A.; Guldenmund, F. Is it possible to influence safety in the building sector? A literature review extending from 1980 until the present. *Saf. Sci.* **2012**, *50*, 1333–1343. [CrossRef]
49. Naveen Kumar, T.; Jagannatha, S.R.; Venkatesha, V.T. Rise in deaths due to fall from height: A 3-year retrospective study. *Med. Leg. Update* **2019**, *19*, 113–116. [CrossRef]
50. Dogan, E.; Yurdusev, M.A.; Yildizel, S.A.; Calis, G. Investigation of scaffolding accident in a construction site: A case study analysis. *Eng. Fail. Anal.* **2021**, *120*, 105108. [CrossRef]
51. Wu, W.; Yang, H.; Li, Q.; Chew, D. An integrated information management model for proactive prevention of struck-by-falling-object accidents on construction sites. *Autom. Constr.* **2013**, *34*, 67–74. [CrossRef]
52. Lipscomb, H.J.; Glazner, J.E.; Bondy, J.; Guarini, K.; Lezotte, D. Injuries from slips and trips in construction. *Appl. Ergon.* **2006**, *37*, 267–274. [CrossRef] [PubMed]
53. Chinniah, Y. Analysis and prevention of serious and fatal accidents related to moving parts of machinery. *Saf. Sci.* **2015**, *75*, 163–173. [CrossRef]

54. Janicak, C.A. Occupational fatalities due to electrocutions in the construction industry. *J. Saf. Res.* **2008**, *39*, 617–621. [CrossRef] [PubMed]
55. Huang, X.; Hinze, J. Analysis of construction worker fall accidents. *J. Constr. Eng. Manag.* **2003**, *129*, 262–271. [CrossRef]
56. Latief, Y.; Suraji, A.; Nugroho, Y.S.; Arifuddin, R. The nature of fall accidents in construction projects: A case of Indonesia. *Int. J. Civ. Environ. Eng.* **2011**, *11*, 80–84.
57. Wu, W.; Yang, H.; Chew, D.A.; Yang, S.-H.; Gibb, A.G.; Li, Q. Towards an autonomous real-time tracking system of near-miss accidents on construction sites. *Autom. Constr.* **2010**, *19*, 134–141. [CrossRef]
58. Wang, X.; Love, P.E.D.; Davis, P.R. BIM + AR: A framework of bringing BIM to construction site. In Proceedings of the Construction Research Congress, Lafayette, Indiana, 21–23 May 2012; pp. 778–786. [CrossRef]
59. Haslam, R.; Hide, S.; Gibb, A.; Gyi, D.; Pavitt, T.; Atkinson, S.; Duff, A. Contributing factors in construction accidents. *Appl. Ergon.* **2005**, *36*, 401–415. [CrossRef] [PubMed]
60. Chi, C.F.; Lin, S.Z. Classification scheme and prevention measures for caught-in-between occupational fatalities. *Appl. Ergon.* **2018**, *68*, 338–348. [CrossRef] [PubMed]
61. Zhao, D.; Thabet, W.; McCoy, A.; Kleiner, B. Electrical deaths in the US construction: An analysis of fatality investigations. *Int. J. Inj. Control. Saf. Promot.* **2014**, *21*, 278–288. [CrossRef]
62. Chi, C.F.; Yang, C.C.; Chen, Z.L. In-depth accident analysis of electrical fatalities in the construction industry. *Int. J. Ind. Ergon.* **2009**, *39*, 635–644. [CrossRef]
63. OSHA. Accidents Data Statistics. 2010–2020. Available online: https://www.osha.gov/data (accessed on 12 January 2021).
64. Ling, F.Y.Y.; Liu, M.; Woo, Y.C. Construction fatalities in Singapore. *Int. J. Proj. Manag.* **2009**, *27*, 717–726. [CrossRef]
65. Cheng, C.W.; Lin, C.C.; Sen Leu, S. Use of association rules to explore cause-effect relationships in occupational accidents in the Taiwan construction industry. *Saf. Sci.* **2010**, *48*, 436–444. [CrossRef]
66. Ali, A.S.; Kamaruzzaman, S.N.; Sing, G.C. A study on causes of accident and prevention in Malaysian construction industry. *J. Des. Built.* **2010**, *3*, 95–113.
67. Abdelhamid, T.S.; Everett, J.G. Identifying root causes of construction accidents. *J. Constr. Eng. Manag.* **2000**, *126*, 52–60. [CrossRef]
68. Osama, J.M.; Hamad, A.A.-I. Effective industrial housekeeping: The supervisor's role. *Prof. Saf.* **1995**, *40*, 30–33.
69. Khankeh, H.R.; Hosseini, S.H.; Farrokhi, M.; Hosseini, M.A.; Amanat, N. Early warning system models and components in emergency and disaster: A systematic literature review protocol. *Syst. Rev.* **2019**, *8*, 315. [CrossRef] [PubMed]
70. Hamid, A.R.A.; Azmi, M.R.A.N.; Aminudin, E.; Jaya, R.P.; Zakaria, R.; Zawawi, A.M.M.; Yahya, K.; Haron, Z.; Yunus, R.; Saar, C.C. Causes of fatal construction accidents in Malaysia. *IOP Conf. Ser Earth Environ. Sci.* **2019**, *220*, 012044. [CrossRef]
71. Cheng, C.-W.; Leu, S.-S.; Lin, C.-C.; Fan, C. Characteristic analysis of occupational accidents at small construction enterprises. *Saf. Sci.* **2010**, *48*, 698–707. [CrossRef]
72. Hsiao, H.; Simeonov, P. Preventing falls from roofs: A critical review. *Ergonomics* **2001**, *44*, 537–561. [CrossRef]
73. Glazner, J.; Bondy, J.; Lezotte, D.C.; Lipscomb, H.; Guarini, K. Factors contributing to construction injury at Denver international airport. *Am. J. Ind. Med.* **2005**, *47*, 27–36. [CrossRef]
74. OSHA. Fatality and Catastrophe Investigation Summaries. 2010–2020. Available online: https://www.osha.gov/pls/imis/accidentsearch.html (accessed on 19 June 2021).
75. Seo, J.; Han, S.; Lee, S.; Kim, H. Computer vision techniques for construction safety and health monitoring. *Adv. Eng. Inform.* **2015**, *29*, 239–251. [CrossRef]
76. Pradhananga, N.; Teizer, J. Automatic spatio-temporal analysis of construction site equipment operations using GPS data. *Autom. Constr.* **2013**, *29*, 107–122. [CrossRef]
77. Zhang, M.; Cao, T.; Zhao, X. Applying sensor-based technology to improve construction safety management. *Sensors* **2017**, *17*, 1841. [CrossRef]
78. Fang, Y.; Cho, Y.K.; Zhang, S.; Perez, E. Case study of BIM and cloud–enabled real-time RFID indoor localization for construction management applications. *J. Constr. Eng. Manag.* **2016**, *142*, 05016003. [CrossRef]
79. Wang, J.; Razavi, S.N. Low false alarm rate model for unsafe-proximity detection in construction. *J. Comput. Civ. Eng.* **2016**, *30*, 04015005. [CrossRef]
80. Saqib, G.; Hassan, M.U.; Zubair, M.U.; Choudhry, R.M. Investigating the acceptance of an electronic incident reporting system in the construction industry: An application of the technology acceptance model. *J. Constr. Eng. Manag.* **2023**, *149*, 04023021. [CrossRef]
81. Choudhry, R.M.; Zahoor, H. Strengths and weaknesses of safety practices to improve safety performance in construction projects in Pakistan. *J. Prof. Issues Eng. Educ. Pract.* **2016**, *142*, 04016011. [CrossRef]

MDPI
St. Alban-Anlage 66
4052 Basel
Switzerland
www.mdpi.com

Buildings Editorial Office
E-mail: buildings@mdpi.com
www.mdpi.com/journal/buildings

www.ingramcontent.com/pod-product-compliance
Lightning Source LLC
Chambersburg PA
CBHW061438210926
51850CB00008B/1546